A Specialist Periodical Report

Aliphatic, Alicyclic, and Saturated Heterocyclic Chemistry
Volume 1
Part I Aliphatic Chemistry

A Review of the Literature Published
during 1970 and 1971

Senior Reporter
W. Parker, Department of Chemistry,
University of Stirling

Reporters for Part I

R. S. Atkinson, University of Leicester
E. W. Colvin, University of Glasgow
F. D. Gunstone, University of St. Andrews

Reporters for Parts II and III

M. S. Baird, University of Newcastle upon Tyne
D. R. Boyd, Queen's University, Belfast
H. Maskill, University of Stirling
J. M. Mellor, University of Southampton
D. G. Morris, University of Glasgow
F. G. Riddell, University of Stirling
B. J. Walker, Queen's University, Belfast

© Copyright 1973

The Chemical Society
Burlington House, London W1V 0BN.

ISBN: 0 85186 502 X
Library of Congress Catalog Card No. 72-83454

Printed in Northern Ireland at The Universities Press, Belfast

Foreword

This first Report surveys the literature over a two-year period (January 1970–December 1971), in order to put in perspective the developments in this wide area of organic chemistry. Subsequent Reports will be annual.

At an advanced production stage it was decided that the length of Volume 1 made it impractical to publish it as one book. It, like Gaul, has therefore been divided into three parts:

I Aliphatic Chemistry
II Three- and four-membered rings (carbocyclic and saturated heterocyclic)
III Five- and six-membered rings: medium sized rings: bridged and caged systems (carbocyclic and saturated heterocyclic)

Aliphatic chemistry is covered in three chapters, two dealing with functional groups and a third devoted, in each Report, to a class of related natural products (in this instance fatty acids). The chapters on functional group chemistry reflect an increasing interest in new synthetic methods, and interconversions. It may well be that this aspect of next year's Report will be extended to give, in effect, an annual comprehensive coverage of General Methods.

It was originally envisaged that the coverage of carbocyclic and saturated heterocyclic chemistry would be dealt with in an integrated fashion. Some 3000 references later it was apparent that this method of tackling the subject matter was unwieldy, and so it was decided to produce separate chapters on carbocyclic and saturated heterocyclic chemistry, with the division based on ring size.

W. P.

April 1973

Contents

Part I Aliphatic Chemistry

Chapter 1 Acetylenes, Allenes, and Olefins   3
By R. S. Atkinson

1	**Acetylenes**	3
	Synthesis	3
	Cycloaddition Reactions	10
	Other Additions to the Acetylenic Bond	16
	Nucleophilic	16
	Electrophilic	19
	Radical Addition	25
	Other Reactions of Acetylenes	27
2	**Allenes**	36
	Synthesis	36
	Cycloaddition Reactions	39
	Addition Reactions	44
	Other Reactions of Allenes	47
3	**Olefins**	52
	Synthesis	53
	Cycloaddition Reactions	61
	Electrophilic Addition	67
	Nucleophilic Addition	71
	Radical Addition	72
	Oxidation and Reduction	73
	Other Reactions and Properties of Olefins	76

Chapter 2 Functional Groups other than Acetylenes, Allenes, and Olefins   79
By E. W. Colvin

1	**Alkanes**	79
	Atomic Carbon	81
2	**Carboxylic Acids**	83
	Preparation and Protection	83
	Absolute Configuration	85
	Reactions	85

3	**Carboxylic Acid Esters**	88
	Preparation	88
	Reactions	93
	α-Anions of Acids and Esters	95
	Reformatsky Reactions	98
	Solid-phase Synthesis	99
	β-Ketoesters	100
	General	103
4	**Carboxylic Acid Amides**	105
5	**Nitriles**	111
	Preparation	111
	Reactions	113
6	**Aldehydes and Ketones**	116
	Preparation	116
	Oxazines	125
	Enone Formation	127
	Conjugate Addition to Enones	128
	Regeneration from Derivatives	129
	Synthons	130
	Alkylation	136
	Reduction	139
	Addition of Grignard Reagents	143
	Halogeno-derivatives	144
	Dimerization	145
	Metal-induced Bond Formation	146
	Aldol Condensation	147
	General	148
7	**Amines**	150
	Preparation	150
	Reactions	155
8	**Alkyl Halides**	156
	Preparation	156
	Reactions	157
9	**Alcohols**	159
	Preparation	159
	Acetals and Thioacetals	165
10	**Sulphur Compounds**	168
11	**Miscellaneous Reactions**	173
12	**Reviews**	175

Chapter 3 Fatty Acids and Related Compounds — 177
By F. D. Gunstone

  1 **Introduction** — 177

  2 **Natural Compounds** — 177
- Long-chain Acids — 177
- Other Long-chain Compounds — 180
- Waxes — 180
- New Lipids — 181

  3 **Synthetic Compounds** — 182
- Fatty Acids and Related Compounds — 182
- Glycerides — 185
- Other Lipids — 186

  4 **Physical Properties** — 186
- Gas–Liquid Chromatography — 186
- Polymorphism — 187
- Infrared and Raman Spectra — 187
- Nuclear Magnetic Resonance Spectroscopy — 187
- Mass Spectrometry — 188

  5 **Chemical Properties** — 188
- Double-bond Migration and Cyclization Reactions — 189
- Dimerization — 190
- Cyclopropane and Cyclopropene Acids — 191
- 1,4-Epoxides — 192
- Oxymercuration Reactions — 192
- Catalytic Hydrogenation — 193
- Other Addition Reactions of Unsaturated Acids or Esters — 193
- Other Reactions — 194

  6 **Biological Reactions** — 196
- Hydrogenation — 196
- Hydration — 196
- Hydroperoxidation — 197
- Biosynthesis and Metabolism — 198
- Prostaglandins — 199

  7 **Reviews** — 200

Author Index — 202

Part I

ALIPHATIC CHEMISTRY

1
Acetylenes, Allenes, and Olefins

BY R. S. ATKINSON

1 Acetylenes

Areas of acetylenic chemistry reviewed recently include the base-catalysed isomerization of acetylenes,[1] nucleophilic additions to acetylenes,[2] additions to activated triple bonds,[3] synthetic and naturally occurring acetylene compounds as drugs,[4] allenic and acetylenic carotenoids,[5] linear polymers from acetylenes,[6] carbonylation of mono-olefinic and monoacetylenic hydrocarbons,[7] and the combustion and oxidation of acetylene.[8] Several books have also appeared.[9]

Synthesis.—The coupling reaction between acetylenic Grignard reagents and allylic halides yields acetylenic olefins which are of value for sterospecific conversion into acyclic isoprenoid polyenes.[10] However, this coupling reaction generally leads to mixtures of allenic and acetylenic products. The lithiated trimethylsilylacetylene has been used to circumvent this problem.[11] Another complementary method[12] is trimethylsilylation of the crude reaction product after treatment with one equivalent of Grignard reagent. Thus the allylic

[1] R. J. Bushby, *Quart. Rev.*, 1970, **24**, 585.
[2] S. I. Miller and R. Tanaka, *Selective Org. Transform.*, 1970, **1**, 143.
[3] E. Winterfeld, *Neure Method. Praep. Org. Chem.*, 1970, **6**, 230.
[4] K. E. Schulte and G. Ruecker, *Fortschr. Arzneim.*, 1970, **14**, 387.
[5] B. C. L. Weedon, *Rev. Pure. Appl. Chem.*, 1970, **20**, 51.
[6] M. J. Benes, M. Janic, and J. Peska, *Chem. listy.*, 1970, **64**, 1094 (*Chem. Abs.*, 1970, **74**, 31 979h).
[7] Ya. T. Eidus, K. V. Puzitskii, A. L. Lapidus, and B. K. Nefedov, *Russ. Chem. Rev.*, 1971, **40**, 429.
[8] A. Williams and D. B. Smith, *Chem. Rev.*, 1970, **70**, 267.
[9] (a) T. F. Rutledge, 'Acetylenic Compounds', Van Nostrand-Reinhold, New York, 1968; (b) T. F. Rutledge, 'Acetylenes and Allenes', Van Nostrand-Reinhold, New York, 1969; (c) 'Chemistry of Acetylenes', ed. H. G. Viehe, Marcel Dekker, New York, 1969.
[10] E. J. Corey, J. A. Katzenellenbogen, N. W. Gilman, S. A. Roman, and B. W. Erickson, *J. Amer. Chem. Soc.*, 1968, **90**, 5618; E. J. Corey, J. A. Katzenellenbogen, and G. H. Posner, *ibid.*, 1967, **89**, 4245.
[11] E. J. Corey and H. A. Kirst, *Tetrahedron Letters*, 1968, 5041.
[12] R. E. Ireland, M. I. Dawson, and C. A. Lipinski, *Tetrahedron Letters*, 1970, 2247.

dibromide (1) gives the trimethylsilylated acetylene (2), from which the enediyne (3) is obtained in 50% overall yield.

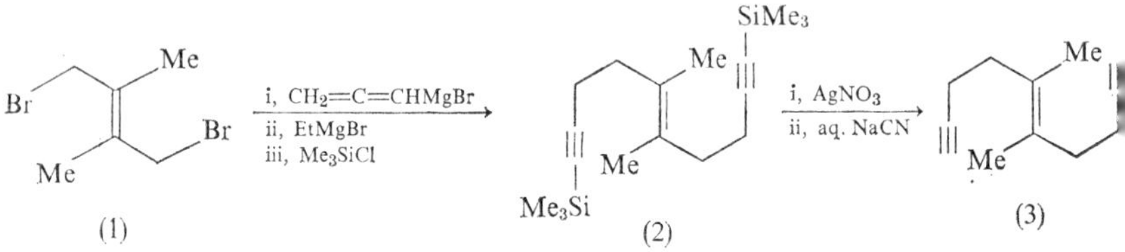

Selenium dioxide oxidation of aryl ketone semicarbazones (4) in acetic acid affords 1,2,3-selenadiazoles (5). Pyrolysis of the latter gives selenium and the arylacetylenes (6) in good yield.[13] The same procedure has been used to prepare cyclo-octyne in 34% yield.[14]

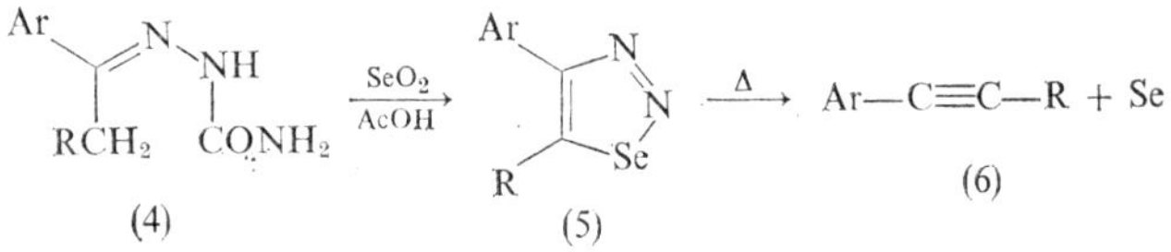

The fragmentation reaction of Eschenmoser,[15] which gives acetylenes from toluene-*p*-sulphonyl hydrazones of $\alpha\beta$-epoxy-ketones (Scheme 1), has been adapted to the synthesis of acetylenes from ketones substituted with leaving groups in the α-position [(7) → (8)].[16]

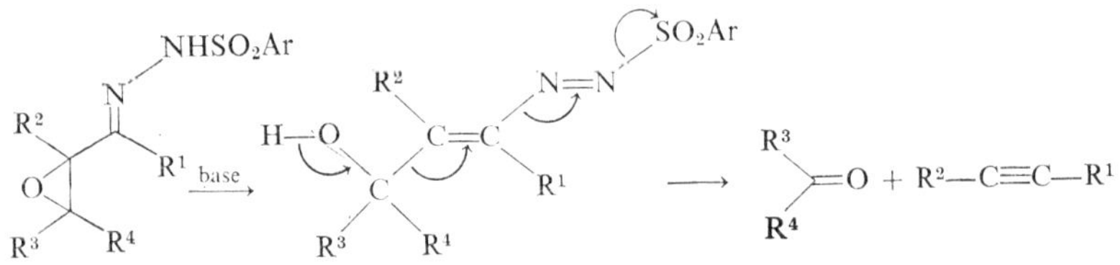

Scheme 1

Cadiot–Chodkiewicz coupling of 1-bromoacetylenes with terminal acetylenes in the presence of cuprous salts and amines is much slower when 1-chloroacetylenes are used.[17] Conditions have been described for the intramolecular coupling of the bromodiyne (9) to (10) in 40% yield using high dilution and complementing the Glaser reaction.[18]

[13] I. Lalezari, A. Shafiee, and M. Yalpani, *Angew. Chem. Internat. Edn.*, 1970, **9**, 464.
[14] H. Meier and I. Menzel, *Chem. Comm.*, 1971, 1059.
[15] A. Eschenmoser, D. Felix, and G. Ohlott, *Helv. Chim. Acta*, 1967, **50**, 708; J. Schreiber, D. Felix, A. Eschenmoser, M. Winter, F. G. Gautschi, K. H. Schulte-Elte, E. Sundt, G. Ohloff, J. Kalvoda, H. Kaufmann, P. Wieland, and G. Anner, *ibid*., p. 2101.
[16] P. Wieland, *Helv. Chim. Acta*, 1970, **53**, 171.
[17] J.-L. Phillipe, W. Chodkiewicz, and P. Cadiot, *Tetrahedron Letters*, 1970, 1795.
[18] G. Eglinton and W. McCrae, *Adv. Org. Chem.*, 1963, **5**, 225.

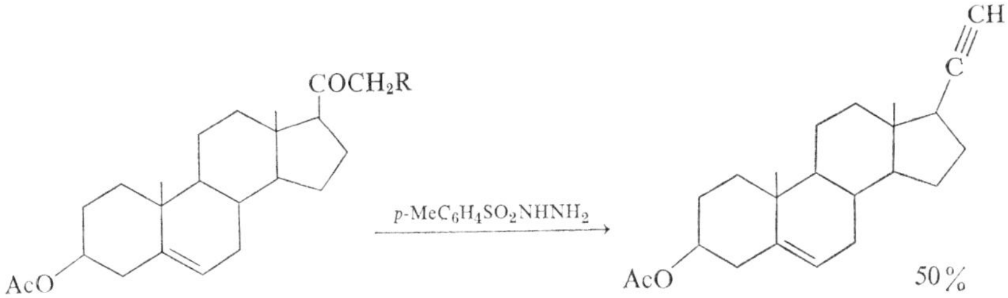

Acetylenes, Allenes, and Olefins

(7) R = F or O₃SMe → (8) 50%

(with p-MeC₆H₄SO₂NHNH₂)

$BrC{\equiv}C(CH_2)_3CONH(CH_2)_3C{\equiv}CH \longrightarrow$ cyclic $(CH_2)_3$–CONH–$(CH_2)_3$–$C{\equiv}C$–$C{\equiv}C$

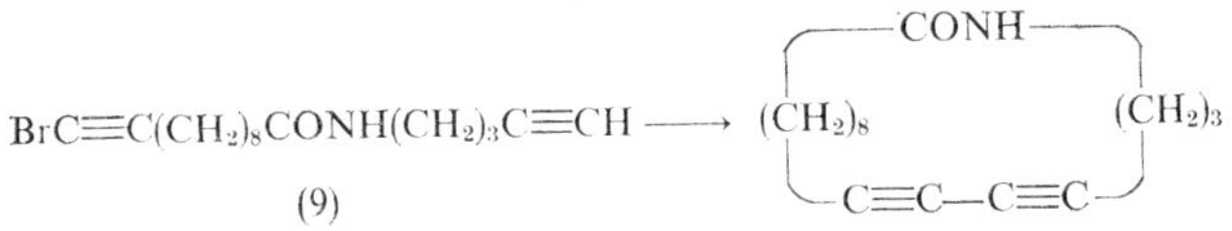

(9) (10)

Labile and unstable acetylenes are conveniently coupled with halogenoacetylenes as their copper(II) salts. The terminal acetylenes required, e.g. (11), can be obtained by deformylation with base of the aldehydes obtained by nickel peroxide oxidation of the corresponding alcohols (12).[19] Iodo-

$$R(C{\equiv}C)_2CH_2OH \xrightarrow[C_6H_6]{NiO_2} R(C{\equiv}C)_2CHO \xrightarrow{NaOH} R(C{\equiv}C)_2H$$
(12) (11)

pyrazoles,[20] iodopyridines,[21] and iodonitrobenzenes[22] couple efficiently with acetylenes in the presence of copper, potassium carbonate, and pyridine. Copper acetylides also react with allylic halides to give ene-ynes (13) and the reaction is accelerated by the presence of halide or cyanide ion.[23] Similarly, reaction with acid chlorides giving acetylenic ketones (14) is catalysed by halide salts and by addition of hexamethylphosphortriamide at a specific time during the reaction.[24]

$$RC{\equiv}CCu + XCH_2C{=}C{-} \longrightarrow RC{\equiv}CCH_2C{=}C{-}$$
(13)

$$R^1{-}C{\equiv}CCu + R^2COCl \longrightarrow R^1C{\equiv}CCOR^2$$
(14)

[19] R. F. Curtis and J. A. Taylor, *J. Chem. Soc. (C)*, 1971, 186.
[20] S. F. Vasilevskii, M. S. Schvartsberg, and I. L. Kotlyarevskii, *Izvest. Akad. Nauk S.S.S.R., Ser. khim*, 1971, **8**, 1764 (*Chem. Abs.*, 1971, **75**, 151 724p).
[21] M. S. Schvartsberg, A. N. Kozhevnikova, and I. L. Kotlyarevskii, *Izvest. Akad. Nauk S.S.S.R., Ser. khim.*, 1971, **8**, 1833 (*Chem. Abs.*, 1971, **75**, 151 645p).
[22] M. S. Schvartsberg, A. A. Moroz, and I. L. Kotlyarevskii, *Izvest. Akad. Nauk S.S.S.R., Ser. khim.*, 1971, **8**, 1306 (*Chem. Abs.*, 1971, **75**, 88 219k).
[23] J. F. Normant, M. Bourgain, and A. M. Rone, *Compt. rend.*, 1970, **270**, C, 354.
[24] J. F. Normant and M. Bourgain, *Tetrahedron Letters*, 1970, 2659.

A synthesis of p,p'-bridged tolans (15) has used the Fritsch–Buttenberg–Wiechell (FBW) rearrangement of *gem*-dihalogeno-olefins with strong base.[25] The effect of diminishing n upon the u.v. spectrum of (15) has been determined and for the o,p'-bridged acetylene and p,p'-bridged diacetylenes.[26] The FBW rearrangement has also been used to prepare dicyclopropylacetylene from (16) with n-butyl-lithium.[27]

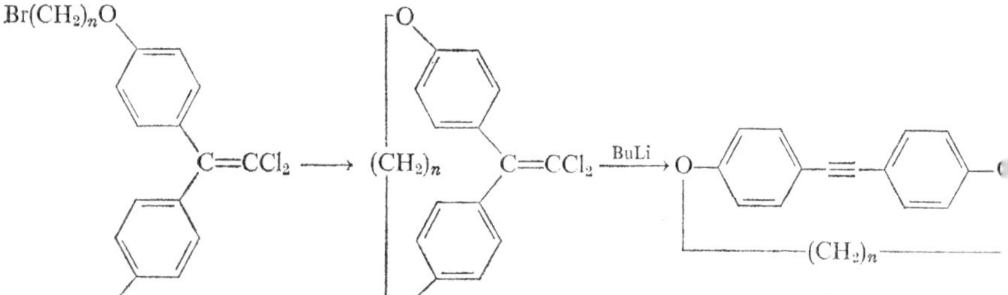

Conversion of terminal olefins into alk-1-ynes by a bromination–dehydrobromination sequence is often easier in theory than in practice. The use of DMSO as solvent and methylsulphinyl carbanion or sodamide as the base gives excellent yields with little isomerization to alk-2-ynes. Moreover, conditions have also been found for transformation of alk-1-ynes into alk-2-ynes.[28]

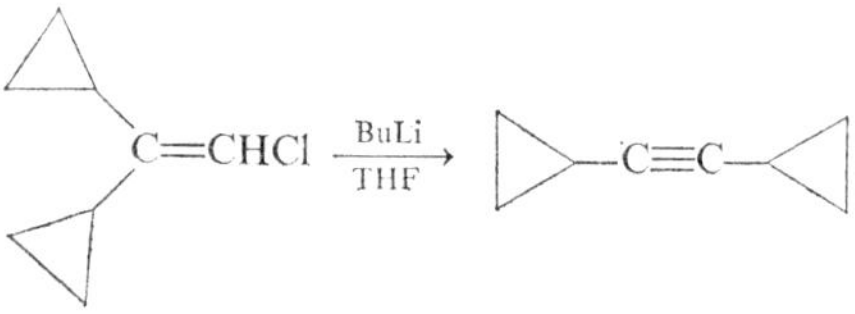

Selective hydroboration of symmetrical conjugated diynes with dialkylboranes provides a route to acetylenic ketones in yields >70%. Protonolysis of the intermediate boranes also gives the corresponding *cis*-ene-ynes in high yield (Scheme 2).[29]

Reaction between acetylenic Grignard reagents and hydroximoyl chlorides (17) is known[30] to give isoxazoles *via* intermediate ketoximes (18). The latter

[25] T. Ando and M. Nakagawa, *Bull. Chem. Soc. Japan*, 1971, **44**, 172.
[26] M. Kataoka, T. Ando, and M. Nakagawa, *Bull. Chem. Soc. Japan*, 1971, **44**, 177; M. Kataoka, T. Ando, and M. Nakagawa, *ibid.*, p. 1909; F. Toda, T. Ando, M. Kataoka, and M. Nakagawa, *ibid.*, p. 1914.
[27] G. Köbrich and D. Merkel, *Angew. Chem. Internat. Edn.*, 1970, **9**, 243.
[28] J. Klein and E. Gurfinkel, *Tetrahedron*, 1970, **26**, 2127.
[29] G. Zweifel and N. L. Polston, *J. Amer. Chem. Soc.*, 1970, **92**, 4068.
[30] G. Palazzo, *Gazzetta*, 1947, **77**, 214.

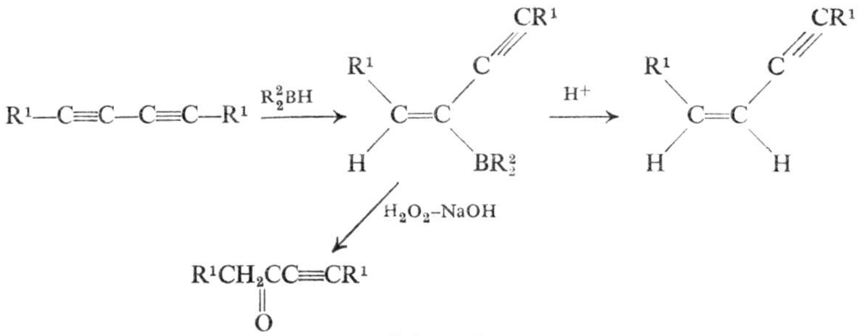

Acetylenes, Allenes, and Olefins

$$R^1-C\equiv C-C\equiv C-R^1 \xrightarrow{R^2_2BH} \begin{array}{c} R^1 \\ \diagdown \\ H \end{array} C=C \begin{array}{c} CR^1 \\ \diagup \\ \diagdown BR^2_2 \end{array} \xrightarrow{H^+} \begin{array}{c} R^1 \\ \diagdown \\ H \end{array} C=C \begin{array}{c} CR^1 \\ \diagup \\ \diagdown H \end{array}$$

$$\downarrow H_2O_2-NaOH$$

$$R^1CH_2\underset{\underset{O}{\|}}{C}C\equiv CR^1$$

Scheme 2

have now been isolated by working at low temperature.[31] In 4,4,4-triethoxybut-1-yne (19), obtained from propargylmagnesium bromide and tetraethyl orthocarbonate, the carboxy-group, masked as an orthoester, is protected from attack by basic reagents, and ready acetylene–allene interconversion is inhibited. Thus the lithio-derivative of (19) is acylated and converted into enol-ether (20) by base-catalysed addition of alcohols. This represents extension of carboxylic acid R by two acetyl units.[32]

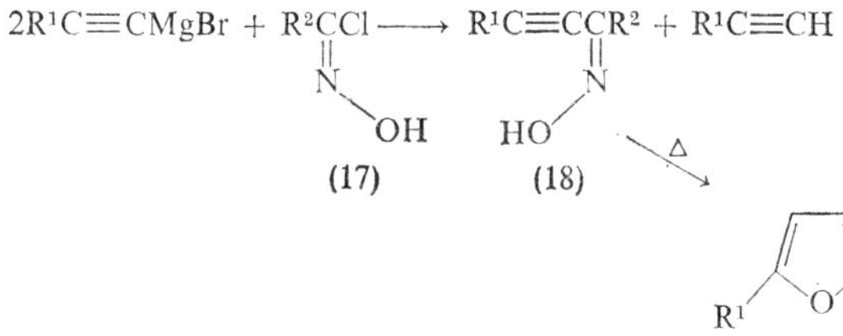

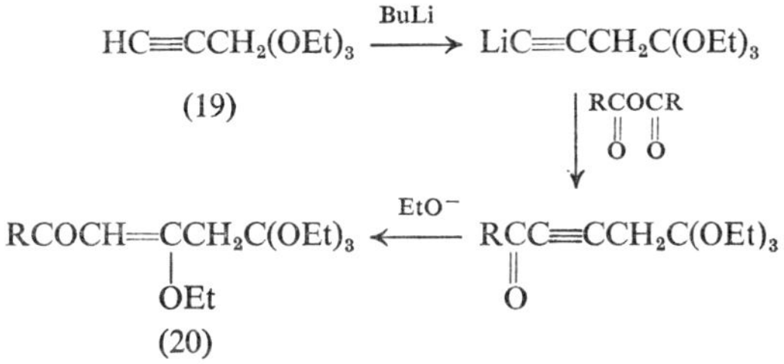

$$2R^1C\equiv CMgBr + R^2CCl \longrightarrow R^1C\equiv CCR^2 + R^1C\equiv CH$$

(17) N–OH, HO–N (18)

$$HC\equiv CCH_2(OEt)_3 \xrightarrow{BuLi} LiC\equiv CCH_2C(OEt)_3$$
(19)

$$\downarrow RCOCR \\ \|\ \| \\ O\ O$$

$$RCOCH=CCH_2C(OEt)_3 \xleftarrow{EtO^-} RCC\equiv CCH_2C(OEt)_3$$
$$\ \ \ \ \ \ \ \ \ |$$
$$\ \ \ \ \ OEt \qquad\qquad\qquad\ \ \ \|$$
$$\ \ \ (20) \qquad\qquad\qquad\quad\ \ O$$

A safe and convenient preparation[33] of dichloroacetylene uses a liquid medium and reduces explosion hazards. All the six homo- and hetero-dihalogenoacetylenes (21) have been obtained pure and characterized by their i.r. and Raman spectra.[34] Chlorofluoroacetylene has also been synthesized by

$$X-C\equiv C-Y \qquad X = Cl, Br, or I$$
$$(21) \qquad\qquad Y = Cl, Br, or I$$

[31] Z. Hamlet, M. Rampersad, and D. J. Shearing, *Tetrahedron Letters*, 1970, 2101; see also S. Morrocchi, A. Ricca, A. Zanarotti, G. Bianchi, R. Gandolfi, and P. Grünanger, *ibid.*, 1969, 3329.
[32] R. Finding and U. Schmidt, *Angew. Chem. Internat. Edn.*, 1970, **9**, 456.
[33] J. Siegel, R. A. Jones, and L. Kurlansik, *J. Org. Chem.*, 1970, **35**, 3199.
[34] E. Kloster-Jenson, *Tetrahedron*, 1971, **27**, 33.

base treatment of 1,1-dichloro-2-fluoroethylene (22). It is a very unstable explosive compound which adds regiospecifically to ethers in the absence of initiators.[35] Trichloroethylene reacts with NN-disubstituted lithium amides

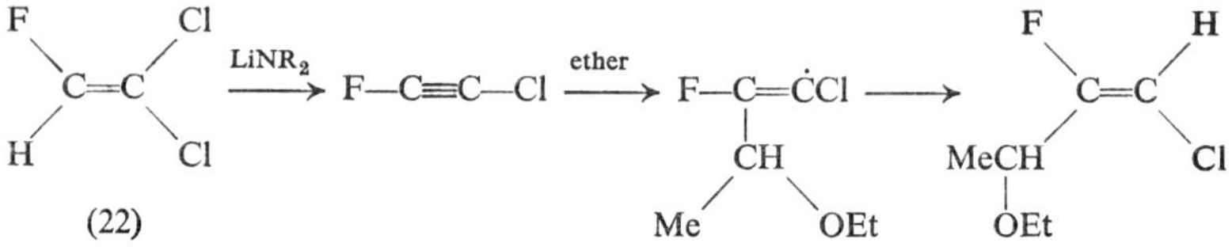

and secondary amines to form chloroketen aminals (23). Upon further treatment with strong base, the latter undergo HCl α-elimination and 'onium' rearrangement to afford yne-diamines (24) in high yields.[36]

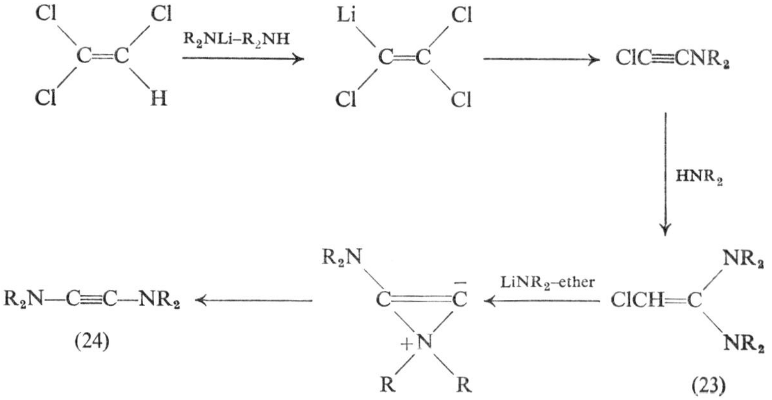

Acetylenic sulphoxides and sulphones are obtained by oxidation of the more readily available acetylenic sulphides using m-chloroperbenzoic acid at $-20\,^\circ$C in chloroform. Alternatively, the sulphoxides may be prepared by dehydrohalogenation of chlorovinylsulphoxides (25).[37] A useful preparative

$$\text{PhC(Cl)=CHSOMe} \xrightarrow[\text{THF}]{\text{KOtBu}} \text{PhC≡CSOMe} \quad (89\%)$$
(25)

route to acetylenic acids (26) uses aroylmalonates as starting materials. Treatment with arenesulphonic anhydrides yields the intermediate enol sulphonates, and decarboxylative elimination of the derived diacids gives (26) in 80% overall yields.[38]

The synthesis of the hexadehydro[18]annulene (27), and hence of [18]-annulene, has been improved.[39] Although cycloheptyne has not been

[35] S. Y. Delavarenne and H. G. Viehe, *Chem. Ber.*, 1970, **103**, 1198.
[36] S. Y. Delavarenne and H. G. Viehe, *Chem. Ber.*, 1970, **103**, 1209.
[37] G. A. Russell and L. A. Ochrymowycz, *J. Org. Chem.*, 1970, **35**, 2106.
[38] I. Fleming and C. R. Owen, *J. Chem. Soc. (C)*, 1971, 2013.
[39] H. P. Figeys and M. Gelbcke, *Tetrahedron Letters*, 1970, 5139.

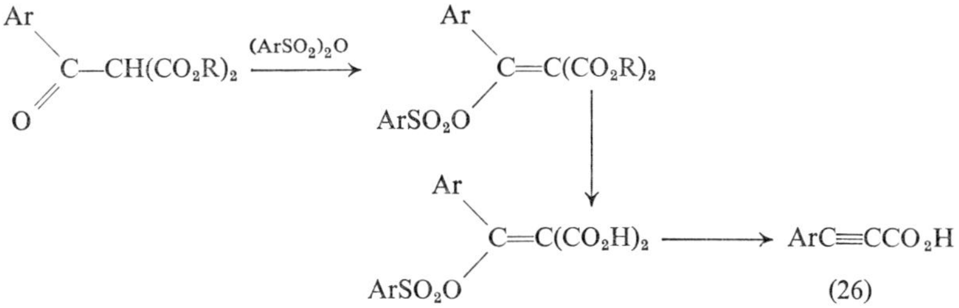

isolated, the thiacycloheptyne (29) has been prepared in low yield by oxidation of bishydrazone (28).[40]

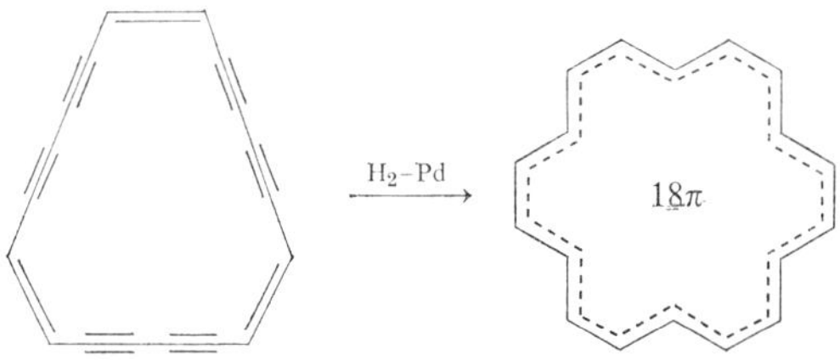

Asymmetric synthesis of acetylenic alcohols is possible by reduction of the corresponding ketones with lithium aluminium hydride complexed with sugar derivatives; the optical yields are 4—7%.[41] The trimethylether of the naturally occurring robustol (30), from the leaves of *Grevillea robusta* A. Cunn., has been synthesized in 55% yield *via* cupric acetate oxidative cyclization of the diacetylene (31).[42]

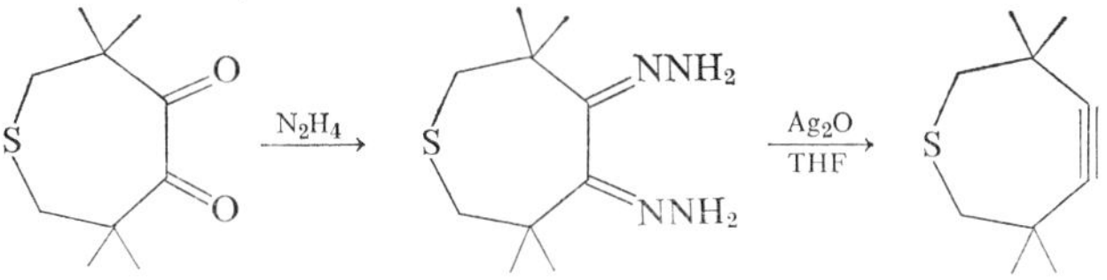

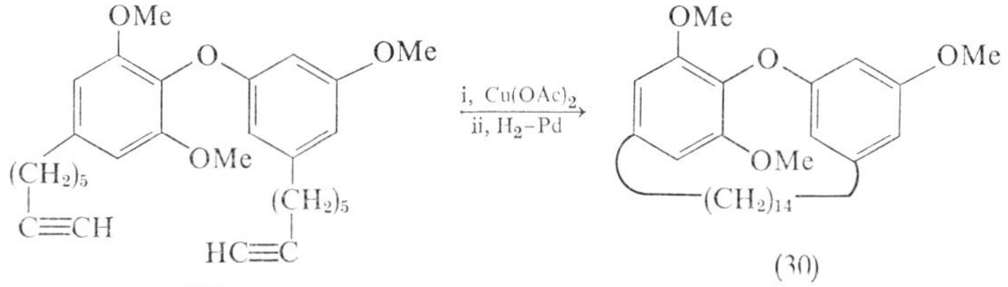

[40] A. Krebs and H. Kimling, *Tetrahedron Letters*, 1970, 761.
[41] S. R. Landor, B. J. Miller, and A. R. Tatchell, *J. Chem. Soc.* (*C*), 1971, 2339.
[42] J. R. Cannon, P. W. Chow, B. W. Metcalf, A. J. Power, and M. W. Fuller, *Tetrahedron Letters*, 1970, 325.

Cycloaddition Reactions.—The use of dimethylacetylene dicarboxylate (DMAD) in cycloaddition reactions with various substrates is illustrated in the Table.

1,2,3-Trimethylindole and DMAD yield the benzazepine (32) and the dienone (33); the suggested mechanism is shown in Scheme 3.[43] A reaction

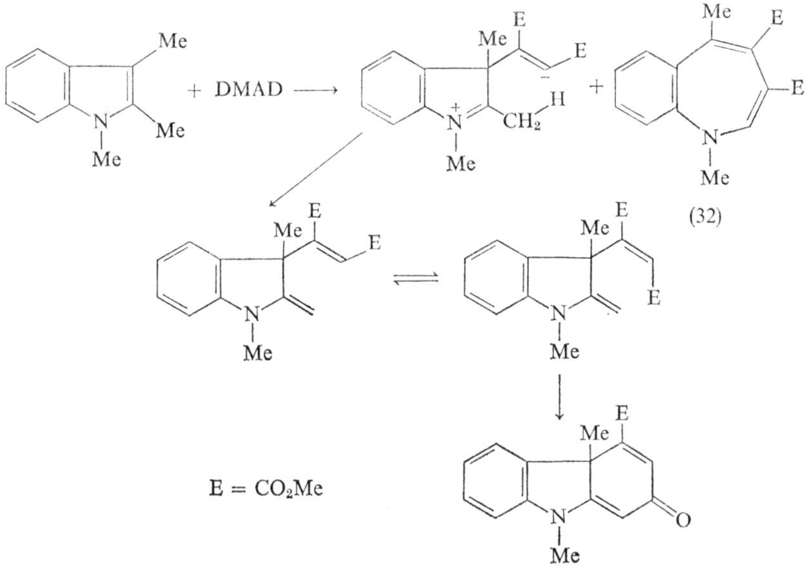

$E = CO_2Me$

Scheme 3

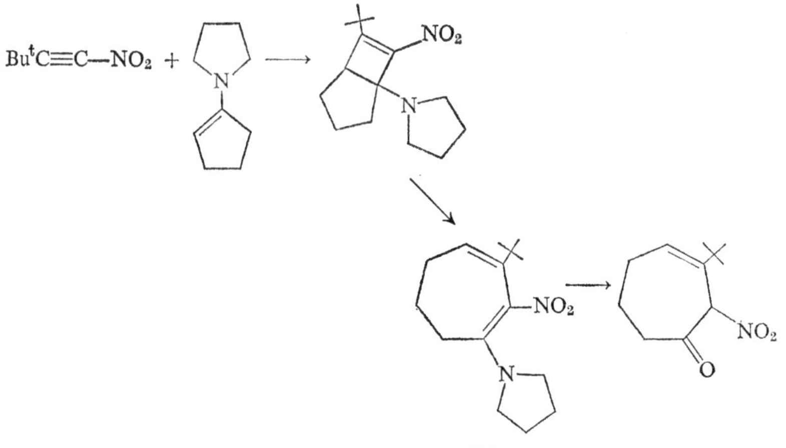

[43] F. Fried, J. B. Taylor, and R. Westwood, *Chem. Comm.*, 1971, 1226.

Acetylenes, Allenes, and Olefins 11

Table *Reactions of dimethylacetylenedicarboxylate with various substrates*

Starting material	Products		Ref.
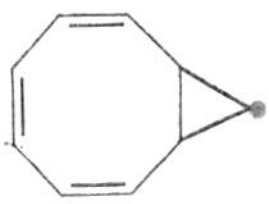	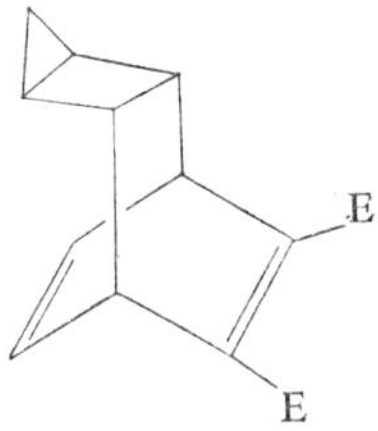		a
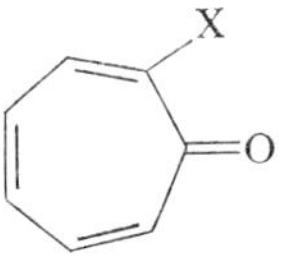	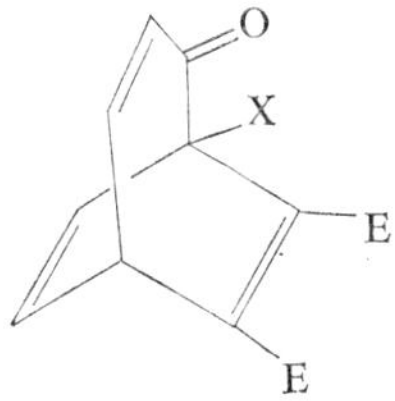	X = H, OMe, or Cl	b
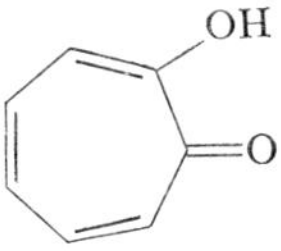	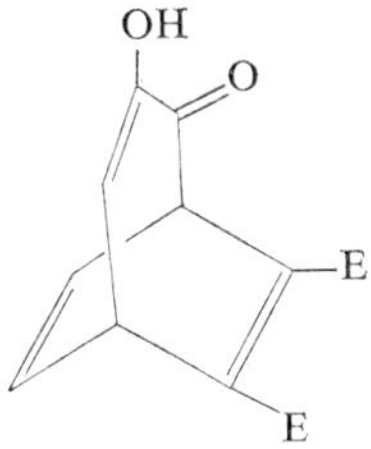		c*
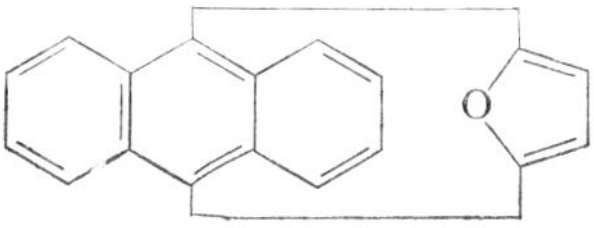	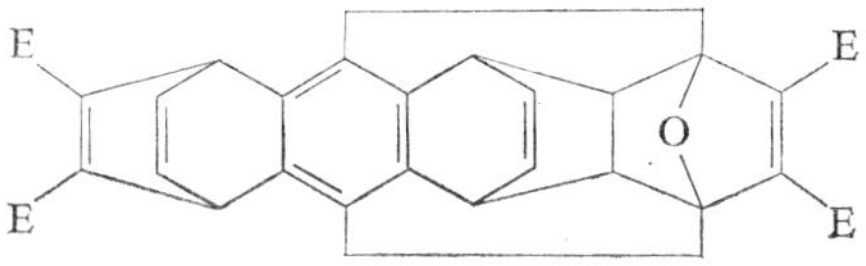		d
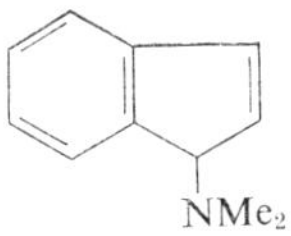	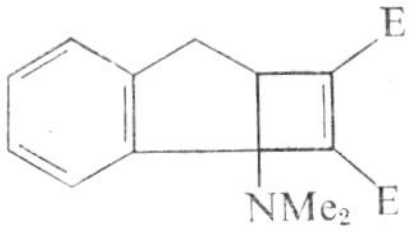		e
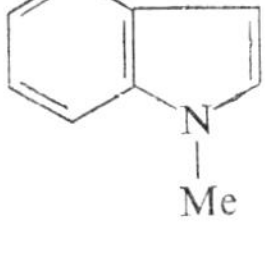	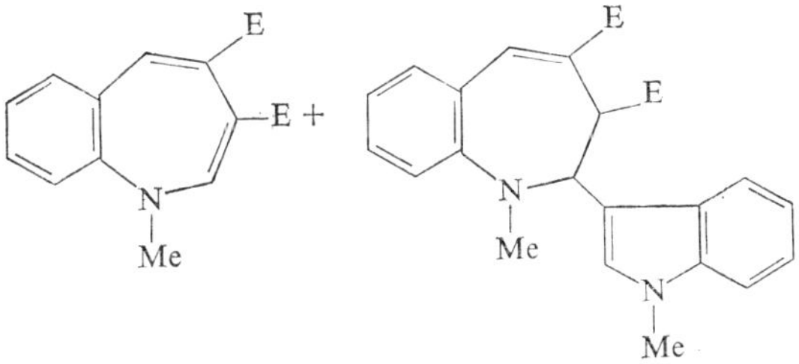		f

Starting material	*Products*	*Ref:*
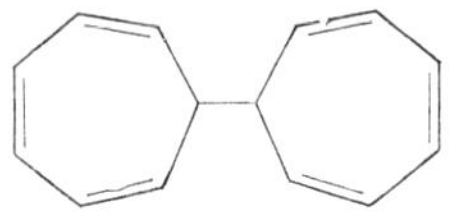	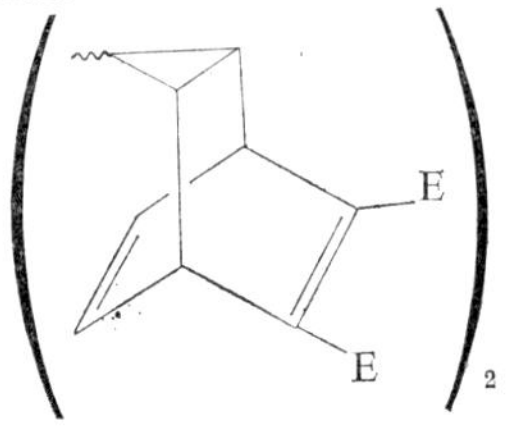	*g*
S=C=S	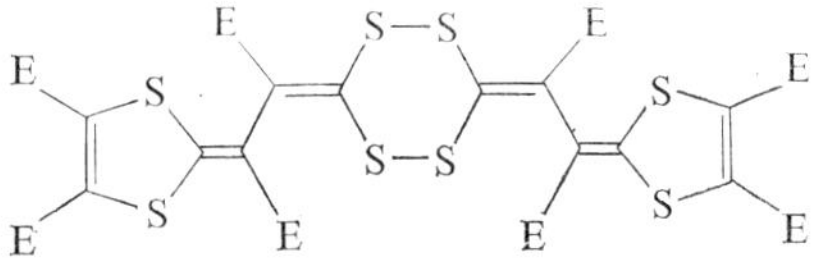	*h*
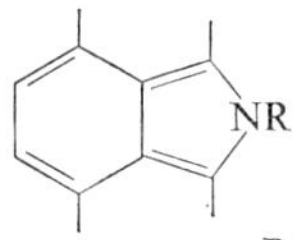 R = H or Me	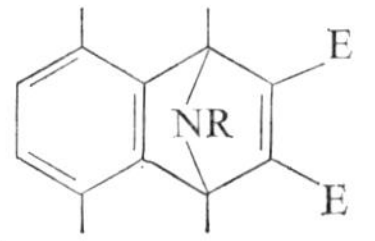 + 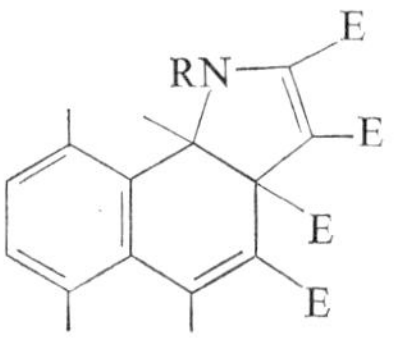	*i*
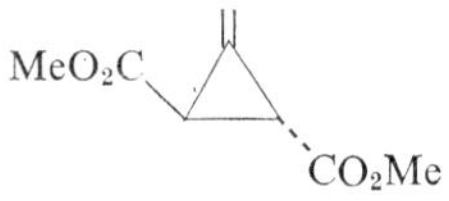	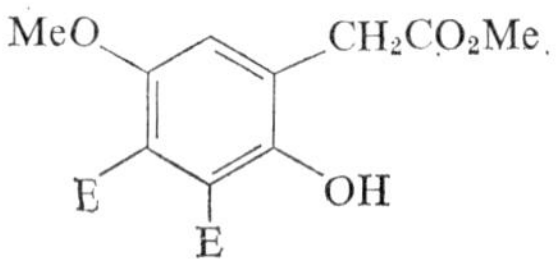	*j*
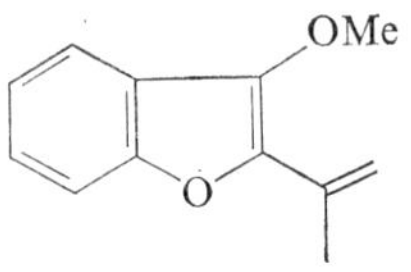	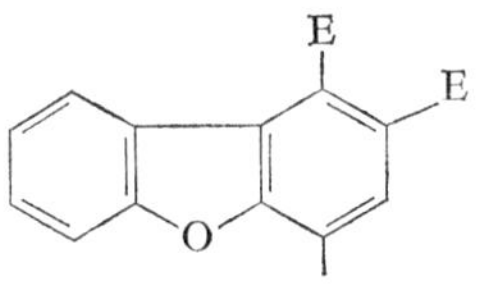	*k*
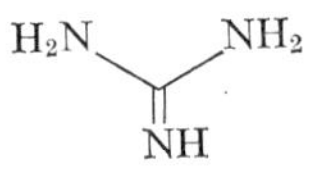	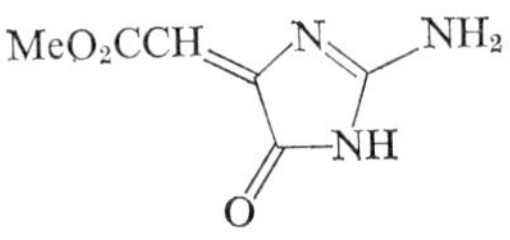	*l*
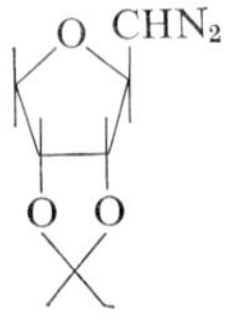	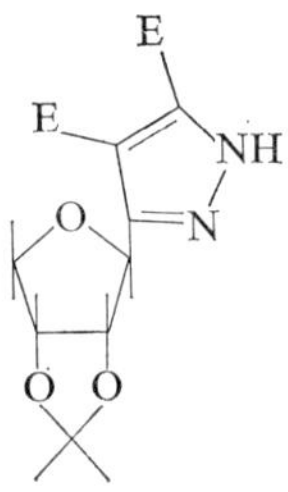	*m*

E = CO_2Me

Acetylenes, Allenes, and Olefins

Starting material   Products   Ref:

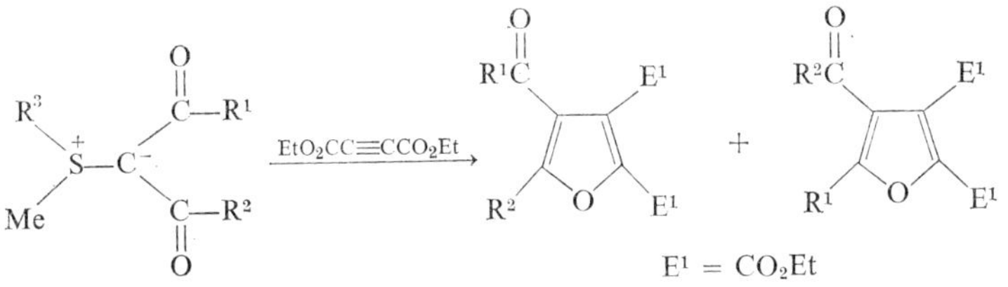

[Structure: R³\S⁺—C(C(=O)R¹)(C(=O)R²) with Me on S, reacting with EtO₂CC≡CCO₂Et to give two furan products with E¹ = CO₂Et] n

* No n.m.r. data reported in abstract
(a) J. E. Baldwin and R. K. Pinschmidt, *Chem. Comm.*, 1971, 820; (b) L.-C. Lin, P.-W. Yang, H.-Y. Huang, and C.-C. Hsieh, *Hua Hsueh*, 1970, (1-2), 1 (*Chem. Abs.*, 1970, **74**, 76 149d); (c) T. Uyehara, M. Funamizu, and Y. Kitahara, *Chem. and Ind.*, 1970, 1500; (d) H. Wynberg and R. Helder, *Tetrahedron Letters*, 1971, 4317; (e) T. W. Doyle, *Canad. J. Chem.*, 1970, **48**, 1633; (f) R. M. Acheson and J. N. Bridson, *Chem. Comm.*, 1971, 1225; see also M.-S. Lin, and V. Snieckus, *J. Org. Chem.*, 1971, **36**, 645; (g) G. H. Wahl and K. Weiss, *J. Org. Chem.*, 1970, **35**, 3902; (h) D. L. Coffen, *Tetrahedron Letters*, 1970, 2633; (i) L. J. Kricka and J. M. Vernon, *Chem. Comm.*, 1971, 942; see also C. O. Bender, R. Bonnett, and R. G. Smith, *J. Chem. Soc. (C)*, 1970, 1251; (j) A. C. Day, C. G. Scales, O. J. R. Hodder, and C. K. Prout, *Chem. Comm.*, 1970, 1228; (k) J. D. Brewer, W. J. Davidson, J. A. Elix, and R. A. Leppik, *Austral. J. Chem.*, 1971, **24**, 1883; (l) A. S. Katner and E. A. Ziege, *Chem. Comm.*, 1971, 864; (m) E. M. Acton, K. J. Ryan, and L. Goodman, *Chem. Comm.*, 1970, 313; (n) Y. Hayasi, M. Kobayasi, and H. Nozaki, *Tetrahedron* 1970, **26**, 4353.

similar to the formation of benzazepines from indoles accounts for cycloaddition of nitroacetylenes to cyclic enamines, affording the ring-opened nitrocycloheptadienylamine (34), which can be hydrolysed to the cycloheptenone with acid.[44]

The product from the reaction of nitroacetylene (35) with ynamine (36) is not the stabilized cyclobutadiene (37)[45] but the isomeric nitrile oxide (38).[44] The nucleophilic ynamines also cycloadd to carbonyl groups and to

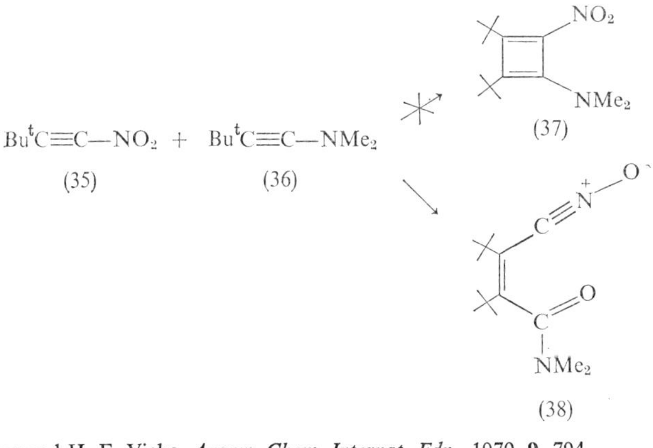

Bu^{t}C≡C—NO₂ + Bu^{t}C≡C—NMe₂

(35)   (36)

[Structures (37) cyclobutadiene with NO₂ and NMe₂ groups, crossed out; and (38) with C≡N⁺—O⁻ and C=O, NMe₂]

[44] V. Jagar and H. E. Viehe, *Angew. Chem. Internat. Edn.*, 1970, **9**, 794.
[45] R. Gompper and G. Seybold, *Angew. Chem. Internat. Edn.*, 1968, **7**, 824; M. Neuenschwander and A. Niederhauser, *Helv. Chim. Acta*, 1970, **53**, 519.

acyclic imines to give acrylamide[46] and acrylamidines,[47] respectively (Scheme 4); cyclic imines give the two-carbon ring expansion.

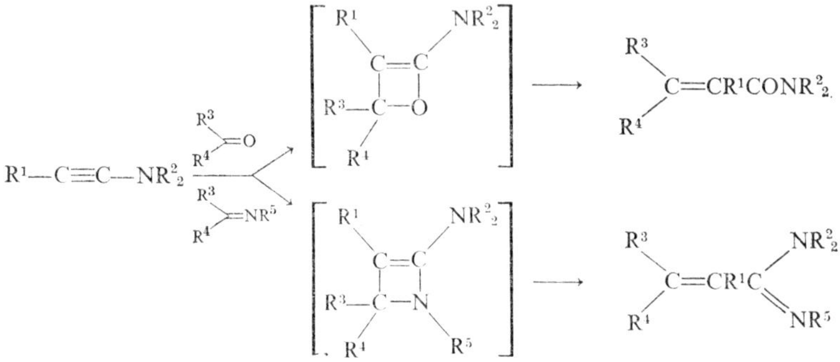

Scheme 4

Cycloaddition of acetylenes with various substituted isocyanates follows a diversity of pathways depending on the substitution of both reactants. Thus cycloaddition of the ethynylogous acid amide (39) with N-carbonylisocyanates gives oxazinones (40).[48] By contrast, one mole of N-benzoylisocyanate reacts with one mole of ethyl propiolate to yield (41), but in a 2:1 molar ratio with 3-methoxypropyne to yield (42).[49] The extremely

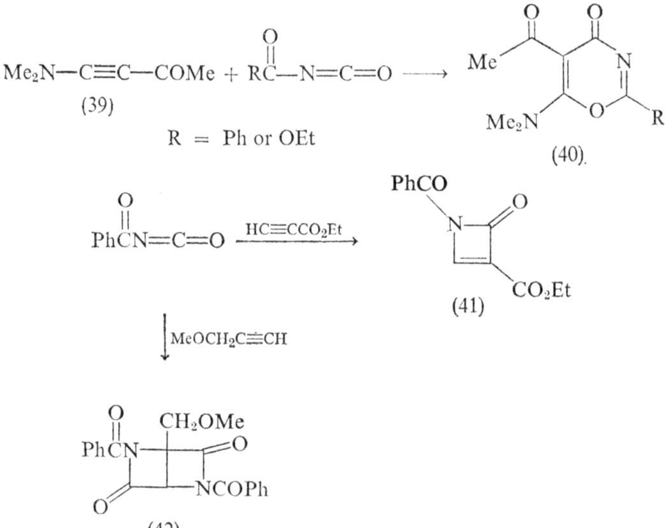

[46] R. Fuks and H. G. Viehe, *Chem. Ber.*, 1970, **103**, 564.
[47] R. Fuks and H. G. Viehe, *Chem. Ber.*, 1970, **103**, 573.
[48] H.-J. Gais and K. Hafner, *Tetrahedron Letters*, 1970, 5101.
[49] B. A. Arbuzov, N. N. Zobova, and F. B. Balabanova, *Izvest. Akad. Nauk S.S.S.R. Ser. khim.*, 1970, **7**, 1570 (*Chem. Abs.*, 1970, **74**, 76 350).

electrophilic halogenosulphonylisocyanates react differently again: (43) reacts with but-2-yne in a 2 : 1 molar ratio to give (44)[50] whereas X-ray analysis shows the structure of the product from chlorosulphonylisocyanate and hex-3-yne to be (45).[51]

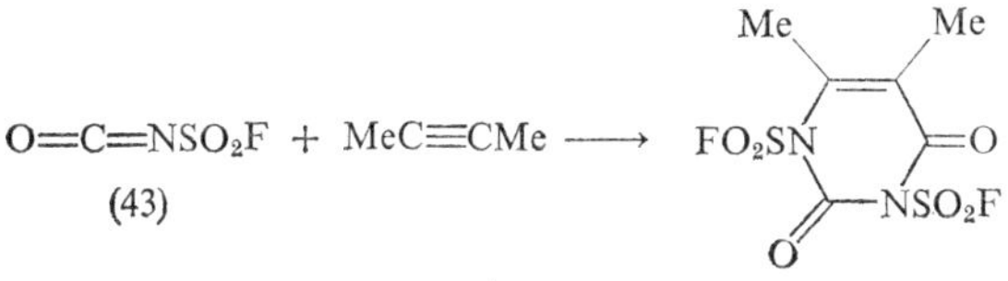

1,3-Dipolar addition of cyanoacetylene to the ylide (46) gives the cyanoindolizine (47), dehydrogenation being effected by the acetylene.[52] The 1,3-dipolar addition of azlactones (48) to DMAD proceeds *via* prior tautomerism to the mesoionic oxazolium 5-oxide (49); pyrroles are the products of subsequent elimination of carbon dioxide.[53]

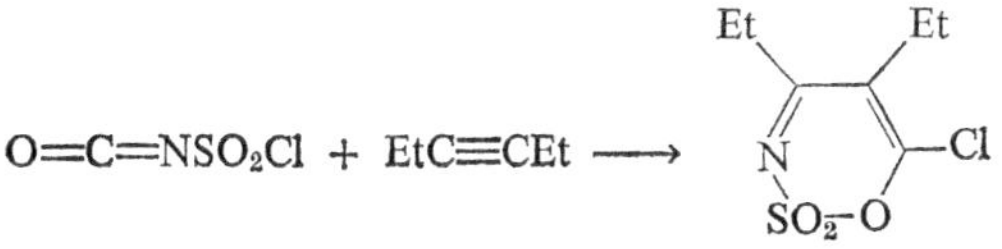

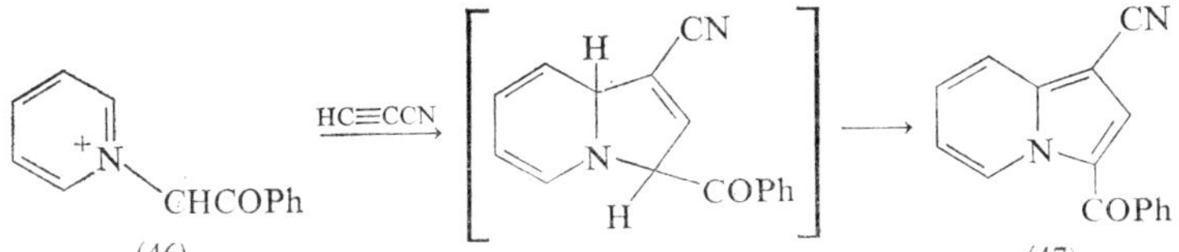

[50] K. Claus and H. Jensen, *Tetrahedron Letters*, 1970, 119.
[51] E. J. Moriconi, J. G. White, R. W. Franck, J. Jansing, J. F. Kelley, R. A. Salomone, and Y. Shimakawa, *Tetrahedron Letters*, 1970, 27.
[52] T. Sasaki, K. Kanematsu, and Y. Yukimoto, *J. Chem. Soc. (C)*, 1970, 481.
[53] H. Gotthardt, R. Huisgen, and H. O. Bayer, *J. Amer. Chem. Soc.*, 1970, **92**, 4340.

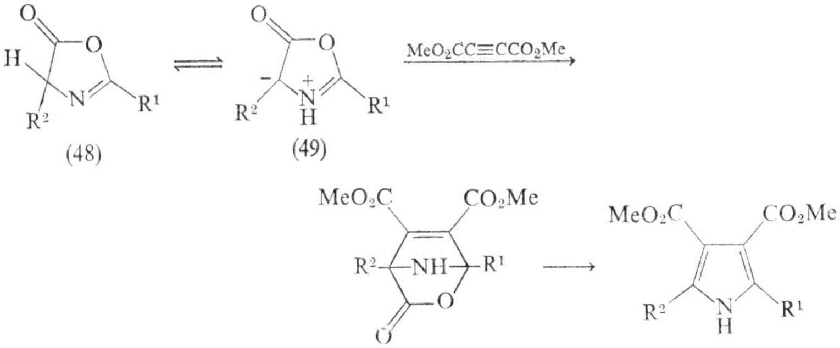

The Diels–Alder addition product (50) from butadiene and 1,2-dibenzoylacetylene is a versatile precursor for preparation of isoindoles, isobenzofurans, and isobenzothiophens.[54]

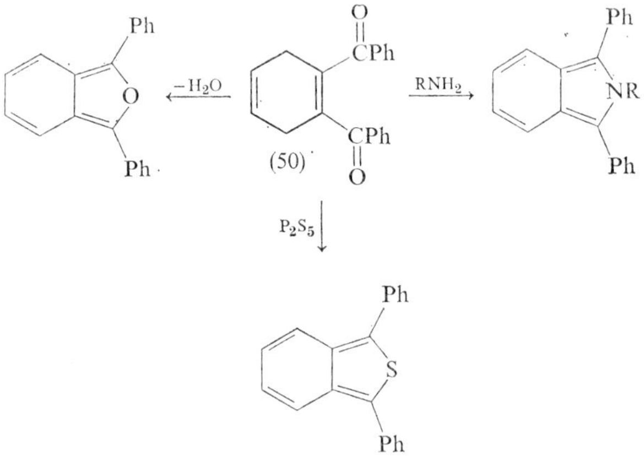

Other Additions to the Acetylenic Bond.—*Nucleophilic.* Acetophenone oxime and DMAD react to yield the 1:1 adduct (51), which undergoes a Claisen-type rearrangement on heating to the pyrrole (52).[55] Similarly, the adduct (53) from phenylamidoxime and DMAD is thermally rearranged to the imidazolone (54).[56] The driving force in both these rearrangements is the making of a C—C bond at the expense of breaking a N—O bond. Adducts

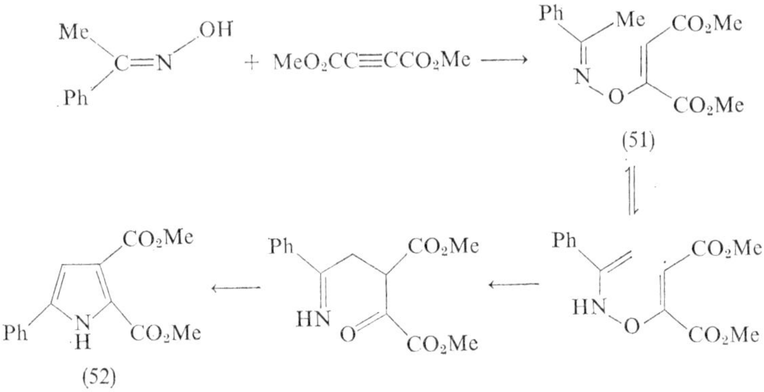

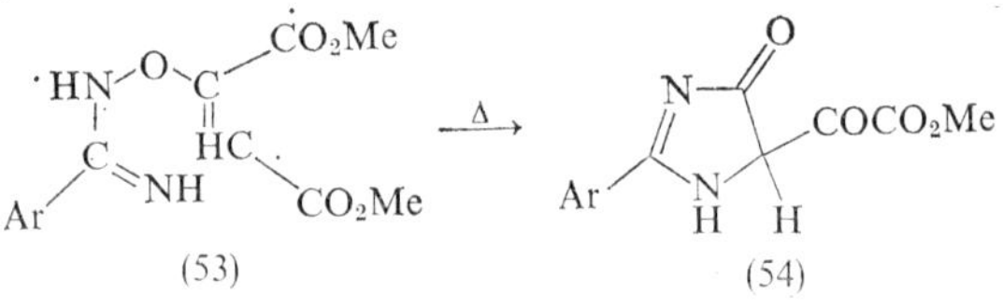

---
[54] J. D. White, M. E. Mann, H. D. Kirschenbaum, and A. Mitra, *J. Org. Chem.*, 1971, **36**, 1048.
[55] T. Sheradsky, *Tetrahedron Letters*, 1970, 25.
[56] N. D. Heindel and M. C. Chun, *Chem. Comm.*, 1971, 664.

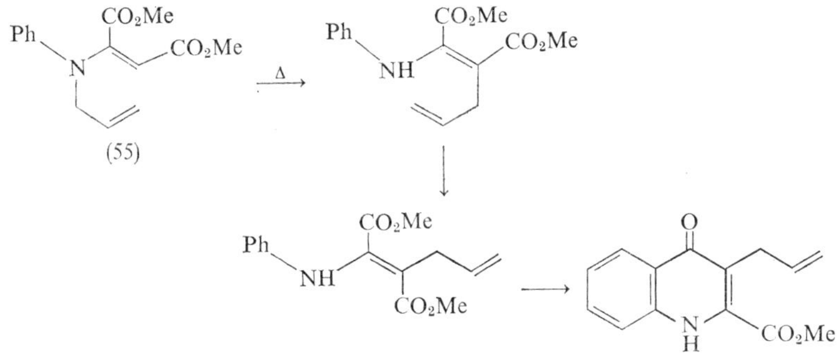

Acetylenes, Allenes, and Olefins

(55) of N-allylanilines with DMAD also undergo a hetero-Cope rearrangement to give anilinofumarates, which are converted into quinolones (56) under the conditions of the reaction.[57]

The stereochemistry of nucleophilic addition to activated triple bonds is related to the nature of the activating groups.[2] In the case of the trifluoromethyl substituted acetylenes where activation is by the inductive effect alone, addition is predominantly *trans*, e.g. methoxide-catalysed addition of methanol to hexafluorobut-2-yne gives (57).[58] *trans*-Addition is also observed in the nucleophilic addition of thiols to ethylsulphonylacetylene[59] to give (58) in a kinetically controlled reaction; isomerization at 100 °C gave the *trans* products (59).

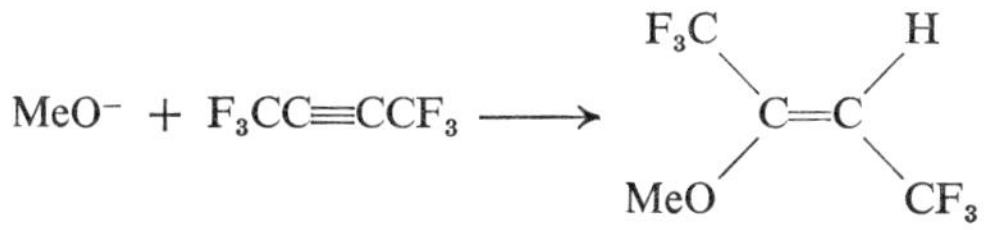

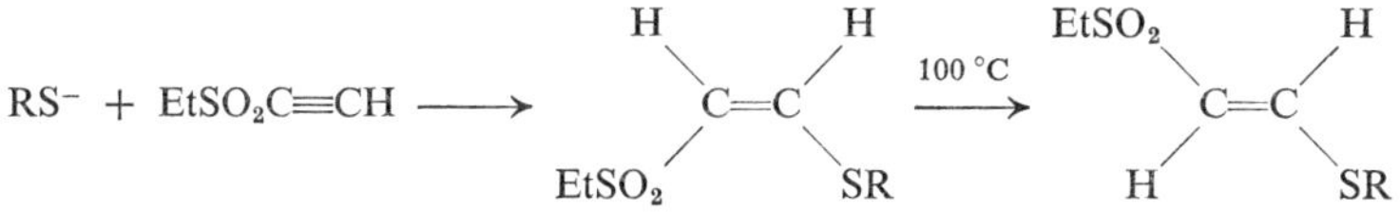

Copper alkyls, prepared from RMgBr and CuBr, react with terminal alkynes to give dienes (60) by two successive sterospecific additions. No metal–hydrogen exchange occurs in ether and the reaction can be limited to the first step; bubbling oxygen into the solution promotes the coupling reaction whose

---

[57] G. Schmidt and E. Winterfeldt, *Chem. Ber.*, 1971, **104**, 2483.
[58] E. K. Raunio and T. G. Frey, *J. Org. Chem.*, 1971, **36**, 345.
[59] E. N. Prilezhaeva, V. I. Laba, V. I. Snegotskii, and R. I. Shekhtman, *Izvest. Akad. Nauk S.S.S.R., Ser. khim.*, 1970, **7**, 1602 (*Chem. Abs.*, 1970, **74**, 3285k).

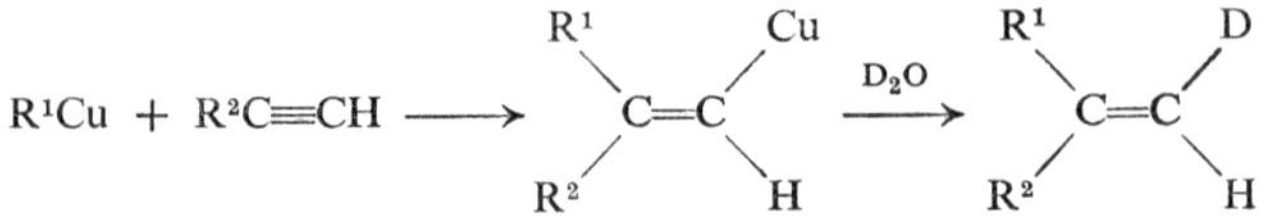

$R^1Cu + R^2C{\equiv}CH \longrightarrow \underset{(61)}{\overset{R^1Cu}{\underset{R^2H}{C{=}C}}} \xrightarrow{D_2O} \underset{(62)}{\overset{R^1D}{\underset{R^2H}{C{=}C}}}$

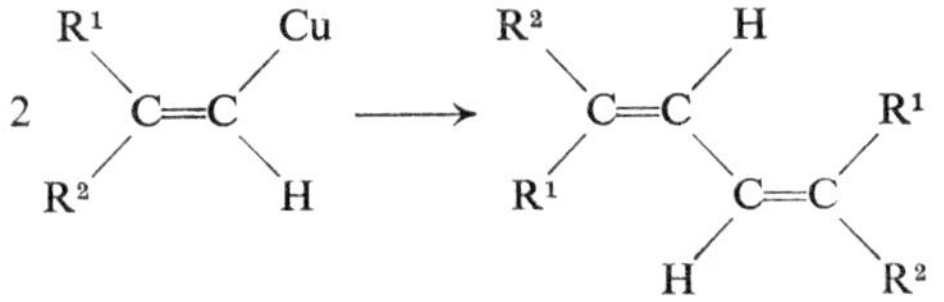

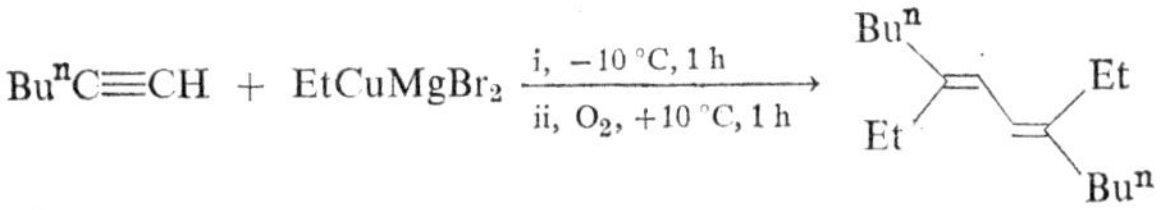

$2\ \underset{(61)}{\overset{R^1Cu}{\underset{R^2H}{C{=}C}}} \longrightarrow \underset{(60)}{\overset{R^2H}{\underset{R^1H}{C{=}C\overset{R^1}{\underset{R^2}{{-}C{=}C}}}}}$

$Bu^nC{\equiv}CH + EtCuMgBr_2 \xrightarrow[\text{ii, }O_2, +10\,°C, 1\,h]{\text{i, }-10\,°C, 1\,h}$ [product] (74%) (1)

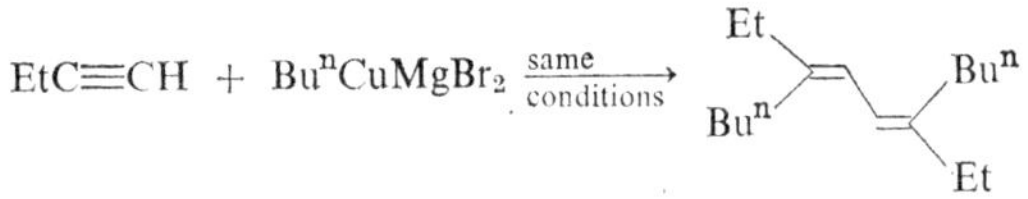

$EtC{\equiv}CH + Bu^nCuMgBr_2 \xrightarrow{\text{same conditions}}$ [product] (67%) (2)

stereospecificity is illustrated by the reactions (1) and (2). The vinylcopper intermediate (61) is identified by hydrolysis with D_2O to give the deuterio-olefin (62) or by iodination to provide iodo-olefins, mono- or di-substituted on C-2 (63) and (64). The more reactive allyl bromide may also add to the intermediate copper complex [reaction (3)].[60]

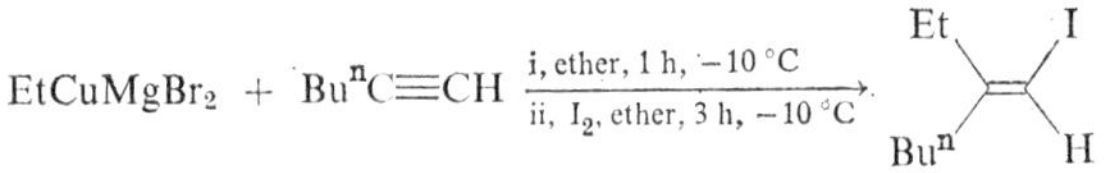

$EtCuMgBr_2 + Bu^nC{\equiv}CH \xrightarrow[\text{ii, }I_2, \text{ether, 3 h, }-10\,°C]{\text{i, ether, 1 h, }-10\,°C}$ (63) 67%

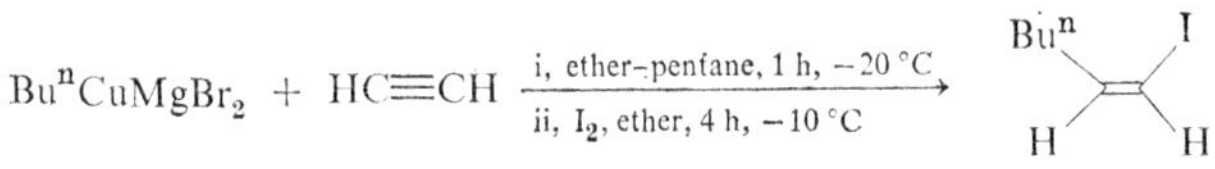

$Bu^nCuMgBr_2 + HC{\equiv}CH \xrightarrow[\text{ii, }I_2, \text{ether, 4 h, }-10\,°C]{\text{i, ether-pentane, 1 h, }-20\,°C}$ (64) 62%

$EtCuMgBr_2 + Bu^nC{\equiv}CH \xrightarrow[\text{ii, HMPT, }\diagdown\!\!\diagup\!\!\diagdown\text{Br, 15 h}]{\text{i, ether, 1 h, }-10\,°C}$ [product] 55% (3)

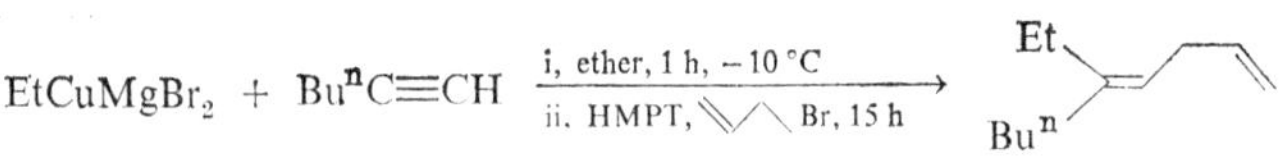

Addition of Grignard reagents to alkenes and alkynes is known to be promoted by the presence of hydroxy-functions near the multiple bonds. The

[60] J. F. Normant and M. Bourgain, *Tetrahedron Letters*, 1971, 2583.

Acetylenes, Allenes, and Olefins

stereochemistry of addition appears to be *trans* [*e.g.* (65)] with *cis*-addition products resulting from a side-reaction—probably addition to an allenol (66) formed by Grignard-promoted isomerization of the alkynol. The *trans*-addition eliminates a mechanism such as (67).[61] Similarly, suitably located tertiary amine functions can cause addition of Grignard reagents to otherwise unreactive multiple bonds, *e.g.* (68) → (69).[62]

PhC≡CCH$_2$NMe$_2$ + H$_2$C=CHCH$_2$MgCl ⟶ PhCH=CCH$_2$NMe$_2$
(68) |
                                                                                    CH$_2$C=CH$_2$
                                                                                            (69)

Electrophilic. Electrophilic addition of bromine to acetylenes has received much less attention than the corresponding reaction with olefins. At low bromine concentration and in the absence of bromide ion, the rate of bromination of ring-substituted phenylacetylenes correlates well with σ^+ values. The ρ value of -5.17 is interpreted in terms of a transition state leading to a vinyl cation intermediate (70) and, in agreement, addition is non-stereospecific.

ArC≡CR $\xrightarrow{\text{Br}_2}{\text{HOAc}}$ [Ar—C$\overset{\delta+}{=}$C$\overset{R}{\underset{\text{Br---Br}}{\diagdown}}^{\delta-}$] $\xrightarrow{-\text{Br}^-}$ ArC$\overset{+}{=}$C$\overset{R}{\diagdown}_{\text{Br}}$ ⟶ ArCBr=CBrR

(70)

Ar = p-XC$_6$H$_4$

In contrast, a cyclic bromonium ion is postulated in alkylacetylene bromination with the isolation of *trans*-dibromides from hex-3-ynes and hex-1-ynes.[63]

A kinetic and product study of the reaction of diphenylacetylenes with triphenylaluminium is interpreted as monomeric aluminium attacking the acetylene electrophilically in the rate-determining step (71); *cis*-Addition products are obtained.[64] Vinylalanes, formed by addition of aluminium

---

[61] F. W. von Rein and H. G. Richey, *Tetrahedron Letters*, 1971, 3777, 3781; H. G. Richey and S. S. Szucs, *ibid.*, p. 3785.
[62] H. G. Richey, W. F. Erickson, and A. S. Heyn, *Tetrahedron Letters*, 1971, 2183.
[63] J. A. Pincock and K. Yates, *Canad. J. Chem.*, 1970, **48**, 3332.
[64] J. J. Eisch and C. K. Hordis, *J. Amer. Chem. Soc.*, 1971, **93**, 2974, 4496.

alkyls to acetylenes, are known to add to disubstituted acetylenes to produce *trans,trans*-dienes, *e.g.* (72), after hydrolysis.[65] Terminal alkynes do not

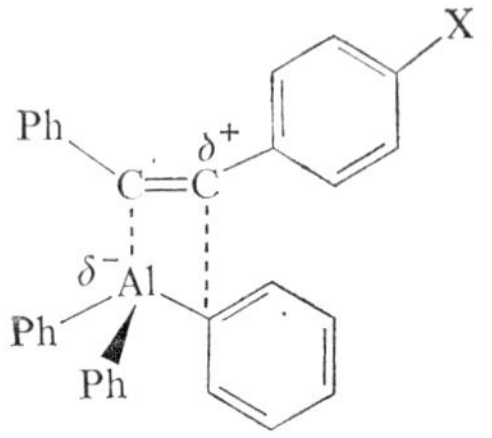

(71) X = Me_2N, MeO, *etc.*

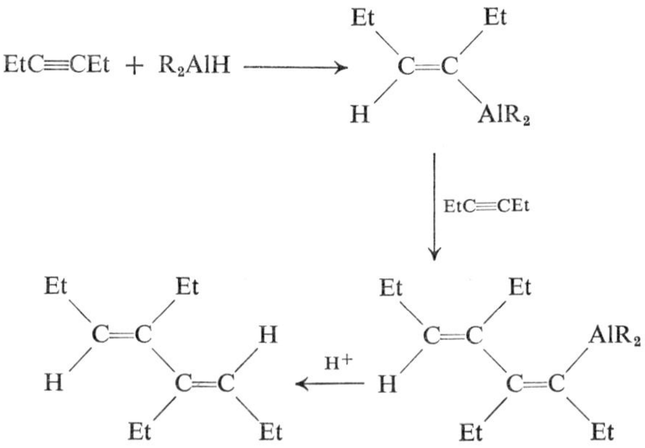

undergo this reaction owing to metalation, but the derived terminal alanes (73) are also converted by cuprous chloride into isomerically pure *trans,trans*-1,3-dienes (74).[66] Hydroalumination has also been used to convert alk-1-ynes into alkylcyclopropanes and *trans*-2-alkyl-1-halogeno-cyclopropanes (75), cleavage of the carbon–aluminium bond occurring with retention of configuration.[67]

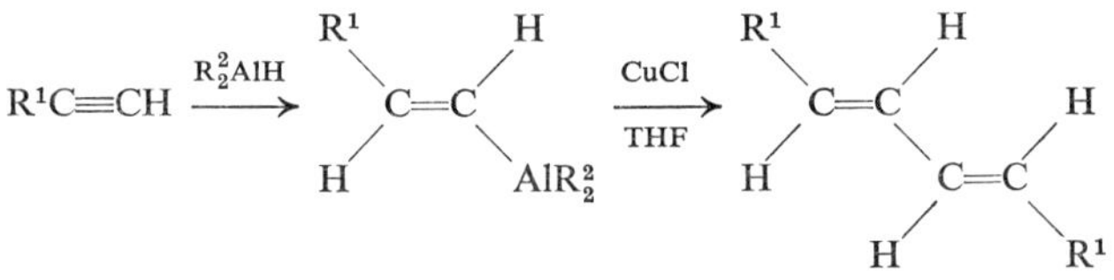

[65] G. Wilke and H. Müller, *Annalen*, 1960, **629**, 222.
[66] G. Zweifel and R. L. Miller, *J. Amer. Chem. Soc.*, 1970, **92**, 6679.
[67] G. Zweifel, G. M. Clark, and C. Whitney, *J. Amer. Chem. Soc.*, 1971, **93**, 1305.

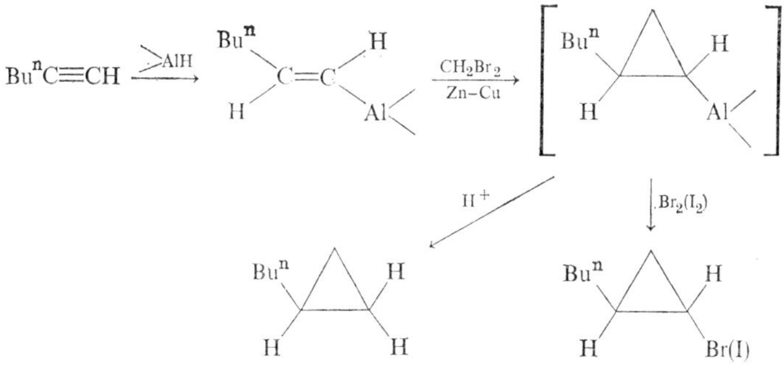

Acetylenes, Allenes, and Olefins

Monohydroboration of terminal alkynes with bulky hydroboranes (76) places the boron exclusively at the terminal position of the triple bond. Addition to unsymmetrically disubstituted alkynes is markedly affected by the triple bond substituent size. Protonolysis or oxidation with H_2O_2 converts the alkynes into alkenes and ketones (aldehydes), respectively.[68] Hydroboration of hex-1-yne with bis(*trans*-2-methylcyclohexyl)borane (77) gives the vinylborane (78). Treatment of (78) successively with 6N-NaOH and iodine gave (79) in 85% yield.[69] Migration of the cyclohexyl group occurs with retention of configuration at the ring. Inversion at the migration terminus is known from previous work.[70] A complementary method for introducing *trans*-olefinic groups on to cycloalkane rings is treatment of α-halogenovinylboranes (80) with sodium methoxide followed by protonolysis.[69]

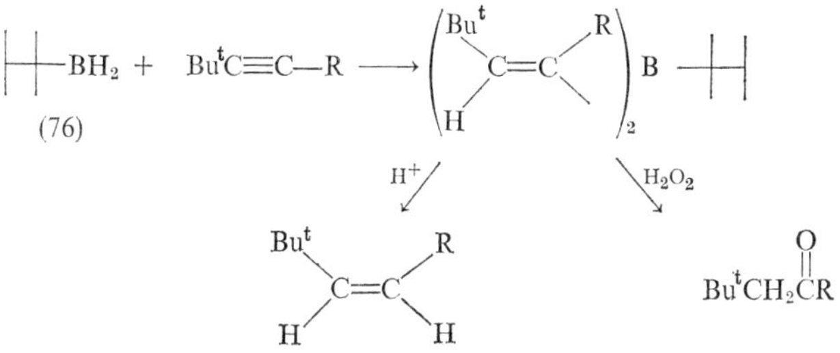

The products of addition of sulphur dichloride to diphenylacetylene are strongly dependent on the solvent. In ether, the unstable vinylsulphenyl

---

[68] G. Zweifel, G. M. Clark, and N. L. Polston, *J. Amer. Chem. Soc.*, 1971, **93**, 3395.
[69] G. Zweifel, R. P. Fischer, J. T. Snow, and C. C. Whitney, *J. Amer. Chem. Soc.*, 1971, **93**, 6309.
[70] G. Zweifel, H. Arzoumanian, and C. C. Whitney, *J. Amer. Chem. Soc.*, 1967, **89**, 3652.

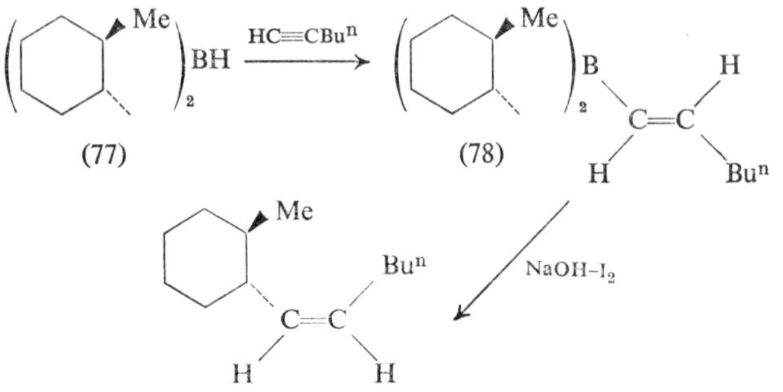

chloride (81) is formed, which readily cyclizes to the benzothiophen (82). In methylene chloride, the divinylsulphide (83) is formed exclusively.[71]

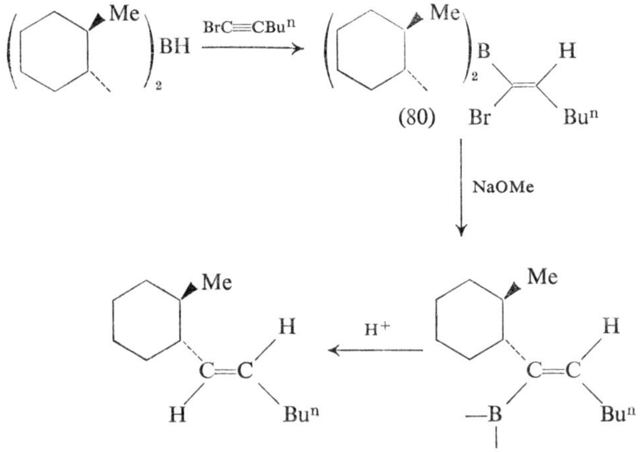

Reaction of 5-halogenopent-1-yne or 6-halogenohex-2-yne with trifluoroacetic acid gives predominantly the 1,4-halogen shifted product (84).[72] Addition of HX to the double bond is *trans* and provides a general method for

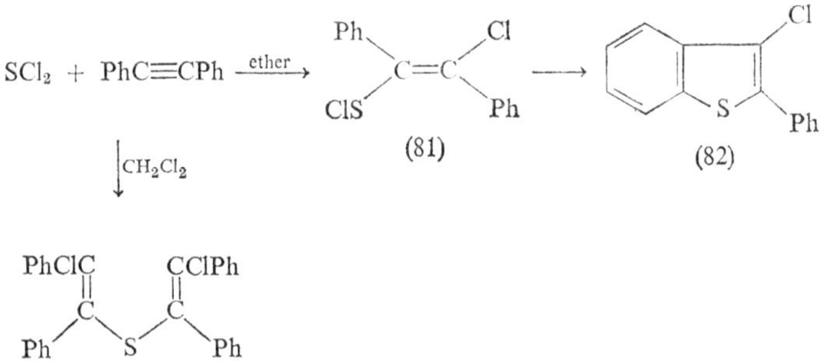

[71] T. J. Barton and R. G. Zika, *J. Org. Chem.*, 1970, **35**, 1729.
[72] P. E. Peterson, R. J. Bopp, and M. A. Ajo, *J. Amer. Chem. Soc.*, 1970, **92**, 2834.

preparation of *trans*-unsymmetrical vinyl halides, which are themselves synthetically useful.[10]

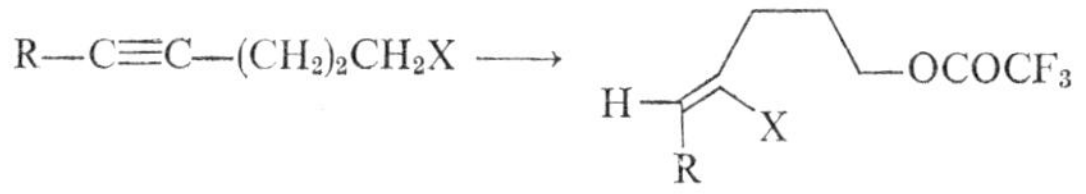

$$R-C\equiv C-(CH_2)_2CH_2X \longrightarrow$$

(84)

R = H or Me
X = halogen

In the presence of superacids, alkynes generate vinyl cations which are trapped by carbon monoxide. Thus a mixture of but-2-yne and carbon monoxide, when bubbled into FSO_3H-SbF_5 in an n.m.r. tube, gave (85), (86), and (87) (Scheme 5); (85) was explained by attack of adventitious water. Curiously, the stereochemistry of (86) implies attack of carbon monoxide upon

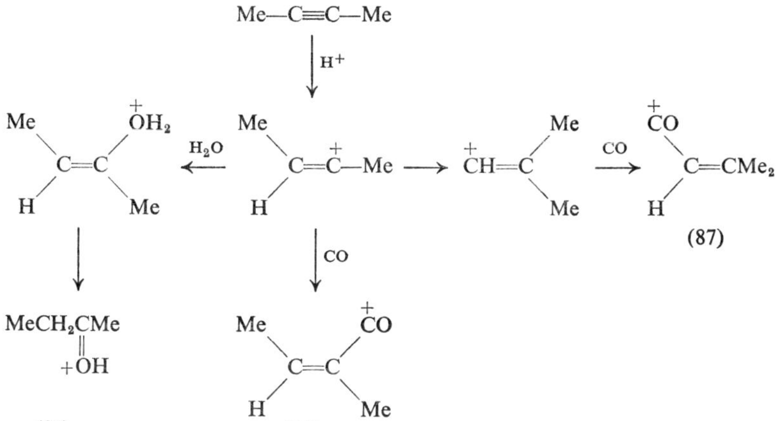

Scheme 5

the more hindered side of the (presumably linear) vinyl cation intermediate. The methyl migration required to explain (87) is also unexpected.[73] Vinyl cations are also involved in the reaction of t-butylacetylene with anhydrous HCl at room temperature (Scheme 6); (88) and (89) are minor products from cyclodimerization of the vinyl cation intermediate.[74]

Acid-catalysed intramolecular cyclization of (90) gives the enone (91). None of the isomeric (92) was obtained, although it is stable under the reaction conditions.[75]

---

[73] H. Hogeveen and C. F. Roobeek, *Tetrahedron Letters*, 1971, 3343.
[74] K. Griesbaum, Z. Rehman, and U-I. Záhorszky, *Angew. Chem. Internat. Edn.*, 1970, 9, 812.
[75] C. E. Harding and H. Hanack, *Tetrahedron Letters*, 1971, 1253.

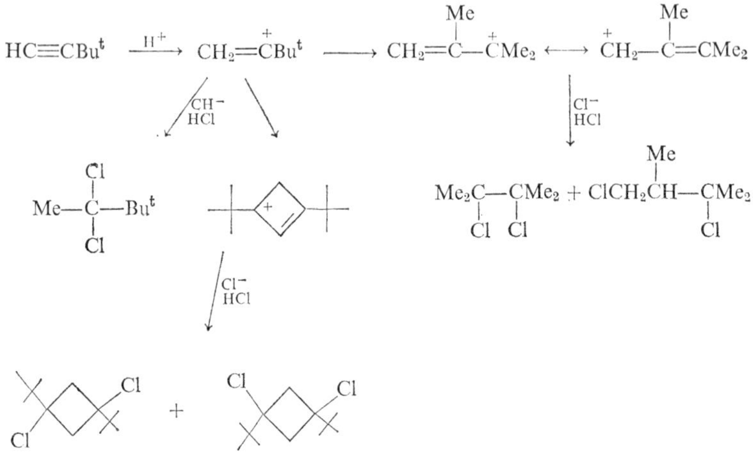

Scheme 6

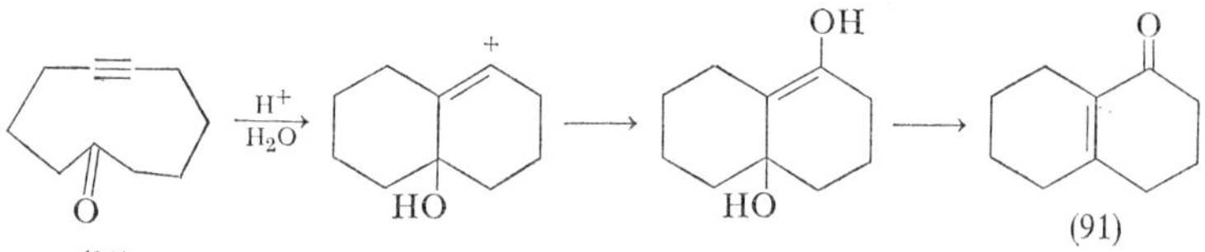

Acetylenic bond participation occurs in the biogenetic-like cyclizations (93) → (94) and (95) → (96), the vinyl cation in the former case being trapped by ethylene carbonate.[76]

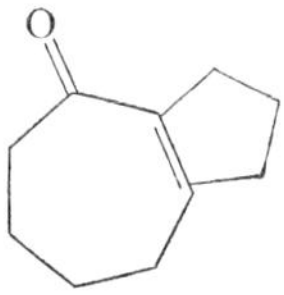

Acetolysis of 1,3-di-t-butylpropargyl tosylate (97) shows an eight-fold rate enhancement by comparison with a model compound (98). The major products were the acetate (99) and the rearranged olefins (100) and (101). None of the other products was allenic: hence the positive charge density resides mostly at C-1 in the propargyl cation.[77]

---

[76] W. S. Johnson, M. B. Gravestock, R. J. Parry, R. F. Myers, T. A. Bryson, and D. H. Miles, *J. Amer. Chem. Soc.*, 1971, **93**, 4332; W. S. Johnson, M. B. Gravestock, and B. E. McCarry, *ibid.*, p. 4334.
[77] R. S. Macomber, *Tetrahedron Letters*, 1970, 4639.

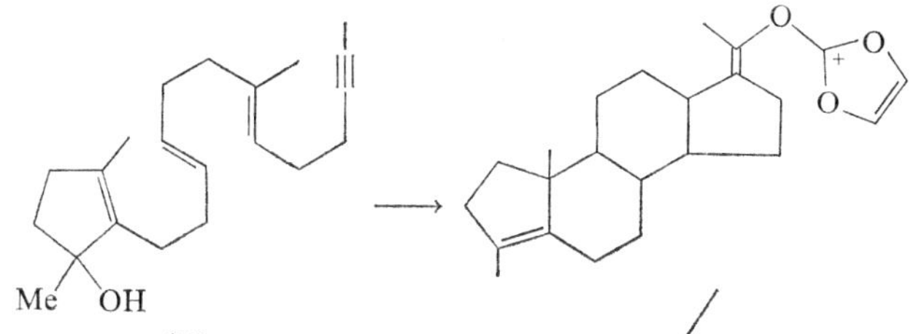

Acetylenes, Allenes, and Olefins

(93)

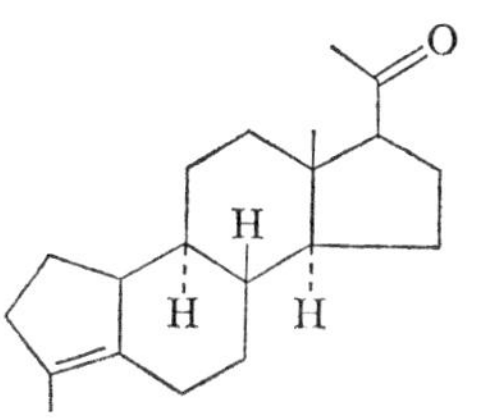

(94)

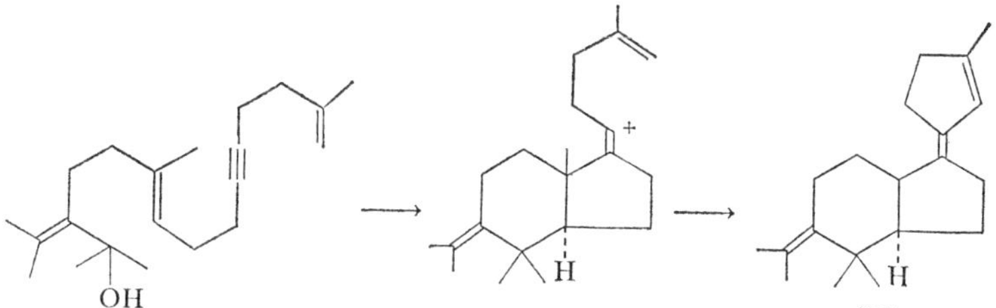

(95)

(96)

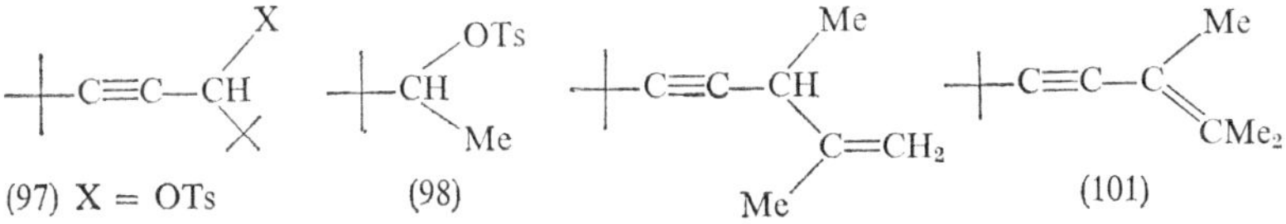

(97) X = OTs
(99) X = OAc

(98)

(100)

(101)

Radical Addition. Free-radical addition of toluene-*p*-sulphonyl iodide or methanesulphonyl iodide to acetylenes proceeds readily and stereoselectively when the two are mixed in ether and illuminated.[78] The high yields of crystalline products (102) obtained imply that in the mechanism (Scheme 7), chain transfer by the sulphonyl iodide (k_0) is much faster than isomerization of the intermediate vinyl radical (k_2). The *trans*-addition was confirmed by zinc-acetic acid reduction to the sulphone, and by X-ray analysis of one of the adducts.

---

[78] W. E. Truce and G. C. Wolf, *J. Org. Chem.*, 1971, **36**, 1727.

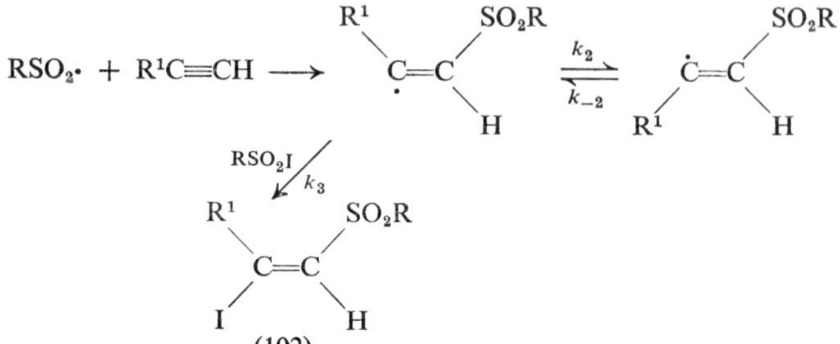

$$RSO_2\cdot + R^1C\equiv CH \longrightarrow \underset{\cdot}{\underset{H}{\overset{R^1}{C}}}=\underset{H}{\overset{SO_2R}{C}} \underset{k_{-2}}{\overset{k_2}{\rightleftharpoons}} \underset{R^1}{\overset{}{\overset{\cdot}{C}}}=\underset{H}{\overset{SO_2R}{C}}$$

$$\downarrow{\scriptstyle RSO_2I,\ k_3}$$

$$\underset{I}{\overset{R^1}{C}}=\underset{H}{\overset{SO_2R}{C}}$$
(102)

Scheme 7

The kinetically controlled radical addition of ethyl mercaptan to ethoxyacetylene is also *trans* and stereospecific at low conversions to give *cis*-1-ethoxy-2-(ethylthio)ethylene (103).[79]

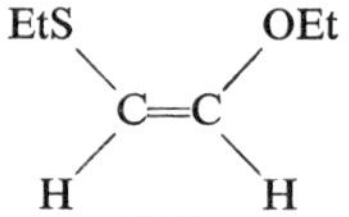

$$\underset{H}{\overset{EtS}{C}}=\underset{H}{\overset{OEt}{C}}$$
(103)

In the peroxide-catalysed reaction of pentamethyldisilane with pentamethyldisilanylacetylene, abstraction of hydrogen from an intermediate vinyl radical (104) must occur to account for acetylene (105) as one of the products.[80] Trialkylboranes are known to undergo facile 1,4-addition to many

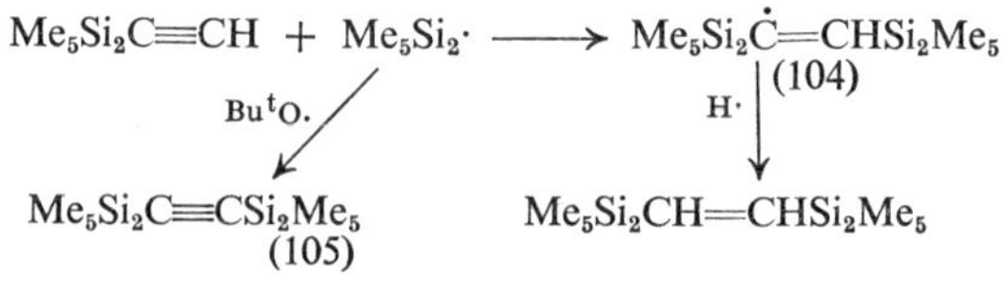

$$Me_5Si_2C\equiv CH + Me_5Si_2\cdot \longrightarrow Me_5Si_2\overset{\cdot}{C}=CHSi_2Me_5$$

$$\downarrow{\scriptstyle Bu^tO\cdot} \qquad\qquad\qquad \downarrow{\scriptstyle H\cdot}\ (104)$$

$$Me_5Si_2C\equiv CSi_2Me_5 \qquad Me_5Si_2CH=CHSi_2Me_5$$
(105)

$\alpha\beta$-olefinic carbonyl compounds. Acetylacetylene does not react spontaneously with trialkylboranes but the radical reaction, as in the case of β-substituted olefinic carbonyl compounds, may be induced by the presence of catalytic amounts of oxygen. Hydrolysis of the intermediate allenic compound (106) produces $\alpha\beta$-unsaturated methyl ketones in good yield.[81]

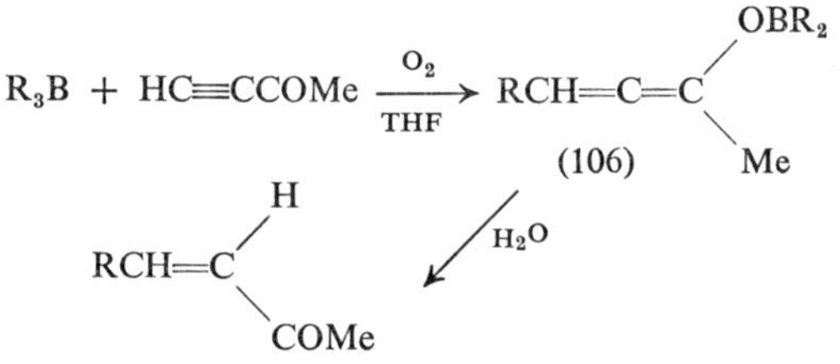

$$R_3B + HC\equiv CCOMe \xrightarrow[THF]{O_2} RCH=C=\underset{Me}{\overset{OBR_2}{C}}$$
(106)

$$RCH=\underset{COMe}{\overset{H}{C}} \xleftarrow{H_2O}$$

[79] D. K. Wedegaertner, R. M. Kopchik, and J. A. Kampmeier, *J. Amer. Chem. Soc.*, 1971, **93**, 6891.
[80] H. Sakurai and M. Yamagata, *Chem. Comm.*, 1970, 1144.
[81] A. Suzuki, S. Nozawa, M. Itoh, H. C. Brown, G. W. Kabalka, and G. W. Holland, *J. Amer. Chem. Soc.*, 1970, **92**, 3503.

Addition of benzyne to tricyclo[4,1,0,0^{2,7}]heptane (107) gives the formal [$_\sigma 2 + _\pi 2 + _\delta 2$] cycloaddition product (108) in 61% yield.[82] The mechanism is believed to involve a biradical intermediate (109) with attack on the C-1—C-7 bond taking place from inside the 'flap' as shown, a feature which is characteristic of these reactions.[83] Dicyanoacetylene reacts with (110) to form (111), and in principle attack takes place here inside the 'flap' at either a or b. The isolation of (111) shows that reaction has occurred at b and that the initial radical addition is fairly sensitive to steric effects.[82]

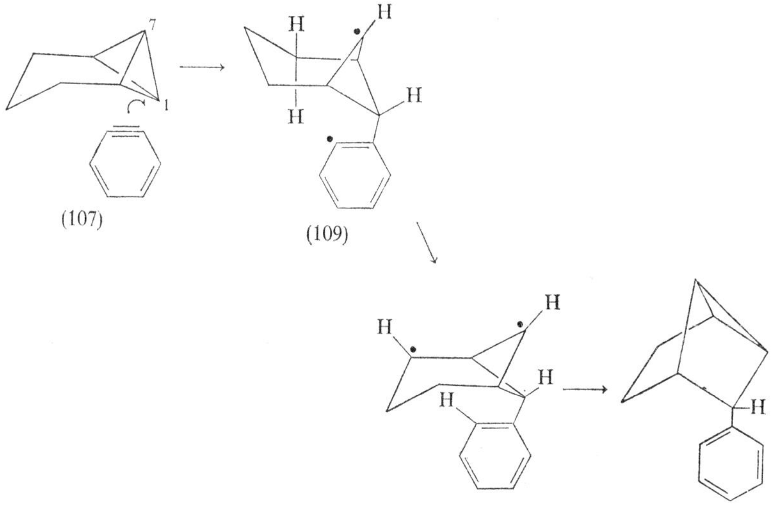

Other Reactions of Acetylenes.—Several recent papers have been concerned with the mechanism of nucleophilic displacement at acetylenic carbon.[84] In the halide displacement from halogenoacetylenes, substitution is feasible by either α-addition and β-elimination [reaction (4)] or by attack on the halogen atom with subsequent attack by the acetylide anion [reaction (5)].

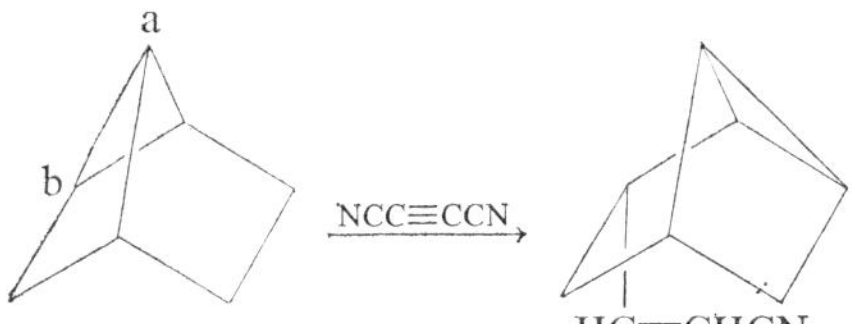

---
[82] P. G. Gassman and G. D. Richmond, *J. Amer. Chem. Soc.*, 1970, **92**, 2090.
[83] P. G. Gassman, *Accounts Chem. Res.*, 1971, **4**, 128.
[84] A. Fujii and S. I. Miller, *J. Amer. Chem. Soc.*, 1971, **93**, 3694.

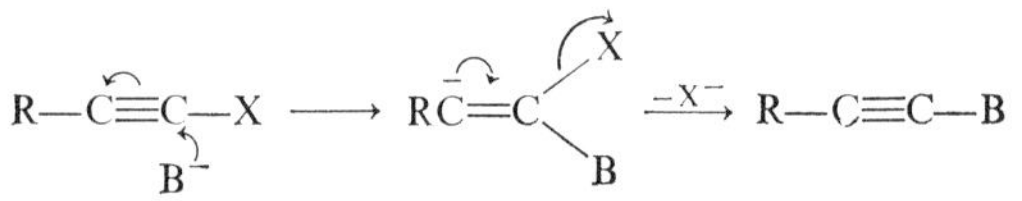

$$R-C\equiv C-X \longrightarrow RC=C\begin{matrix}X\\B\end{matrix} \xrightarrow{-X^-} R-C\equiv C-B \quad (4)$$

$$R-C\equiv C-X \longrightarrow RC\equiv C^- + BX \xrightarrow{-X^-} R-C\equiv C-B \quad (5)$$

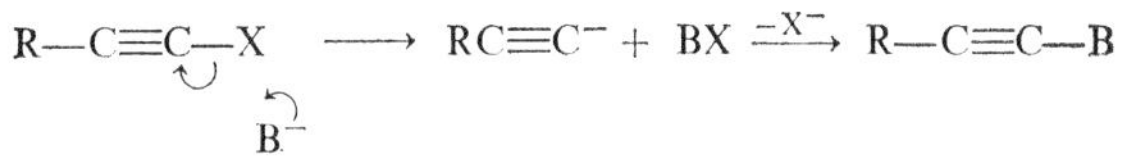

Viehe and Delavarenne have suggested[36,85] a third possibility which includes β-addition, α-elimination, and 'onium' rearrangement [reaction (6)] similar to the mechanism of the Frisch–Buttenberg–Wiechell rearrangement. Their

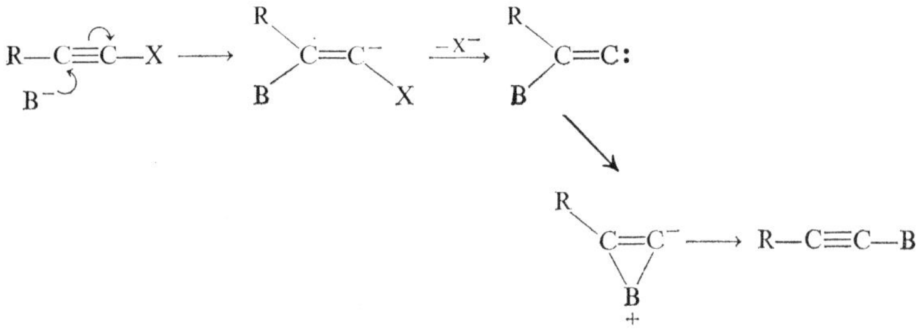

$$R-C\equiv C-X \longrightarrow \begin{matrix}R\\B\end{matrix}C=C\begin{matrix}\\X\end{matrix} \xrightarrow{-X^-} \begin{matrix}R\\B\end{matrix}C=C: $$

$$\searrow \begin{matrix}R\\B^+\end{matrix}C=C^- \longrightarrow R-C\equiv C-B \quad (6)$$

evidence is that addition of thiophenolate anion in protonic solvents to t-butylchloroacetylene gives (112) and (113). The latter are thermodynamically unstable with respect to (114) but all three isomers yield the acetylene (115) on treatment with lithium diethylamide. This acetylene is also obtained on

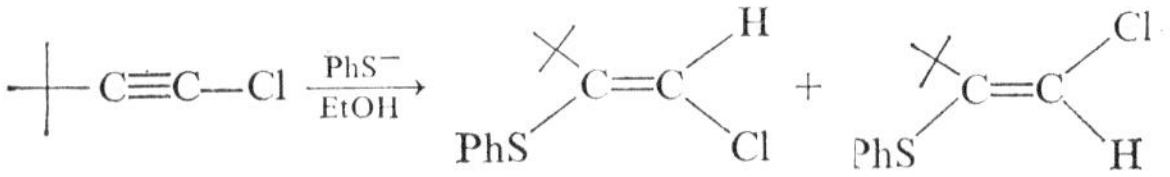

$$\text{t-Bu}-C\equiv C-Cl \xrightarrow[\text{EtOH}]{\text{PhS}^-} \begin{matrix}\text{t-Bu}\\\text{PhS}\end{matrix}C=C\begin{matrix}H\\Cl\end{matrix} + \begin{matrix}\text{t-Bu}\\\text{PhS}\end{matrix}C=C\begin{matrix}Cl\\H\end{matrix}$$

(112)   (113)

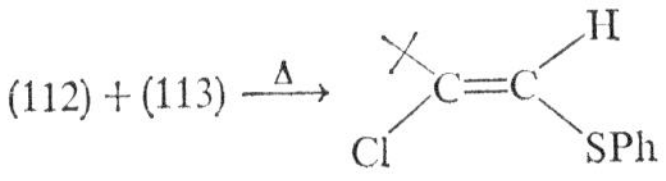

$$(112) + (113) \xrightarrow{\Delta} \begin{matrix}\text{t-Bu}\\Cl\end{matrix}C=C\begin{matrix}H\\\text{SPh}\end{matrix}$$

(114)

$$(112), (113), \text{ and } (114) \xrightarrow{\text{LiNEt}_2} \text{t-Bu}-C\equiv C-\text{SPh}$$

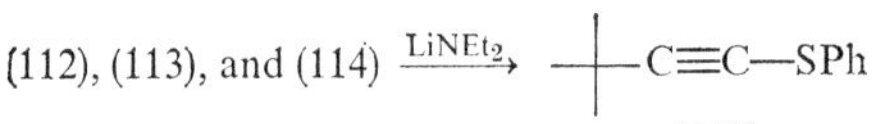

(115)

addition of thiophenolate anion to t-butylchloroacetylene in aprotic solvents, and the inference is that the mechanism in reaction (7) is operating, assuming that a drastic change in mechanism does not occur with absence of proton-donating solvents. The validity and generality of this mechanism remains to be proved.

[85] H. G. Viehe and S. Y. Delavarenne, *Chem. Ber.*, 1970, **103**, 1216.

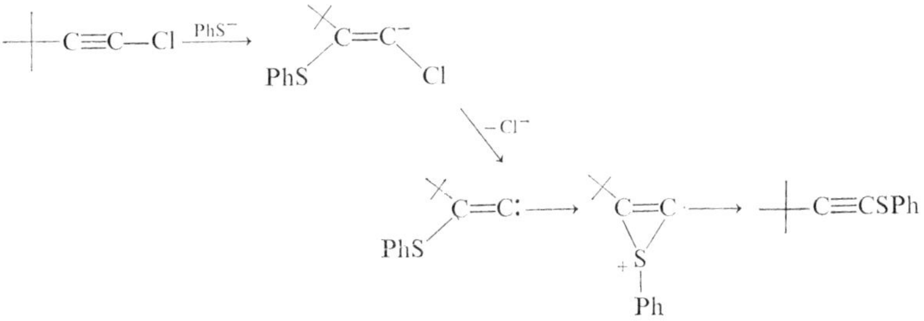

Scheme (top)

$$\text{—C≡C—Cl} \xrightarrow{\text{PhS}^-} \underset{\text{PhS}}{\overset{\times}{>}}\text{C=C}\overset{-}{<}_{\text{Cl}}$$

$$\xrightarrow{-\text{Cl}^-} \underset{\text{PhS}}{\overset{\times}{>}}\text{C=C:} \longrightarrow \overset{\times}{>}\text{C=C} \underset{\overset{|}{\text{Ph}}}{\overset{+}{\text{S}}} \longrightarrow \text{—C≡CSPh}$$

The reaction of phenylhalogenoacetylenes with trialkyl phosphite gives the acetylenic phosphonate by an Arbuzov reaction. Halogenoacetylenes are more reactive than the corresponding alkyl, aryl, or vinyl halides towards triethylphosphite, and the mechanism has been studied by using ethanol as a trapping agent.[84,86] The results were interpreted as a nucleophilic attack by phosphite on halogen [as in reaction (5)], the acetylide anion being trapped by added ethanol (Scheme 8). At least one other mechanism is operating and this

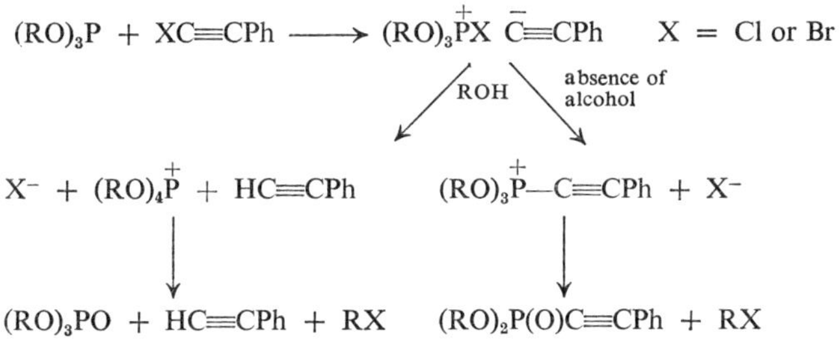

$$(RO)_3P + XC≡CPh \longrightarrow (RO)_3\overset{+}{P}X \ \overset{-}{C}≡CPh \quad X = Cl \text{ or } Br$$

with ROH / absence of alcohol branching to:

$$X^- + (RO)_4\overset{+}{P} + HC≡CPh \qquad (RO)_3\overset{+}{P}\text{—C≡CPh} + X^-$$

$$\downarrow \qquad\qquad\qquad\qquad \downarrow$$

$$(RO)_3PO + HC≡CPh + RX \qquad (RO)_2P(O)C≡CPh + RX$$

Scheme 8

is the major pathway in the case of phenylchloroacetylene in dilute THF solution. This is believed to be an attack on the α-carbon and subsequent β-elimination [reaction (4)].

Phenylethynyl ethers may be obtained from phenylchloroacetylene by treatment with alkoxide ion in aprotic solvents.[87] A detailed study of the reaction of phenylchloroacetylene has shown the latter to be triphilic towards methoxide ion with 99% of the attack occurring at the carbon bearing the chlorine. With phenylbromoacetylene the corresponding figure is 83%.[88]

Acetylenes generally react with carbenes to form cyclopropenes. Whether this is a reaction *via* a singlet carbene or a two-step process *via* a biradical cannot be determined from the product, as is the case in the trapping with olefins. Isolation of indenes (116) from the reaction of acetylenes with diphenylcarbene generated photochemically from the diazo-compound, however, was considered to be evidence for the intermediacy of a triplet carbene.

---

[86] P. Simpson and D. W. Burt, *Tetrahedron Letters*, 1970, 4799; D. W. Burt and P. Simpson, *J. Chem. Soc.* (C), 1971, 2872.
[87] R. Tanaka and S. I. Miller, *Tetrahedron Letters*, 1971, 1753.
[88] R. Tanaka, M. Rodgers, R. Simonaitis, and S. I. Miller, *Tetrahedron*, 1971, 27, 2651.

The cyclopropenes (117) were shown not to be intermediates in the reaction.[89]

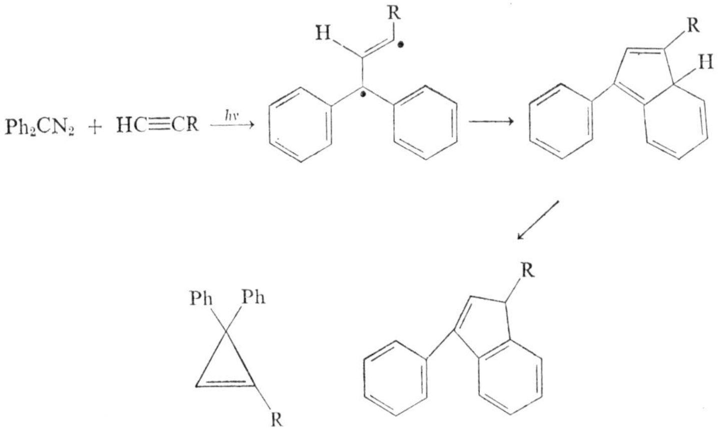

Among Cope rearrangements involving acetylenes is the thermolysis of hex-5-en-1-yn-3-ol. The products were all considered to be derived by further reaction of the primary oxy-Cope rearrangement product (118). The Arrhenius energy of 30 ± 2 kcal and ΔS^* of −14 e.u. are indicative of a concerted reaction *via* a cyclic transition state and suggest that participation of the triple bonds in electrocyclic reactions leads to increased rates by comparison with the corresponding olefins.[90] Similarly, thermolysis of β-hydroxyacetylenes (119) gives carbonyl and allenic products resulting from a retro-ene reaction (1,5-sigmatropic hydrogen shift).[91]

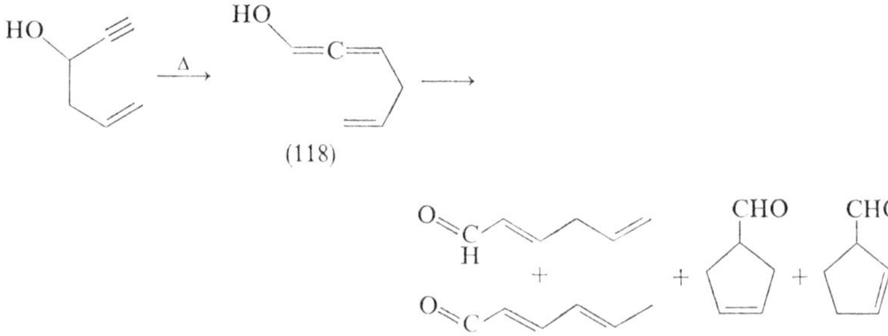

---

[89] M. E. Hendrick, W. J. Baron, and M. Jones, *J. Amer. Chem. Soc.*, 1971, **93**, 1554.
[90] A. Viola and J. H. MacMillan, *J. Amer. Chem. Soc.*, 1971, **93**, 2404.
[91] A. Viola, J. H. MacMillan, R. J. Proverb, and B. L. Yates, *J. Amer. Chem. Soc.*, 1971, **93**, 6967.

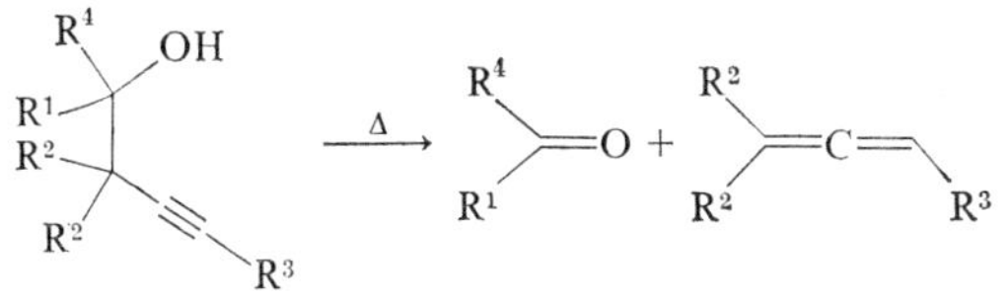

(119)

Cyclization of the aromatic acetylenic aldehyde (120) with acid gives the rearranged allenyldihydroisoquinoline (121).[92]

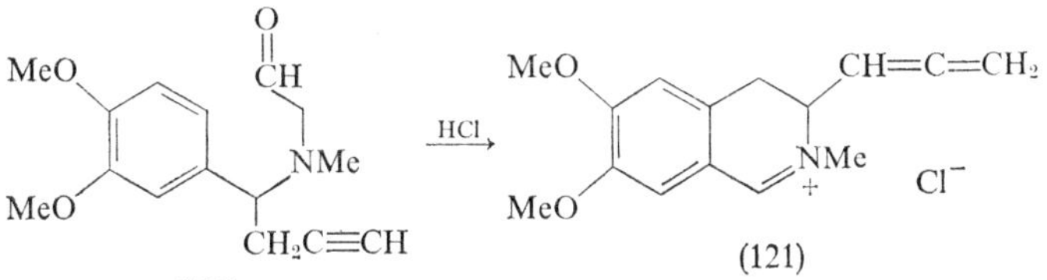

Allylboranes react with monosubstituted acetylenes and protonolysis of the resulting vinylboranes is a general method for synthesis of 1,4-substituted pentadienes (122).[93] Using alkoxyacetylenes (R = OAlk) the resulting alkoxypentadienes can be hydrolysed to the non-conjugated enones.

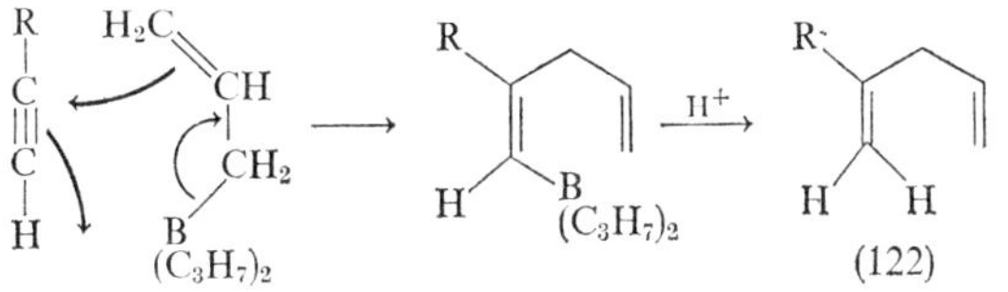

Oxidation of acetylenes with N-bromosuccinimide in DMSO gives α-dicarbonyl compounds in some cases. The yield using diphenylacetylene is practically quantitative.[94] Oxidation with peracid has been studied using di-t-butylacetylene, and the products obtained compared with those obtained by decomposition of the related diazoketone (123). The primary products from the former were the αβ-unsaturated ketone (124) and the cyclopropylketone (125). An almost identical ratio of products (124) and (125) was

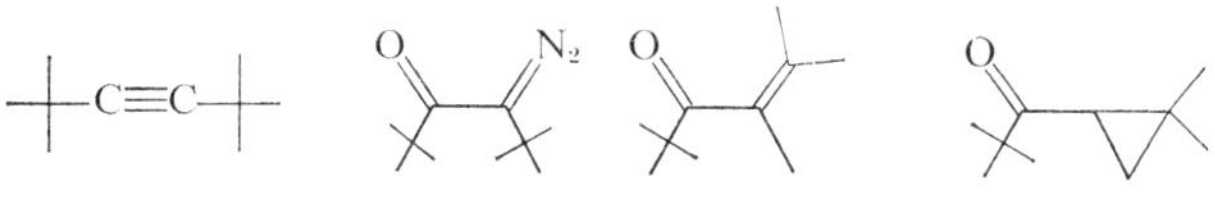

[92] M. Sainsbury, S. F. Dyke, D. W. Brown, and R. G. Kinsman, *Tetrahedron*, 1970, **26**, 5265.
[93] B. M. Mikhailov, Yu. N. Bubnov, S. A. Korobeinikova, and S. I. Frolov, *J. Organometallic Chem.*, 1971, **27**, 165.
[94] S. Wolfe, W. R. Pilgrim, T. F. Garrard, and P. Chamberlain, *Canad. J. Chem.*, 1971, **49**, 1099.

obtained from decomposition of the diazoketone, implying a common intermediate ketocarbene for both reactions. Although similar products were obtained from analogous reactions using (126) and cyclodecyne (127), the distributions were different.[95]

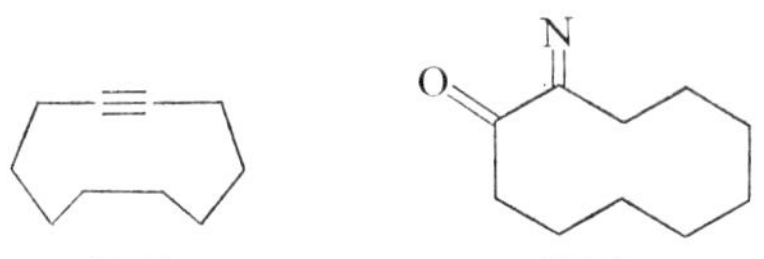

(127)   (126)

Oxidation of secondary–tertiary acetylenic glycols (128) provides a route to furanones (129).[96]

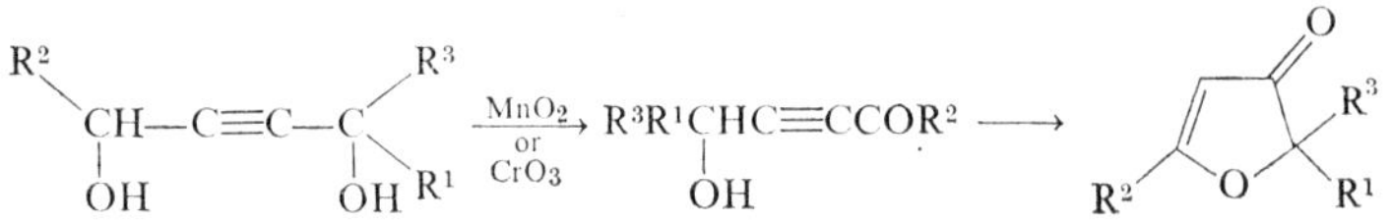

(128) → $R^3R^1CHC{\equiv}CCOR^2$ → (129)

An attempt to convert the aldehyde (130) into the corresponding ester (131) by the method of Corey[97] gave, in addition, the ester (132) and its transformation products. Ester (132) was also obtained in the absence of MnO_2, and the mechanism via (133) is supported by studies using NaCN in MeOD.[98]

$Ph-C{\equiv}C-CH{=}CHCHO \xrightarrow[MnO_2]{MeOH-CN^-} Ph-C{\equiv}C-CH{=}CHCO_2Et$
(130)   (131)

+ $Ph-CH{=}C{=}CHCH_2CO_2Me$
(132)

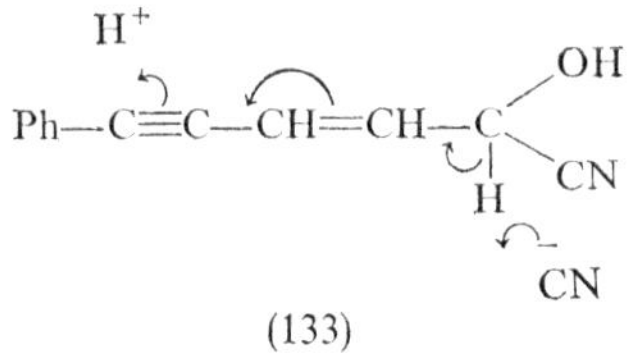

(133)

Oxidation of internal acetylenes with ruthenium tetroxide gives the corresponding diketones with some of the acids resulting from cleavage.[99]

DMAD reacts with ethylene trithiocarbonate (134) to give ethylene and the 1,3-dithiole-2-thione (135).[100] Further study has indicated that acetylenes

[95] J. Ciabattoni, R. A. Campbell, C. A. Renner, and P. W. Concannon, *J. Amer. Chem. Soc.*, 1970, **92**, 3826.
[96] M. F. Shostakovskii, T. A. Favorskaya, A. S. Medvedeva, L. P. Safronova, and V. K. Voranov, *Zhur. org. khim.*, 1970, **6**, 2377 (*Chem. Abs.*, 1970, **74**, 64 134g).
[97] E. J. Corey, N. W. Gilman, and B. E. Ganem, *J. Amer. Chem. Soc.*, 1968, **90**, 5616.
[98] P.-H. Bonnet and F. Bohlmann, *Chem. Ber.*, 1971, **104**, 1616.
[99] H. Gopal and A. J. Gordon, *Tetrahedron Letters*, 1971, 2941.
[100] D. B. J. Easton and D. Leaver, *Chem. Comm.*, 1965, 585.

Acetylenes, Allenes, and Olefins

bearing electron-withdrawing groups and the —C(=S)S— unit are necessary for the success of this reaction. Thus (136) reacts to give (137), but the SS'-ethylene dithiocarbonate was unreactive. Bromocyanoacetylene reacts anomalously to give (138). No mechanisms have been advanced for these reactions.[101]

$$\underset{X}{\overset{S}{\diagup}}\!\!=\!Y + MeO_2CC\!\equiv\!CCO_2Me \longrightarrow \underset{MeO_2C}{\overset{MeO_2C}{\diagup}}\!\!\underset{X}{\overset{S}{\diagdown}}\!\!=\!Y + C_2H_4$$

(134) X = Y = S (135) X = Y = S
(136) X = O, Y = S (137) X = S, Y = O

(138)

t-Butyl isocyanide reacts with acetylenes in the presence of nickel acetate, and the pyrrole derivatives (139) are formed in high yield.[102]

$$R^1C\!\equiv\!CR^2 + Me_3CNC \xrightarrow{Ni(OAc)_2}$$

(139)

It is often difficult to induce selective addition to double bonds in the presence of triple bonds because of the reactivity of the latter. Dicobalt octacarbonyl selectively adds to a triple bond in the presence of a double bond and allows selective transformation of the non-co-ordinated olefinic bond.[103] Removal of the protecting metal is simple. Thus vinylacetylenes when treated with strong acids usually form the products of hydration of the triple bond, and the ene-yne-ol (140) reacts with fluoroboric–acetic acid at 25 °C for 24 h to form an intractable mixture. Its complex (141) reacted at 0 °C for 15 min to give 91 % of (142) on work-up, which implies a metal-stabilized carbonium ion intermediate. Oxidative degradation with $Fe(NO_3)_3$ generates the acetylene (143) in excellent yield.

On treatment with hydrobromic–acetic acids the diol (144) gives the dibromotetraene (145). The corresponding methoxylated analogue (146), however, yields the benzofuranyl acetylene (147) after chromatography over alumina. Allene (148) was shown to be an intermediate in this reaction and was converted into (147) by alumina.[104]

---
[101] B. R. O'Connor and F. N. Jones, *J. Org. Chem.*, 1970, **35**, 2002.
[102] M. Jautelat and K. Ley, *Synthesis*, 1970, 593.
[103] K. M. Nichols and R. Pettit, *Tetrahedron Letters*, 1971, 3475.
[104] S. Kobayashi, M. Shinya, and H. Taniguchi, *Tetrahedron Letters*, 1971, 71.

A study of the metallation of 1-phenylpropyne (149) has shown that the dianion produced is different from that prepared from 3-phenylpropyne and

$$\text{PhC}\equiv\text{CCH}_3 \xrightarrow{\text{BuLi}} \text{PhC}\equiv\text{CCH}^{2-} \rightleftharpoons \text{Ph}\bar{\text{C}}\text{HC}\equiv\bar{\text{C}} \underset{\text{BuLi}}{\overset{\text{PhCH}_2\text{C}\equiv\text{CH}}{\diagdown}} \text{PhCH}=\text{C}=\text{CH}_2$$
(149)   (150)

phenylallene. Protonation of (150) with D_2O yielded (151), hence a hydrogen transfer takes place at some stage of the reaction. This transfer does not take place either at the dianion or at the final product stage and must therefore occur at the monoanion step. It is attributed to the rearrangement (152) → (153), an allowed intramolecular 1,3-hydrogen shift in a propargylic anion.[105]

(150) ⟶ PhCHDC≡CD   PhCD=C=$\bar{\text{C}}$H ⟶ PhCHDC≡C⁻ ⟶ (151)
         (151)   (152)   (153)

All the hydrogens in penta-1,3-diyne can be removed by n-butyl-lithium in the presence of $NNN'N'$-tetramethylethylenediamine to give C_5Li_4 which, on treatment with water, gives penta-1,3-diyne, penta-1,4-diyne, and penta-1,2-diene-4-yne.[106]

The hindered acetylene (154) has been synthesized, and its radical anion and dianion were studied by e.s.r. spectroscopy. A comparison was made with dimesitylacetylene, in which complete coplanarity between the aromatic rings is possible.[107]

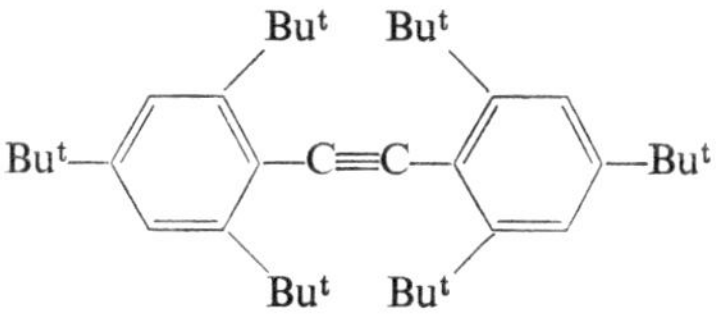

(154)

Studies of electronic charge shift in acetylenes have been made using ^{13}C n.m.r. spectra.[108] The isolation, characterization, and synthesis of many naturally occurring acetylenes and polyacetylenes have been reported by Bohlmann and his group.[109]

---

[105] J. Klein and S. Brenner, *Tetrahedron*, 1970, **26**, 2345.
[106] R. West and T. L. Chwang, *Chem. Comm.*, 1971, 813.
[107] H. E. Zimmermann and J. R. Dodd, *J. Amer. Chem. Soc.*, 1970, **92**, 6507.
[108] D. Rosenberg and W. Drenth, *Tetrahedron*, 1971, **27**, 3893.
[109] F. Bohlmann *et al.*, *Chem. Ber.*, 1970, **103**, 561, 834, 1879, 1886, 2095, 2100, 2245, 2327, 2853, 2856, 2860, 3414; *ibid.*, 1971, **104**, 11, 954, 958, 961, 1322, 1329, 1362, 1962, 2030, 2354.

2 Allenes

Synthesis.—A usually reliable method of preparing allenes is by the insertion of cyclopropylidene carbenes (carbenoids). However, treatment of tetrasubstituted *gem*-dibromocyclopropanes with methyl-lithium gives bicyclobutanes with allenes as minor products.[110] The dibromocyclopropane (155) yields only the bicyclobutane (156) and no allene is detectable, whereas the isomeric dibromocyclopropane (157) yields, in addition to the bicyclobutane (158), appreciable amounts of allene (159). This is attributed to a lengthening of the C-2—C-3 bond in an unsymmetrical transition state with positive charge better accommodated on a carbon bearing two phenyl groups.

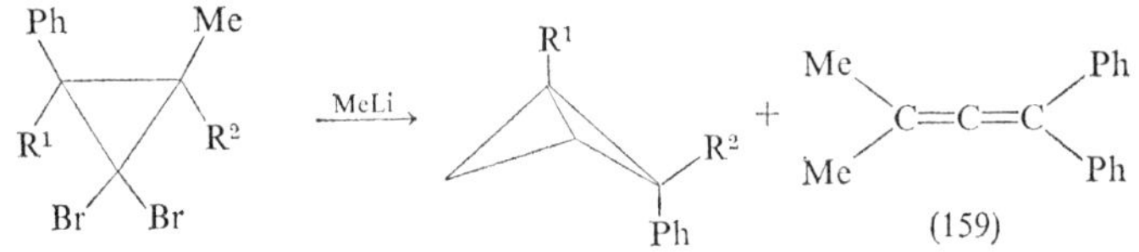

(155) $R^1 = Me, R^2 = Ph$
(157) $R^1 = Ph, R^2 = Me$

(156) $R^1 = Ph, R^2 = Me$
(158) $R^1 = Me, R^2 = Ph$

The acetylene (160), derived from phenyldibromocyclopropane (161) with excess methyl-lithium, is now believed to be the product of the reaction of methyl bromide with the dianion of the allene generated *in situ*.[111] The acetylene (160) appears to be the kinetically formed product on neutralization but may be converted into the allene (162) by shaking with powdered potassium hydroxide in hexane.

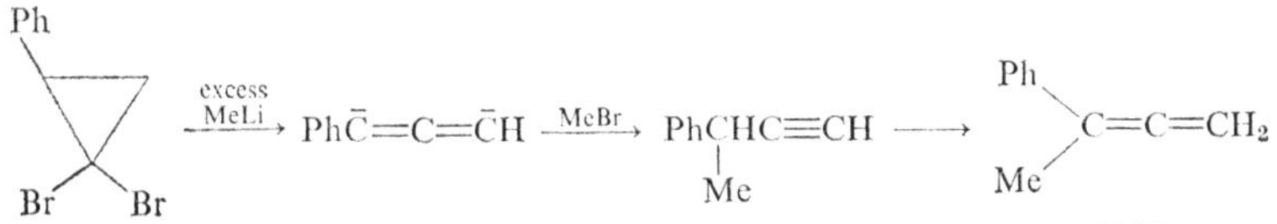

$$Ph\bar{C}=C=\bar{C}H \xrightarrow{MeBr} PhCHC\equiv CH \longrightarrow$$

Decomposition of the *N*-nitroso-oxazolidone (163) with base in a solution of ethoxyacetylene gives the allene diethylacetal (164); the carbene intermediate in the suggested mechanism is supported by previous work.[112]

An attempt was made to prepare diphenyldiazoallene (165) by base decomposition of the dinitrosourethane (166). The major product was the

---

[110] W. R. Moore and J. B. Hill, *Tetrahedron Letters*, 1970, 4553.
[111] E. V. Dehmlow and G. C. Ezimora, *Tetrahedron Letters*, 1971, 563; E. V. Dehmlow and G. C. Ezimora, *ibid.*, p. 1599.
[112] M. S. Newman and C. D. Beard, *J. Org. Chem.*, 1970, **35**, 2412; M. S. Newman and A. O. M. Okorodudu, *ibid.*, 1969, **34**, 1220.

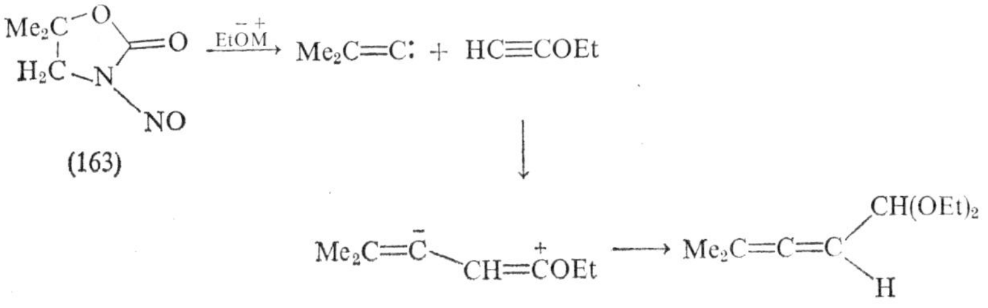

$$\underset{(163)}{\overset{Me_2C-O}{\underset{H_2C-N}{|}}\diagdown\underset{NO}{N}}{=}O \xrightarrow{\text{EtOM}} Me_2C=\overset{-\ +}{C:} + HC{\equiv}COEt$$

$$Me_2C{=}\bar{C}{\diagdown}_{CH=\overset{+}{C}OEt} \longrightarrow Me_2C{=}C{=}C\diagup^{CH(OEt)_2}_{H}$$
$$(164)$$

diazopropanone (167), believed to be derived from an oxadiazole and confirming that at least the first steps of the anticipated reaction had occurred.[113] The presence of at least a small quantity of the evidently unstable (165) as one of the reaction products was indicated by the trapping of the derived carbene, giving the adduct (168) with tetramethylethylene.

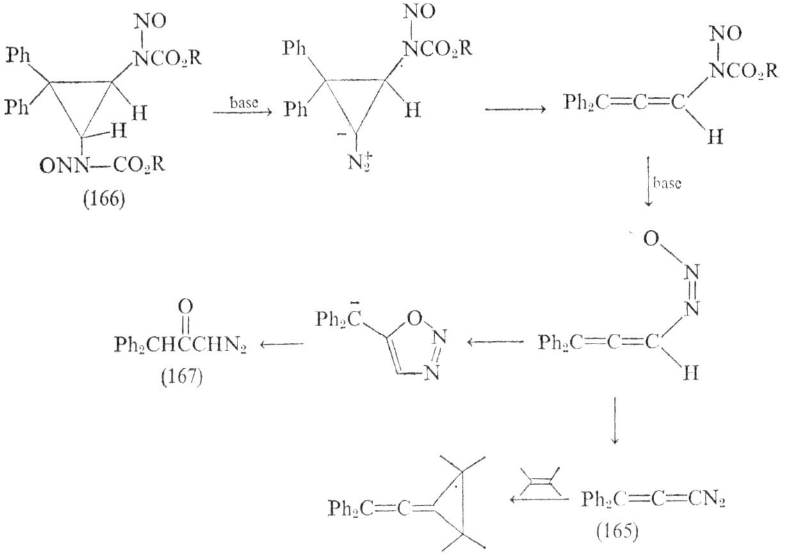

By analogy with the preparation of ketens from acid chlorides, the γ-chlorovinylaldehydes (169)—vinylogous acid chlorides—were treated with triethylamine and yielded the allenic aldehydes (170).[114]

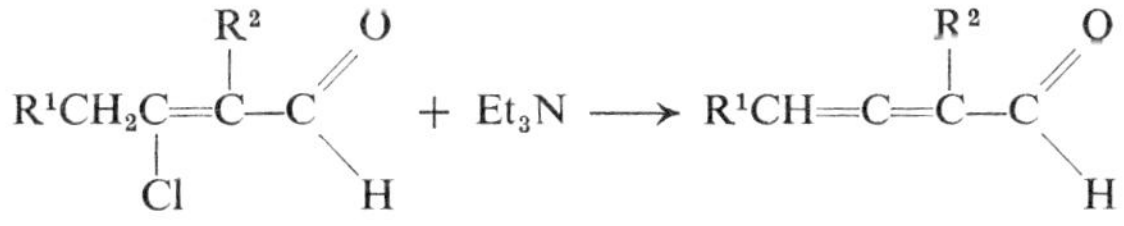

$$\underset{(169)}{R^1CH_2\underset{Cl}{\overset{R^2}{\underset{|}{C}}}{=}C{-}C\diagup^{O}_{H}} + Et_3N \longrightarrow \underset{(170)}{R^1CH{=}C{=}\overset{R^2}{\underset{|}{C}}{-}C\diagup^{O}_{H}}$$

[113] D. J. Northington and W. M. Jones, *Tetrahedron Letters*, 1971, 317.
[114] E. Schelhorn, H. Frischleder, and S. Hauptmann, *Tetrahedron Letters*, 1970, 4315.

Allenes of the type (171) have been prepared by cycloaddition of ynamines with carbon dioxide.[115] The reaction with diethylaminopropyne is complete in one hour at $-60\,^\circ$C and the only contaminant is a small quantity of the aminocyclobutenone (172).

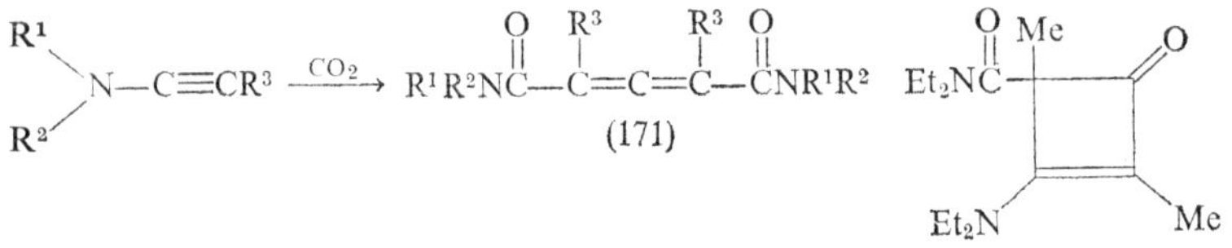

$$R^1R^2N-C\equiv CR^3 \xrightarrow{CO_2} R^1R^2N\overset{O}{\underset{\|}{C}}-\overset{R^3}{\underset{|}{C}}=C=\overset{R^3}{\underset{|}{C}}-\overset{O}{\underset{\|}{C}}NR^1R^2$$
(171)

(172)

1,1-Dialkyl-3-iodoallenes are efficiently prepared from 1,1-dialkylprop-2-yn-1-ols [*e.g.* (173) → (174)] using a two-phase system, the product being extracted into petroleum.[116]

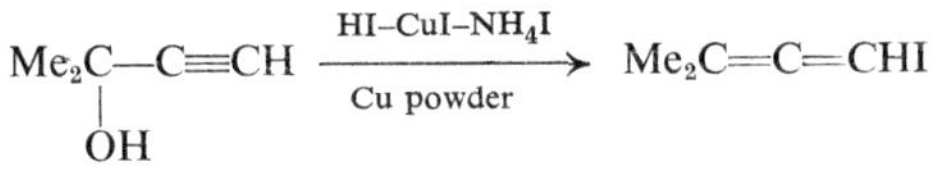

$$Me_2\underset{\underset{OH}{|}}{C}-C\equiv CH \xrightarrow[\text{Cu powder}]{HI-CuI-NH_4I} Me_2C=C=CHI$$

(173)   (174)

Dibrominated diarylallenes (175) are readily derivatized *via* the carbenoid formed with butyl-lithium. The acid (176) appears to be formed in the work-up from the bromide (177), which seems to be readily solvolysed.[117]

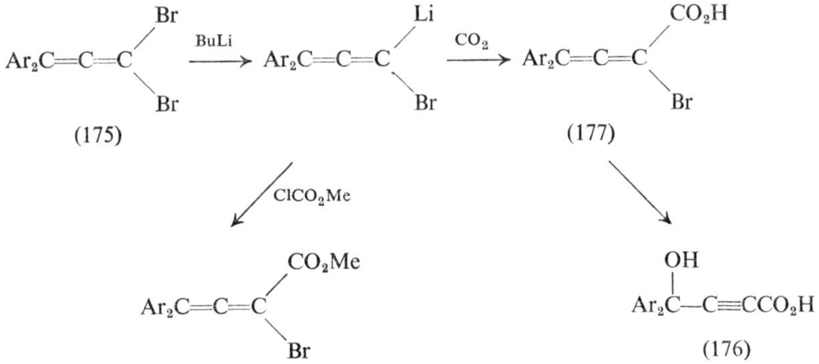

$$Ar_2C=C=C\underset{Br}{\overset{Br}{\diagup}} \xrightarrow{BuLi} Ar_2C=C=C\underset{Br}{\overset{Li}{\diagup}} \xrightarrow{CO_2} Ar_2C=C=C\underset{Br}{\overset{CO_2H}{\diagup}}$$

(175)   (177)

↓ ClCO₂Me

$$Ar_2C=C=C\underset{Br}{\overset{CO_2Me}{\diagup}}$$

$$Ar_2\underset{\underset{OH}{|}}{C}-C\equiv CCO_2H$$
(176)

Other allene syntheses employing acetylenic starting materials include the preparation of β-allenic esters (178) *via* an intramolecular Claisen rearrangement;[118] the rearrangement of acetylenic esters of sulphinic acid (179) to

---

[115] J. Ficini and J. Pouliquen, *J. Amer. Chem. Soc.*, 1971, **93**, 3295.
[116] P. M. Greaves, M. Kalli, P. D. Landor, and S. R. Landor, *J. Chem. Soc.* (C), 1971, 667; see also J. Gore and M.-L. Roumestant, *Tetrahedron Letters*, 1971, 1027.
[117] G. Köbrich and E. Wagner, *Angew. Chem. Internat. Edn.*, 1970, **9**, 524.
[118] J. K. Crandall and G. L. Tindell, *Chem. Comm.*, 1970, 1411.

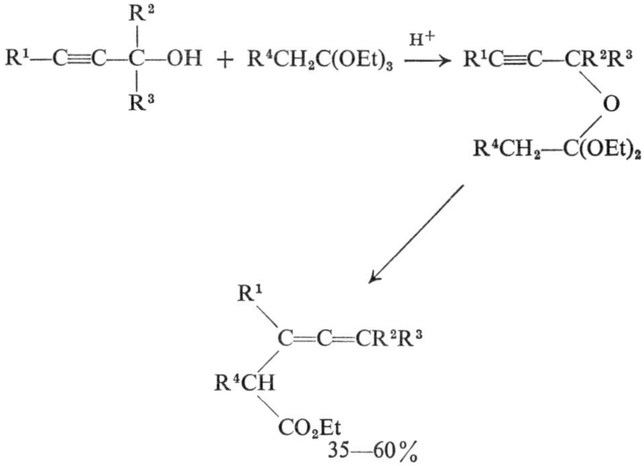

$$R^1-C\equiv C-\underset{R^3}{\overset{R^2}{C}}-OH + R^4CH_2C(OEt)_3 \xrightarrow{H^+} R^1C\equiv C-CR^2R^3$$

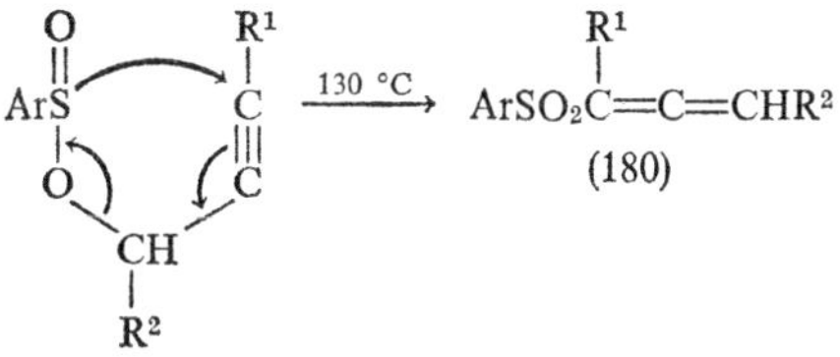

$$\downarrow \quad\quad\quad O-C(OEt)_2-CH_2R^4$$

$$\underset{R^4CH}{\overset{R^1}{\diagdown}}C=C=CR^2R^3$$
$$\overset{\diagdown}{CO_2Et}$$
35—60%
(178)

$$\text{(179)} \xrightarrow{130\,°C} \underset{(180)}{ArSO_2\overset{R^1}{\underset{|}{C}}=C=CHR^2}$$

(179) structure: ArS(O)(O)–C(R¹)=C–CH(R²) cyclic

allenyl sulphones (180) involves a similar cyclic rearrangement. When optically active acetylene is used, the absolute configuration of the product, deduced from the polarizability sequence of substituents on the allene, is in agreement with a cyclic mechanism. Rearrangement of the corresponding sulphenate ester to sulphoxide (180; SO instead of SO_2) is much faster but appears to follow a similar pathway.[119]

The potentially useful allenic amines (181) are prepared from $\alpha\beta$-unsaturated ketones by the route shown.[120]

Propargyl chlorides are converted in good yields *via* a hydroboration procedure into terminal allenes (182).[121]

1,2-Cyclononadiene has been partially resolved by crystallization of a diastereoisomeric mixture of its platinum complexes (183). Optically purer samples of both enantiomers were prepared from optically active *trans*-cyclo-octenes *via* the dibromocarbene adduct.[122]

Cycloaddition Reactions.—The theoretical predictions of Woodward and Hoffmann to the effect that [2 + 2] cycloadditions of allenes (and ketens)

---

[119] G. Smith and C. J. M. Stirling, *J. Chem. Soc.* (C), 1971, 1530.
[120] J.-P. Dulcere, M. Santelli, and M. Bertrand, *Compt. rend.*, 1970, **271**, C, 585.
[121] G. Zweifel, A. Horng, and J. T. Snow, *J. Amer. Chem. Soc.*, 1970, **92**, 1427.
[122] A. C. Cope, W. R. Moore, R. D. Bach, and H. J. S. Winkler, *J. Amer. Chem. Soc.*, 1970, **92**, 1243.

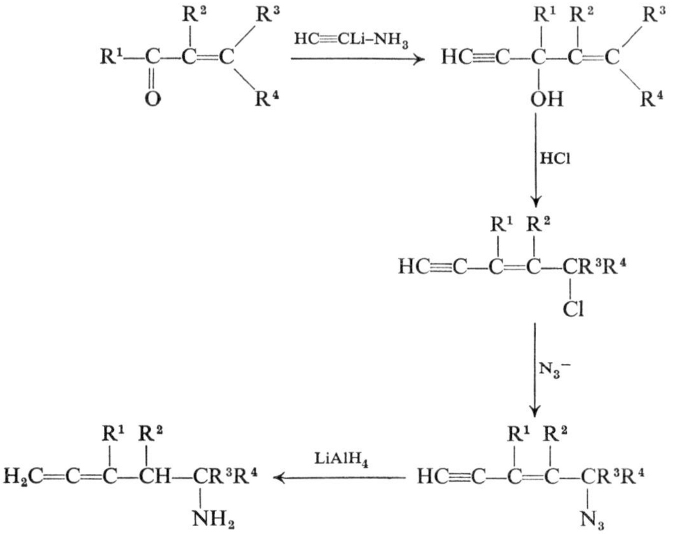

[Scheme showing:]

R¹—C(=O)—C(R²)=C(R³)(R⁴) →(HC≡CLi-NH₃)→ HC≡C—C(R¹)(OH)—C(R²)=C(R³)(R⁴)

↓ HCl

HC≡C—C(R¹)=C(R²)—CR³R⁴(Cl)

↓ N₃⁻

HC≡C—C(R¹)=C(R²)—CR³R⁴(N₃)

→(LiAlH₄)→ H₂C=C=C(R¹)—CH(R²)—CR³R⁴(NH₂)
(181)

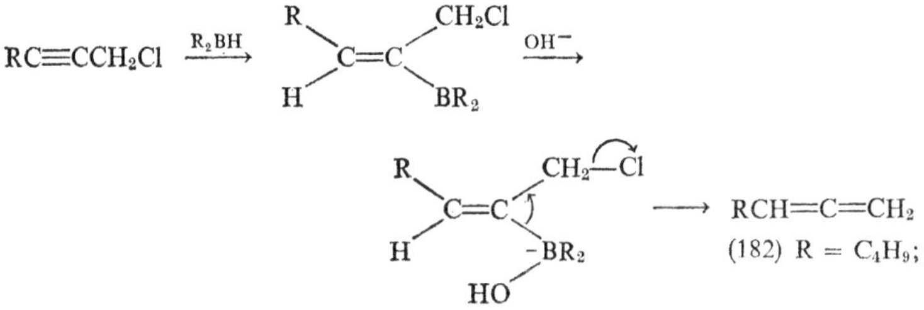

RC≡CCH₂Cl →(R₂BH)→ (R)(H)C=C(CH₂Cl)(BR₂) →(OH⁻)→

(R)(H)C=C(CH₂—Cl)(−BR₂)(HO) → RCH=C=CH₂
(182) R = C₄H₉; 83%

[(±)-C₉H₁₄(PtCl₂)Am*] Am* = optically active α-methylbenzylamine
(183)   or p-nitro-α-methylbenzylamine.

may be concerted have led to a detailed investigation of the dimerization of allenes. Dimerization of equimolar mixtures of tetradeuterioallene and allene leads to a statistical distribution of [²H₀], [²H₄], and [²H₈] dimeric products (1,2-dimethylenecyclobutanes). When 1,1-dideuterioallene is dimerized under similar conditions, an intramolecular kinetic isotope effect of k_H/k_D = 1.14 ± 0.02 is observed with an excess of deuterium located in the vinyl positions (184). These results are reconciled by postulating the presence of

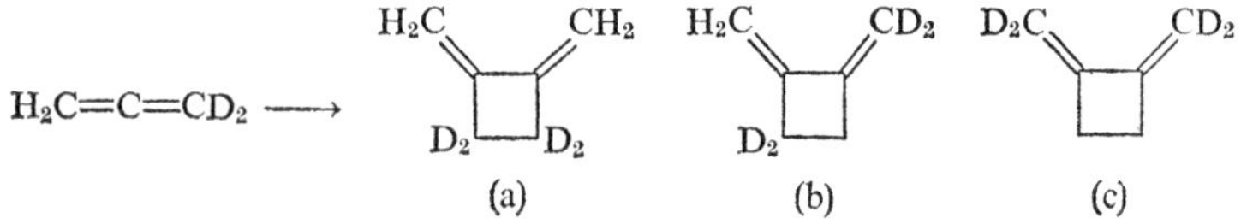

H₂C=C=CD₂ →

(a) cyclobutane with H₂C= and =CH₂ exocyclic, ring D₂/D₂
(b) cyclobutane with H₂C= and =CD₂ exocyclic, ring D₂
(c) cyclobutane with D₂C= and =CD₂ exocyclic

(184)

more than one energy barrier in the mechanistic pathway, with the rate-determining step not identical to the product-determining step.[123] Dimerization of optically active 2,3-pentadiene gives six stereoisomeric 1,2-diethylidene-3,4-dimethylcyclobutanes (185), which were isolated and identified by their 220 MHz spectra. The results did not distinguish between a concerted cycloaddition and a biradical mechanism.[124]

(185) (186) (187)

X-Ray structure determinations of the major dimer from 3-chloro-1,1-diphenylallene and the symmetrical dimer from 3-chloro-1-mesityl-allene shows them to have structures (186) and (187), respectively.[125]

The stereochemistry of the dimerization products of 1-adamantyl-3-chloroallene has been assigned.[126] Although the dimerization of optically active cyclononadiene gives results which strongly support a $[_\pi 2_s + _\pi 2_a]$ concerted cycloaddition,[127] the possibility that an intermediate biradical is produced and reacts stereospecifically must also be considered.

Cycloaddition of 1,1-dimethylallene with tetrafluoroethylene is believed to be a non-synchronous process.[128] The biradical (188) was suggested as an intermediate in formation of the 1:1 adducts isolated [(189a) and (189b)], rotation of the CH_2 group out of the allyl plane occurring more easily than that of the heavier hyperconjugated CMe_2 group and accounting for the 3.6:1 ratio of (189a):(189b).

(188) (189a) (189b)

It appears that the 1,2-addition of an electron-deficient olefin to an allene resembles the analogous 1,2-addition of olefins to 1,3-dienes, accepted to be a two-step mechanism.[129] Benzyne reacts with allenes to give benzocyclobutenes;

---

[123] W. R. Dolbier and S.-H. Dai, *J. Amer. Chem. Soc.*, 1970, **92**, 1774.
[124] J. J. Gajewski and W. A. Black, *Tetrahedron Letters*, 1970, 899.
[125] S. R. Byrn, E. Maverick, O. J. Muscio, K. N. Trueblood, and T. L. Jacobs, *J. Amer. Chem. Soc.*, 1971, **93**, 6680.
[126] T. L. Jacobs and O. J. Muscio, *Tetrahedron Letters*, 1970, 4829.
[127] W. R. Moore, R. D. Bach, and T. M. Ozretich, *J. Amer. Chem. Soc.*, 1969, **91**, 5918.
[128] D. R. Taylor, M. R. Warburton, and D. B. Wright, *J. Chem. Soc.* (C), 1971, 385; D. R. Taylor and D. B. Wright, *ibid.*, p. 391.
[129] P. D. Bartlett, *Quart. Rev.*, 1970, **24**, 473.

1,3-dienes and acetylenes, products of an ene reaction, are also obtained (Scheme 9). Benzocyclobutene formation is favoured where the allene is

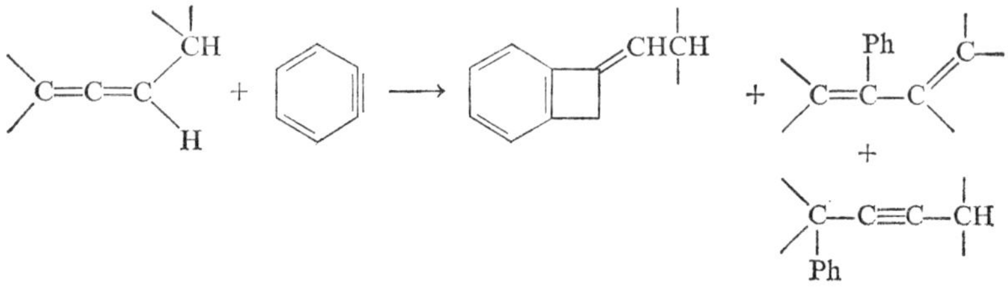

Scheme 9

substituted with electron-rich groups.[130] Phenylallene undergoes a new type of dimerization on heating to give the naphthalene (190), for which a self Diels–Alder adduct is the postulated intermediate.[131]

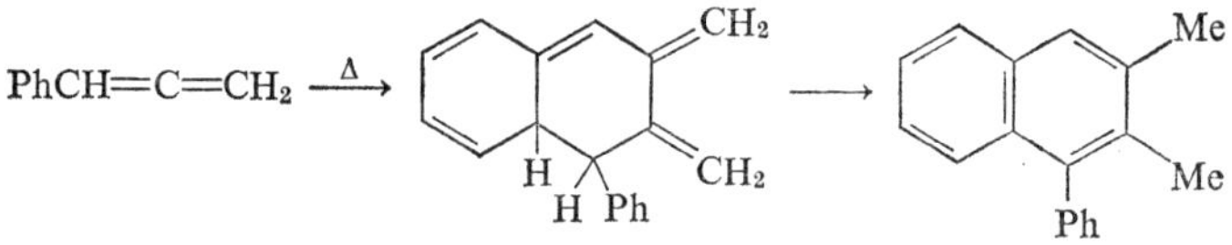

Tetramethoxyallene and tetracyanoethylene combine to yield an equilibrium mixture of the cyclobutane (191) and the stabilized ring-opened dipolar species (192).[132] At $-95\,°C$ the signals due to both (191) and (192) are recognizable in the n.m.r. spectrum. A similar situation is observed in the adduct from benzenesulphonyl isocyanate and tetrathioalkylallenes,[133] the equilibrium between (193) and (194) being established.

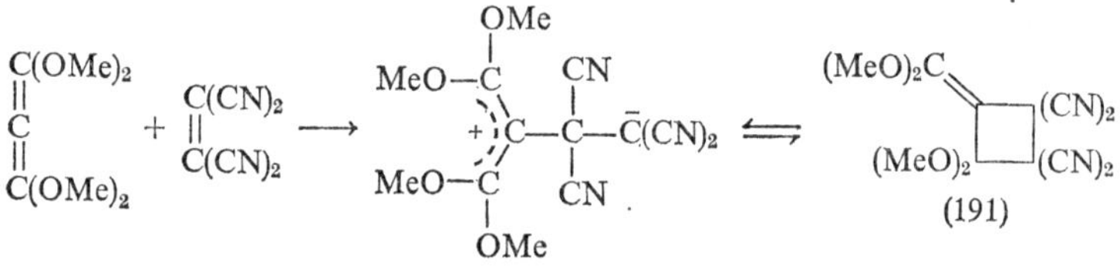

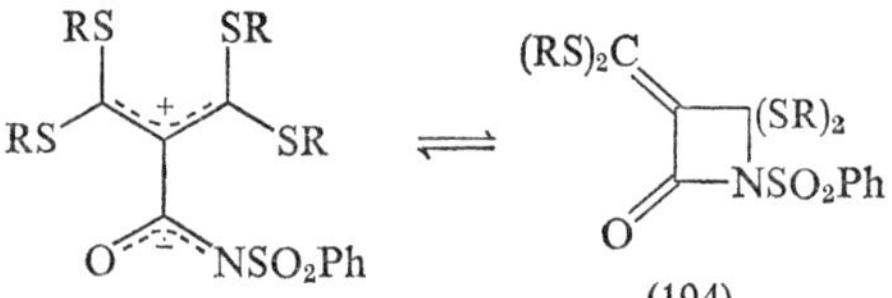

Earlier work in cycloadditions with allene suggested that intramolecular isotope effects with values of $k_H/k_D > 1.00$ derive from non-concerted, probably biradical processes and that when $k_H/k_D < 1.00$ a concerted mechanism

[130] H. H. Wasserman and L. S. Keller, *Chem. Comm.*, 1970, 1483.
[131] J. E. Baldwin and L. E. Walker, *J. Org. Chem.*, 1971, **36**, 1440.
[132] R. W. Hoffmann and W. Schäfer, *Angew. Chem. Internat. Edn.*, 1970, **9**, 733.
[133] R. Gompper and D. Lack, *Angew. Chem. Internat. Edn.*, 1971, **10**, 70.

is likely. The isotope effect $k_H/k_D = 0.93$ was considered diagnostic of a concerted reaction between tetracyanoethylene oxide and dideuterioallene (Scheme 10).[134] A similar conclusion was reached in a study of tetracyanoethylene oxide addition to deuteriated styrenes.[135]

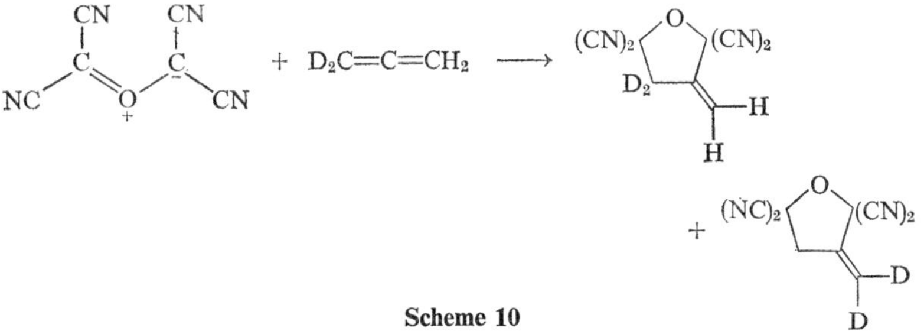

Scheme 10

Intramolecular nitrone–allene cycloaddition has been used to prepare the bicyclic isoxazolidine (195).[136]

Addition of azomethine oxides to allenes does not lead to the anticipated 1,3-dipolar addition product (196) but to the 3-pyrrolidinones (197).[137]

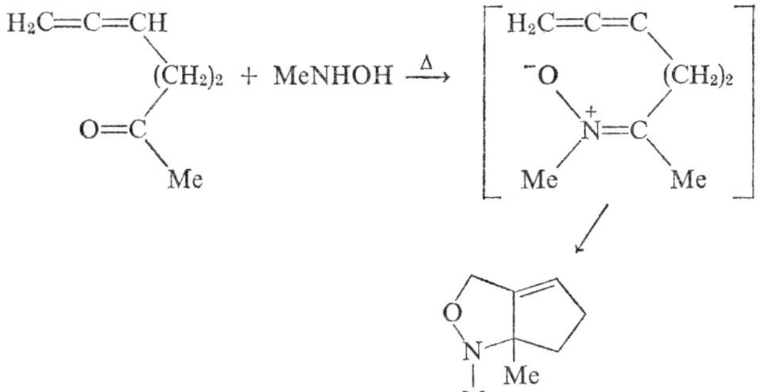

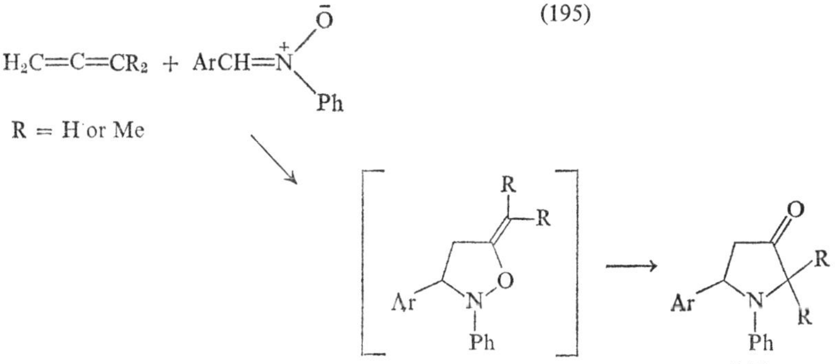

[134] W. R. Dolbier and S. H. Dai, *Tetrahedron Letters*, 1970, 4645.
[135] W. F. Bayne and E. I. Snyder, *Tetrahedron Letters*, 1970, 2263.
[136] N. A. Lebel and E. Banucci, *J. Amer. Chem. Soc.*, 1970, **92**, 5278.
[137] M. C. Aversa, G. Cum, and N. Uccella, *Chem. Comm.*, 1971, 156.

The rearrangement is formulated as passing through a heterolytic cleavage of the N—O bond, but a radical mechanism seems more likely.

The cycloaddition of alkenylidenecyclopropanes (198) with the triazoline (199) involves participation of the strained σ-bond of the cyclopropane. A concerted reaction is believed to be involved since single isomers of (200) and (201) are obtained which are the least thermodynamically stable.[138]

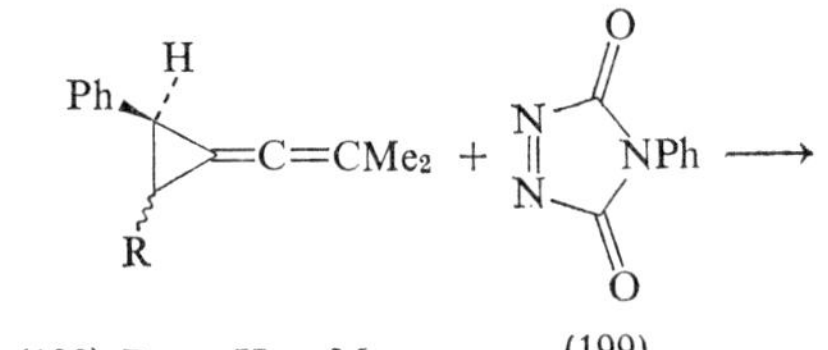

(198) R = H or Me (199)

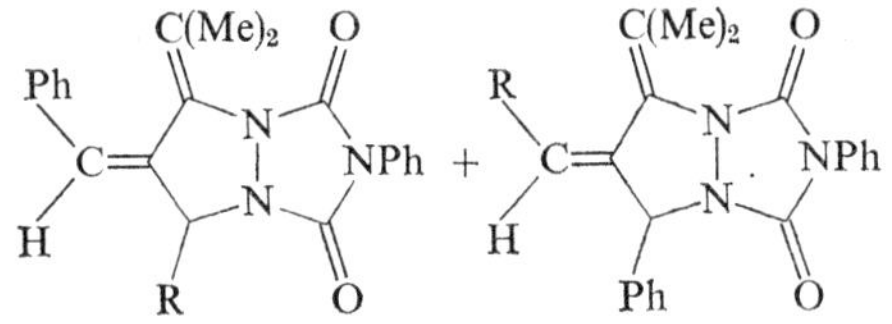

(200) (201)

Allenic esters react predictably with benzonitrile oxide to give isoxazoles (202), but spiro-isoxazolines (203) result when aromatization of the intermediate is difficult.[139]

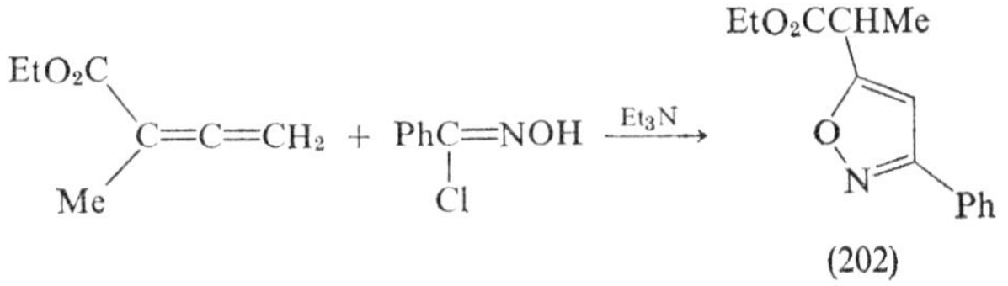

(202)

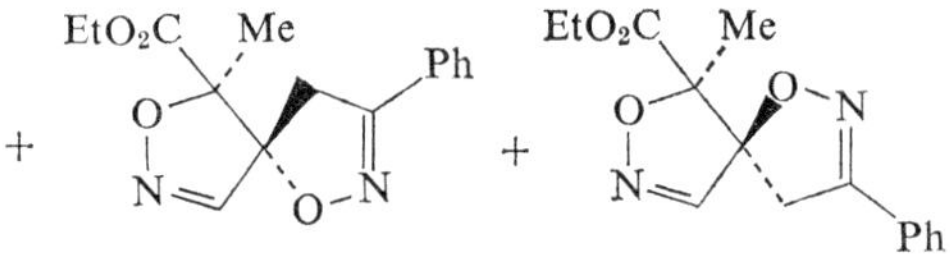

(203)

Addition Reactions.[139a]—A detailed study[140] of the free-radical chlorination of penta-2,3-diene with t-butyl hypochlorite shows that substitution products are obtained by both allylic and allenic hydrogen-abstraction; addition

[138] D. J. Pasto and A. Chen, *J. Amer. Chem. Soc.*, 1971, **93**, 2562.
[139] P. Battioni, L. Vo-Quang, J.-C. Raymond, and Y. Vo-Quang, *Compt. rend.*, 1970, **271**, C, 1468.
[139a] M. C. Caserio, *Selective Org. Transform.*, 1970, **1**, 239.
[140] L. R. Byrd and M. C. Caserio, *J. Amer. Chem. Soc.*, 1970, **92**, 5422.

products are also obtained. None of the products isolated using optically active penta-2,3-diene are optically active. Free-radical addition of hydrogen bromide to allenes yields olefinic products almost exclusively by addition of a bromine atom to the central carbon of the allenic system.[141] Addition of hypochlorous acid to allenic hydrocarbons, an electrophilic attack, leads, in all cases, to the placement of the chlorine on the central carbon atom and the OH group on the more substituted of the allenyl carbons.[142] Electrophilic hydrochlorination of phenylallenes is believed to proceed through a transition state more closely related to the localized allylic cation (204) than to the delocalized cation (205).[143]

$$Ph\overset{+}{C}HCH=CH_2 \qquad Ph\overset{\frown}{CH=CH=CH_2}\overset{+}{}$$
$$(204) \qquad\qquad (205)$$

Optically active penta-2,3-diene reacts with iodine, iodine bromide, or iodine chloride in methanol to give optically active *trans*-3-iodo-4-methoxy pent-2-ene (206). The optical purity of the product was found to vary with the nature of the iodinating agent in the order ICl > IBr > I$_2$. Iodine produced in the reaction is responsible for the racemization, which is explained below (Scheme 11).[144]

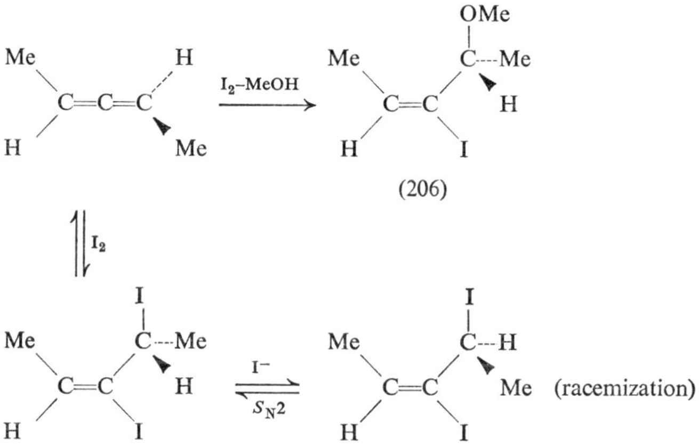

Scheme 11

Oxymercuration studies on optically active penta-2,3-diene using mercuric acetate in methanol suggest that formation of the unsymmetrical mercurinium

[141] R. Y. Tien and P. I. Abell, *J. Org. Chem.*, 1970, **35**, 956.
[142] J.-P. Bianchini and M. Cocordano, *Tetrahedron*, 1970, **26**, 3401.
[143] T. Okuyama, K. Izawa, and T. Fueno, *Tetrahedron Letters*, 1970, 3295.
[144] M. C. Findlay, W. L. Waters, and M. C. Caserio, *J. Org. Chem.*, 1971, **36**, 275.

ion is the rate-determining step in the formation of the chiral products (207). However, on using mercuric chloride the open ion (208) is apparently formed, leading to racemic products.[145]

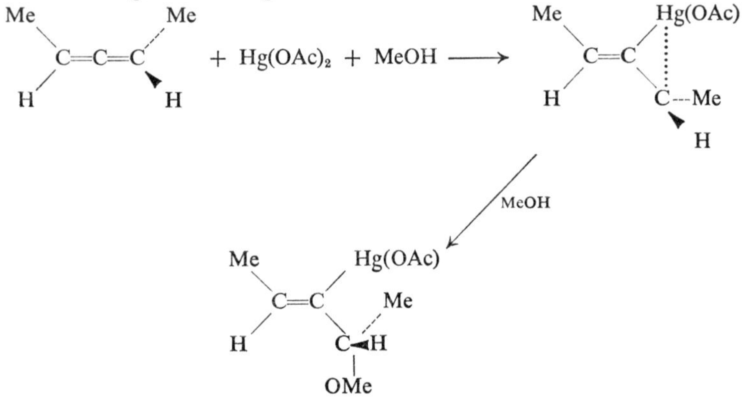

$$\underset{H}{\overset{Me}{>}}C=C=C\underset{H}{\overset{Me}{<}} + Hg(OAc)_2 + MeOH \longrightarrow \underset{H}{\overset{Me}{>}}C=C\underset{\underset{H}{\overset{|}{C}}\text{---}Me}{\overset{Hg(OAc)}{<}}$$

$$\downarrow MeOH$$

$$\underset{H}{\overset{Me}{>}}C=C\underset{\underset{OMe}{\overset{|}{C}\text{-}H}}{\overset{Hg(OAc)}{<}}\overset{Me}{}$$

(207) (+cis-isomer)

$$\underset{|}{\overset{HgCl}{\overset{|}{C}}}$$
$$\underset{|}{C}\cdots\overset{+}{\cdots}\underset{|}{C}$$

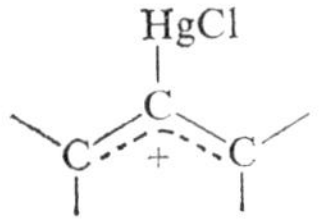

(208)

Reduction of organomercurials formed from cyclic allenes (10—14-membered rings) with sodium borohydride yields an increasing ratio of *trans*:*cis* monosubstituted olefins as the ring size is increased.[146] The stereochemistry and mechanism of the reduction of cyclic allenes using di-imide[147] and sodium in liquid ammonia[148] have been investigated. A stereospecific reduction with sodium in liquid ammonia of the intermediate cyclopropylallene (209) has been used to synthesize *trans*-chrysanthemic acid (210).[149]

Dehydrohalogenation of 1-halogenocyclohexenes with potassium t-butoxide in t-butanol–DMSO has been studied using the deuteriated substrate (211). Two competitive pathways seem to be operating to give cyclohexa-1,2-diene and cyclohexyne (Scheme 12).[150] The proposal of Wittig in 1966 that the C_{12} hydrocarbon (212), also produced in the above reaction, was derived by dimerization of cyclohexa-1,2-diene has been vindicated. When the deuteriated substrate (211) is used the dimer (212) has no protium at vinylic positions.[151]

[145] W. S. Linn, W. L. Waters, and M. C. Caserio, *J. Amer. Chem. Soc.*, 1970, **92**, 4018; see also R. D. Bach, *J. Amer. Chem. Soc.*, 1969, **98**, 1771.
[146] R. Vaidyanathaswamy, D. Devaprabhakara, and V. V. Rao, *Tetrahedron Letters*, 1971, 915.
[147] G. Nagendrappa and D. Devaprabhakara, *Tetrahedron Letters*, 1970, 4243.
[148] R. Vaidyanathaswamy, G. C. Joshi, and D. Devaprabhakara, *Tetrahedron Letters*, 1971, 2075.
[149] R. W. Mills, R. D. H. Murray, and R. A. Raphael, *Chem. Comm.*, 1971, 555.
[150] A. T. Bottini, F. P. Corson, R. Fitzgerald, and K. A. Frost, *Tetrahedron Letters*, 1970, 4753.
[151] A. T. Bottini, F. P. Corson, R. Fitzgerald, and K. A. Frost, *Tetrahedron Letters*, 1970, 4757.

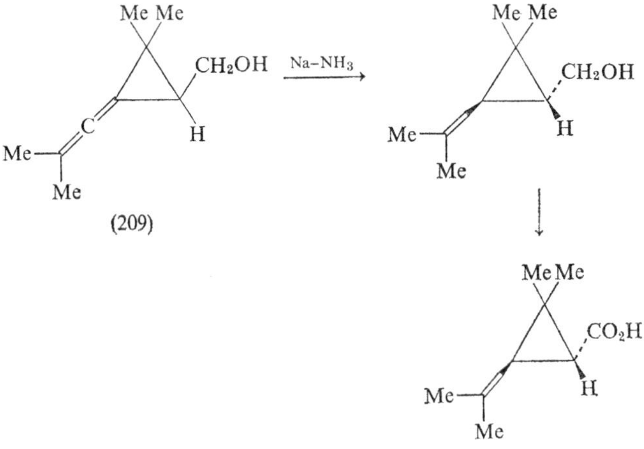

[Structures (209) and (210) with Na-NH$_3$ reduction shown]

Other Reactions of Allenes.—Allylic halides undergo solvolysis at rates greater than those of their saturated analogues owing to charge delocalization through the π-system. The vinyl cation,[152] it may be anticipated, could be

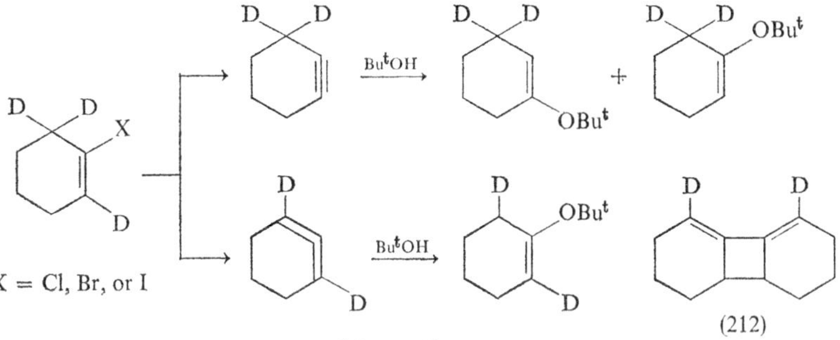

Scheme 12

similarly stabilized by such a neighbouring double bond. This is demonstrated[153] in solvolyses of the allenic chlorides (213), which proceed readily in

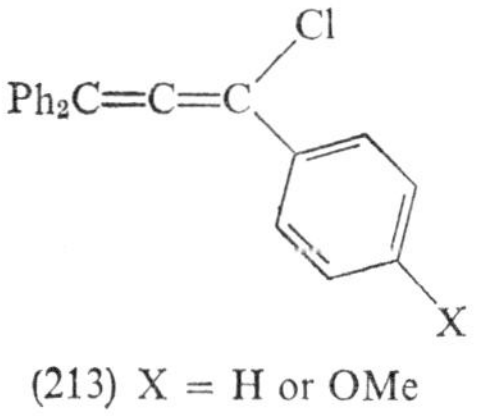

(213) X = H or OMe

[152] M. Hanack, *Accounts Chem. Res.*, 1970, **3**, 209.
[153] M. D. Shiavelli, S. H. Hixon, and H. W. Moran, *J. Amer. Chem. Soc.*, 1970, **92**, 1082; M. D. Shiavelli, S. H. Hixon, H. W. Moran, and C. J. Boswell, *ibid.*, 1971, **93**, 6989.

aqueous acetone at 25 °C at rates far in excess of those of the corresponding aryl vinyl halides.[154] Allenyl bromide itself solvolyses in aqueous ethanol at a rate 4×10^3 slower than propargyl bromide.[155] This is attributable to a lower ground-state energy in allenyl bromide as a result of the stronger sp^2 hybridized carbon–halogen bond relative to the sp^3 bond in propargyl bromide as well as to differences in transition-state geometries.

Homoallenic participation has been measured using (214), which is believed to solvolyse through the same intermediate ion (Scheme 13) as the isomeric tosylate (215)[156] on the basis of virtually identical product yields. The tosylate (214) solvolyses approximately 40 times as fast as its saturated analogue

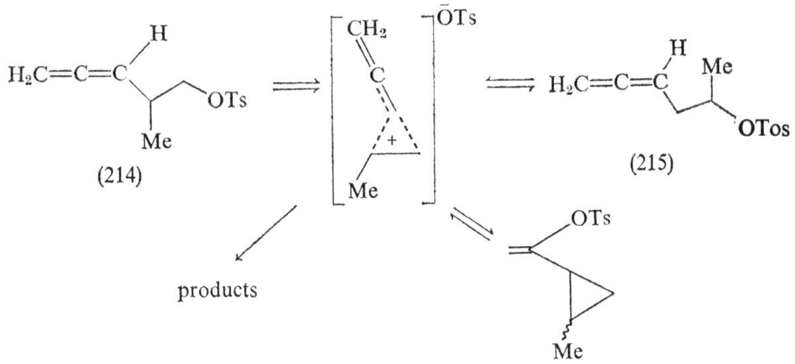

Scheme 13

2-methyl-1-pentyltosylate. Similarly, the bishomoallenic tosylate (216) solvolyses at a rate which suggests that participation is occurring, and the cyclopentene derivatives (217) and (218) are among the products.[157] This

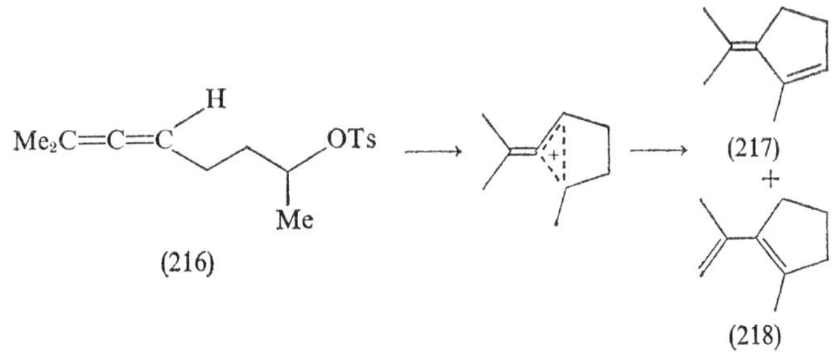

result suggests that intramolecular participation by an allenyl group is superior to that of a similarly located ethylene.[158]

---

[154] Z. Rappoport and A. Gal, *J. Amer. Chem. Soc.*, 1969, **91**, 5246.
[155] C. V. Lee, R. J. Hargrove, T. E. Dueber, and P. J. Stang, *Tetrahedron Letters*, 1971, 2519.
[156] R. S. Macomber, *J. Amer. Chem. Soc.*, 1970, **92**, 7101.
[157] B. Ragonnet, M. Santelli, and M. Bertrand, *Tetrahedron Letters*, 1971, 955.
[158] P. D. Bartlett, W. D. Closson, and T. J. Cogdell, *J. Amer. Chem. Soc.*, 1965, **87**, 1308.

The allenic dianion (219) is obtained from the acetylene (220) by treatment with two moles of butyl-lithium.[159] Alkylation can be carried out successively using two different reagents to yield the tetrasubstituted allenes (221); hydrolysis of the monoalkylated products yields $\alpha\beta$-unsaturated aldehydes (222).

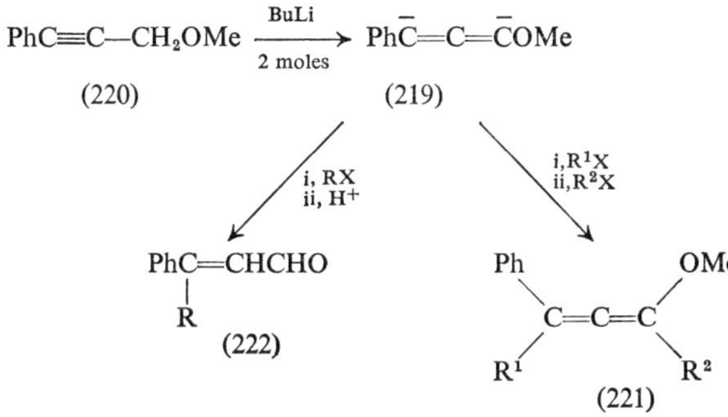

$$PhC\equiv C-CH_2OMe \xrightarrow[\text{2 moles}]{BuLi} Ph\bar{C}=C=\bar{C}OMe$$
$$(220) \qquad\qquad (219)$$

(219) → i, RX; ii, H⁺ → PhC=CHCHO with R substituent (222)

(219) → i, R¹X; ii, R²X → tetrasubstituted allene

$$\begin{array}{c}Ph\\R^1\end{array}C=C=C\begin{array}{c}OMe\\R^2\end{array}$$
(221)

Adsorption of MeC≡CD on zinc oxide gives rise to an OH band at 3515 cm⁻¹ in the i.r. spectrum and no band at 2598 cm⁻¹ (OD) is observed, suggesting that the following process is occurring:

$$CH_3C\equiv CD + ZnO \longrightarrow \underset{\underset{Zn-O-H}{|}}{CH_2-C\equiv CD}$$

The spectrum is identical to that of adsorbed allene, implying that the adsorbed species is the propargyl anion. In agreement with this conclusion is the isomerization of allene to methylacetylene when the former is circulated over the activated zinc oxide employed.[160]

Epoxidation of vinylallenes (223) leads to cyclopentenones (224) or allenylepoxides (225) in amounts which depend upon the substitution pattern of (223).[161]

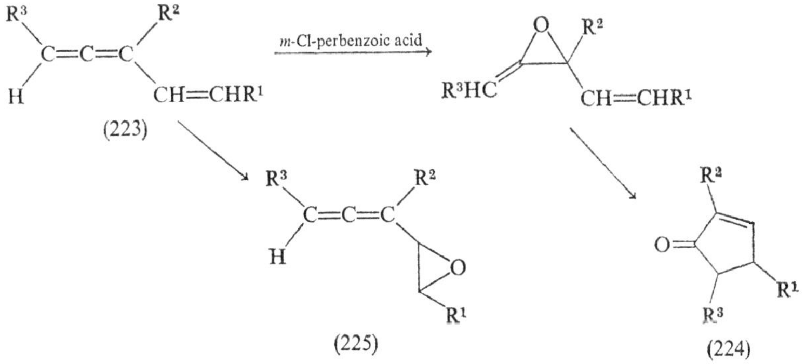

$$\begin{array}{c}R^3\\H\end{array}C=C=C\begin{array}{c}R^2\\CH=CHR^1\end{array}$$
(223)

m-Cl-perbenzoic acid →

$$R^3HC\underset{O}{\overset{R^2}{\diagup\diagdown}}CH=CHR^1$$

→ (225): $\begin{array}{c}R^3\\H\end{array}C=C=C\begin{array}{c}R^2\\\end{array}$ with epoxide bearing R¹

→ (224): cyclopentenone with R², R¹, R³ substituents

[159] Y. Leroux and R. Mantione, *Tetrahedron Letters*, 1971, 591; Y. Leroux and R. Mantione, *J. Organometallic Chem.*, 1971, **30**, 295.
[160] C. C. Chang and R. J. Kokes, *J. Amer. Chem. Soc.*, 1970, **92**, 7517.
[161] J. Grimaldi and M. Bertrand, *Bull. Soc. chim. France*, 1971, 957.

Isomerization of the acetylenic dimethylsulphonium bromide (226) occurs readily in ethanol to yield the allene (227). Addition of β-ketoester, β-diketone, or β-ketosulphone gives the corresponding furans (228) in high yield.[162]

Thermolysis of the allenic diamide (229) yielded the α-pyrone (230), where the keten (231) was a presumed intermediate formed by a reverse ene reaction.[163]

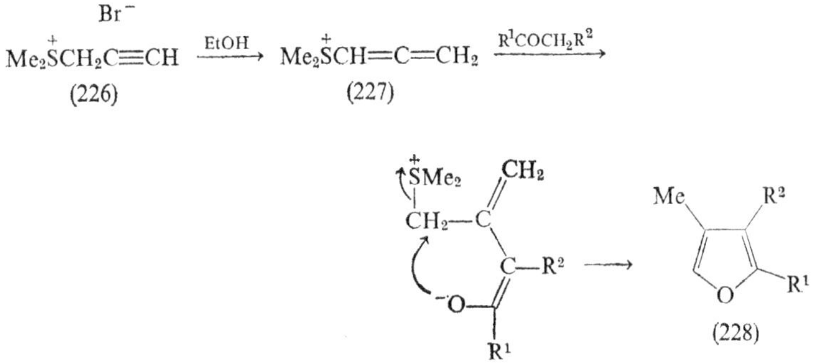

The absolute configuration of (−)-cyclonona-1,2-diene has been established as (S) by an o.r.d.–c.d. correlation[164] and by the extended chemical correlation shown (Scheme 14).[165] The absolute configuration of the (+)-isomer of (232) was established by X-ray diffraction.

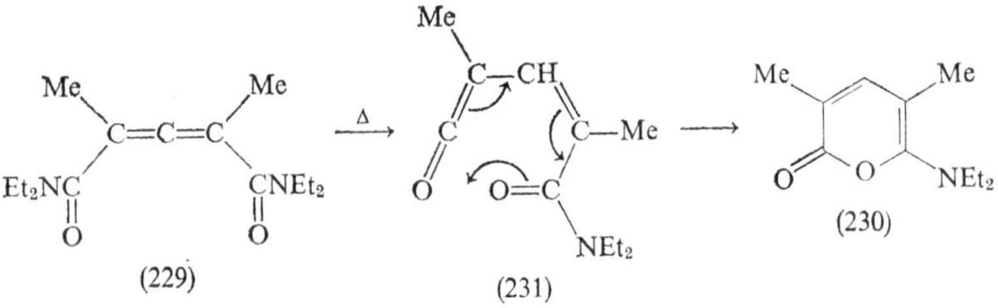

Investigations into ^{13}C n.m.r. spectra of allenes have shown that the central allenic carbon (C-β) has an extremely low chemical shift. This technique is useful for studying charge distribution in allenes, and for a given alkyl substituent there is a fair linear relationship between the number of substituents and δ(C-β).[166] Among studies of the long-range spin–spin coupling over five

---

[162] J. W. Batty, P. D. Howes, and C. J. M. Stirling, *Chem. Comm.*, 1971, 534.
[163] J. Ficini, J. Pouliquen, and J.-P. Paulme, *Tetrahedron Letters*, 1971, 2483.
[164] W. R. Moore, H. W. Anderson, S. D. Clark, and T. M. Ozretich, *J. Amer. Chem. Soc.*, 1971, **93**, 4932.
[165] R. D. Bach, U. Mazur, R. N. Brummel, and L.-H. Lin, *J. Amer. Chem. Soc.*, 1971, **93**, 7120.
[166] R. Steur, J. P. C. M. van Dongen, M. J. A. de Bie, W. Drenth, J. W. de Haan, and L. J. M. van de Ven, *Tetrahedron Letters*, 1971, 3307.

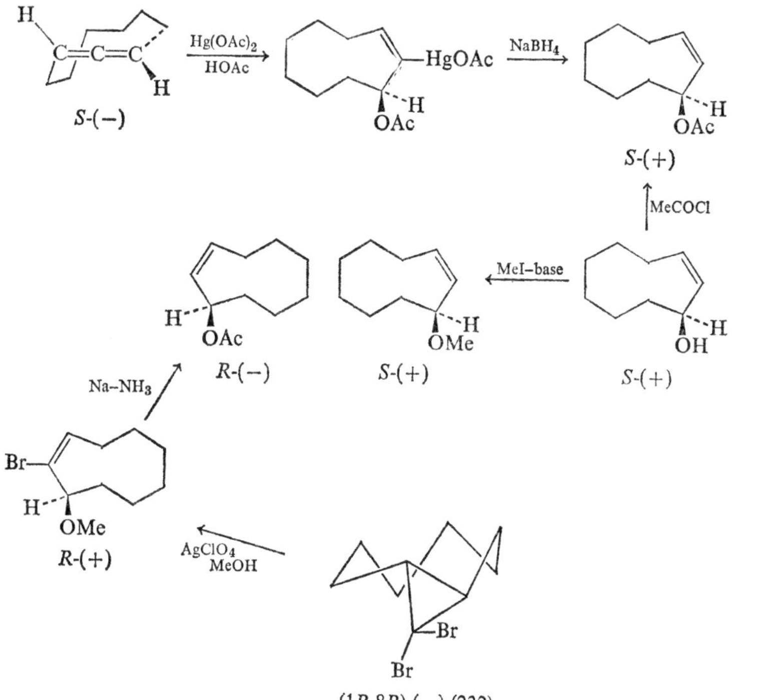

Scheme 14

bonds in the ^{1}H n.m.r. of allenes is an attempt to relate the value of ^5J to the dihedral angle ϕ between the two planes involved (233).[167]

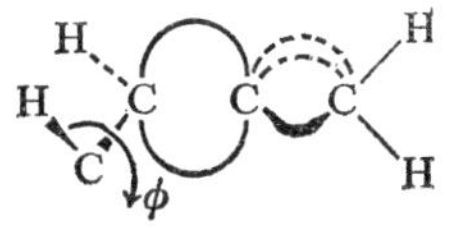

(233)

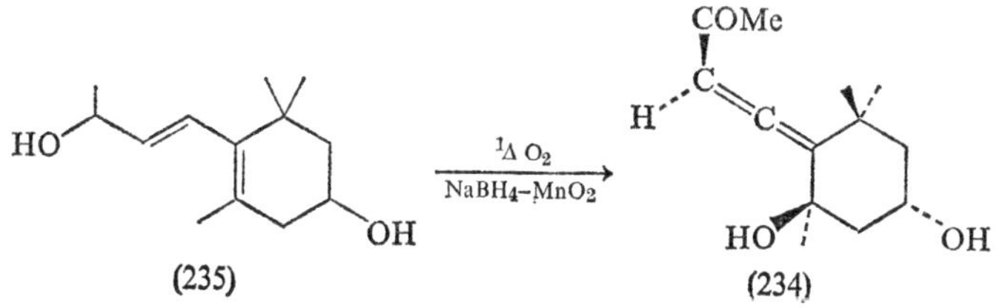

(235) (234)

An X-ray elucidation of the p-bromobenzoate shows (234) to be the structure of the allenic ketone formed by photo-oxidation and manganese dioxide oxidation of (235). It is concluded that the naturally occurring allenic ketone

[167] M. Santelli, *Chem. Comm.*, 1971, 938.

isolated from the large flightless grasshopper differs from (234) in configuration at the allenic carbon and supports the view that the latter is derived by *in vivo* degradation of the carotenoid neoxanthin.[168]

The propargyl–allenyl free radical (236) has been studied by e.s.r. by reaction of methylacetylene and allene with photochemically generated t-butoxyl radicals. 1,1-Dimethylallene reacts by abstraction of either an allenyl

$$\dot{C}H_2-C\equiv CH \longleftrightarrow CH_2=C=\dot{C}H$$
(236)

or a methyl hydrogen to give (237) and (238), distinguishable by their e.s.r. spectra.

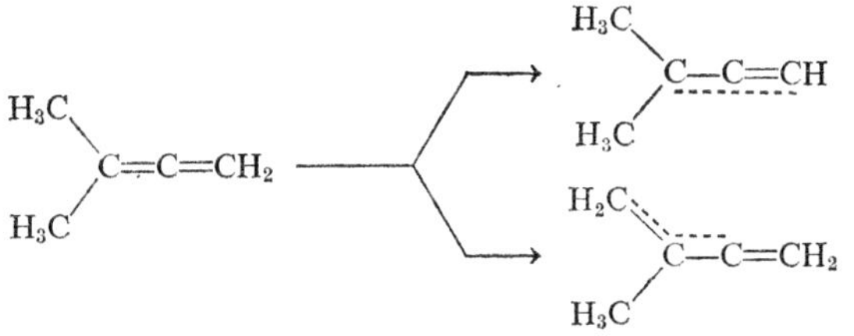

$$\begin{array}{c} H_3C \\ \diagdown \\ H_3C \end{array} C=C=CH_2 \longrightarrow \begin{array}{c} H_3C \\ \diagdown \\ H_3\dot{C} \end{array} C-C\equiv CH \quad (237)$$

$$\longrightarrow \begin{array}{c} H_2\dot{C} \\ \diagdown \\ H_3C \end{array} C-C=CH_2 \quad (238)$$

The radical (238) may be considered in two valence tautomeric forms [(239) and (240)] which differ by a 90° rotation about the C-3—C-4 axis. E.s.r. and calculations favour (240), containing the unpaired electron in a delocalized orbital (shaded), over (239) with the electron in a localized orbital.[169]

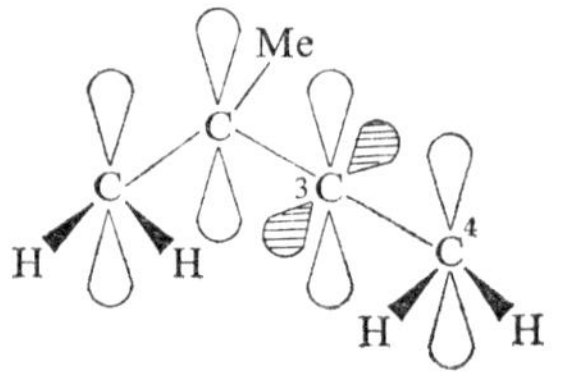

(239) 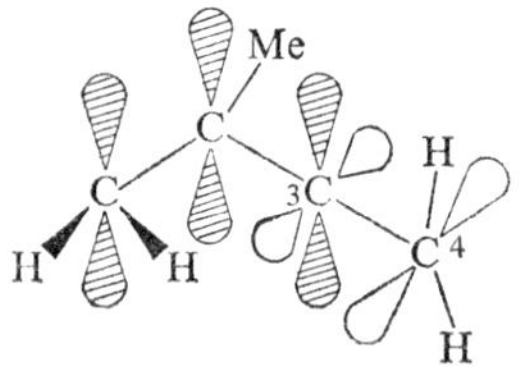   (240)

3 Olefins

Reviews on olefins which have appeared during the past two years include the addition–elimination reactions of palladium compounds with olefins;[170] elimination reactions leading to olefins;[171,172] the stereochemistry of the Wittig reaction;[173] stereoselective and stereospecific olefin synthesis;[174,175] steric

---

[168] T. E. DeVille, J. Hora, M. B. Hursthouse, T. P. Toube, and B. C. L. Weedon, *Chem. Comm.*, 1970, 1231.
[169] J. K. Kochi and P. J. Krusic, *J. Amer. Chem. Soc.*, 1970, **92**, 4110.
[170] R. F. Heck, *Fortschr. Chem. Forsch.*, 1971, **16**, 221.
[171] J. Sicher, *Pure Appl. Chem.*, 1971, **25**, 655.
[172] A. J. Parker, *Chem. Technol.*, 1971, 297.
[173] M. Schlosser, *Topics Stereochem.*, 1970, **5**, 1.
[174] J. Reucroft and P. G. Sammes, *Quart. Rev.*, 1971, **25**, 135.
[175] J. D. Faulkner, *Synthesis*, 1971, 175.

Acetylenes, Allenes, and Olefins

selectivity in addition of carbenes to olefins;[176] addition of halogen azides,[177] pseudo-halogens,[178] or sulphenyl halides[179] to olefins; and prototropic isomerization of olefins with functional groups.[180] In addition, a book has also appeared.[181]

Synthesis.—Much effort has been expended recently in devising new methods for stereospecific syntheses of substituted olefins[174,175] and, in particular, their application to the synthesis of biologically important terpenoids. Studies of C_{17} and C_{18} Cecropia juvenile hormone necessitated a stereospecific synthesis of the heptenol (241), which was prepared[182] as shown. Conversion of the

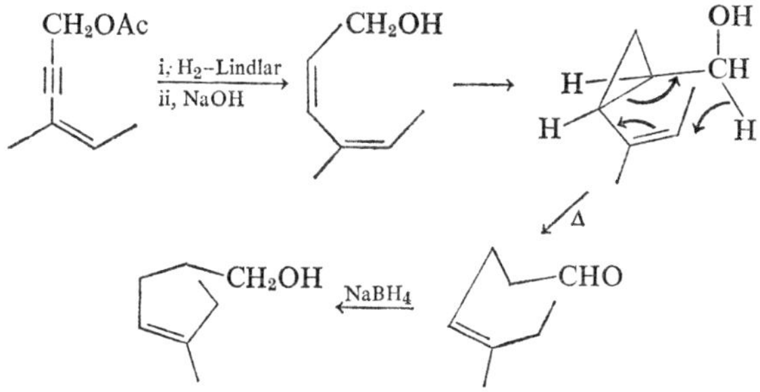

phosphonium salt derived from (241) into the trisubstituted olefin (242) uses another stereospecific reaction, the addition of formaldehyde to the β-oxidoylide (243). Olefin (242) is converted in several steps into (±)-C_{17}

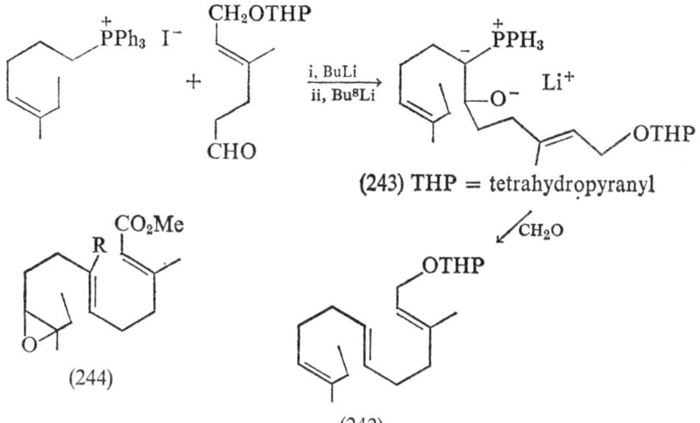

(243) THP = tetrahydropyranyl

[176] R. A. Moss, *Selective Org. Transform.*, 1970, **1**, 35.
[177] A. Hassner, *Accounts Chem. Res.*, 1971, **4**, 9.
[178] D. Swern, *Amer. Chem. Soc. Div. Petrol. Chem.*, 1970, **15**, E39.
[179] D. R. Hogg, *Mech. Reactions Sulfur Compounds*, 1970, **5**, 87.
[180] L. A. Yanovskaya and Kh. Shakhidayatov, *Russ. Chem. Rev.*, 1970, **39**, 859.
[181] 'The Chemistry of Alkenes', ed. J. Zabicky, Interscience, London, 1970.
[182] E. J. Corey, H. Yamamoto, D. K. Herron, and K. Achiwa, *J. Amer. Chem. Soc.*, 1970, **92**, 6635.

(244; R = Me) or (±)-C_{18} (244; R = Et) juvenile hormone.[183] The β-oxidoylides may also be converted to chloro-olefins by halogenation.[184]

Reaction of lithium dialkylcopper complexes with acyclic allylic acetates proceeds by two predictable pathways depending on the nature of X, Y, and Z (Scheme 15).[185] Thus the allylic acetate (245), on treatment with

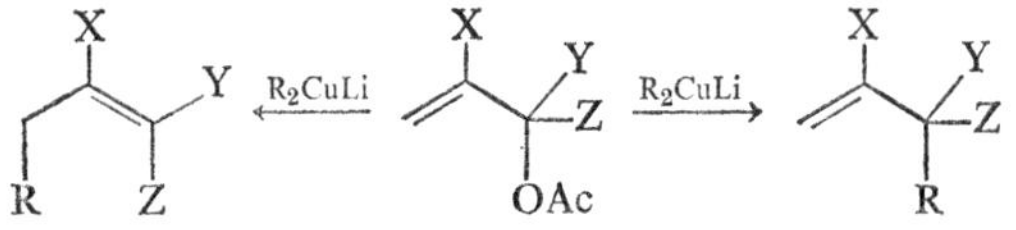

X, Y, Z = H or alkyl

Scheme 15

lithium dimethylcuprate, stereoselectively gave (246), which contains the juvenile hormone skeleton.

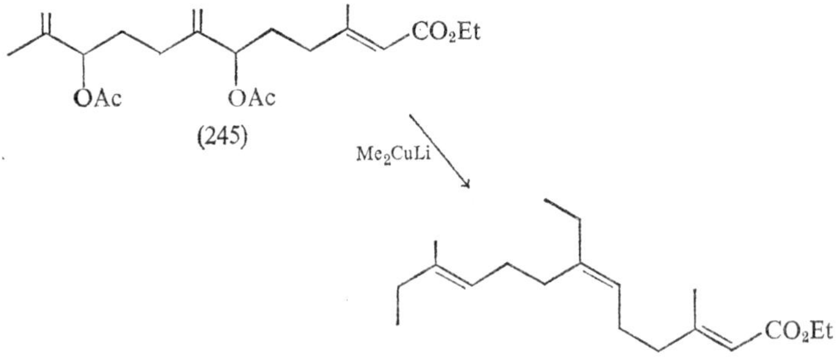

The dienol derivative (247), when heated, rearranges stereospecifically by a 1,5-sigmatropic hydrogen shift to (248), from which the aldehyde is obtained on hydrolysis.[186]

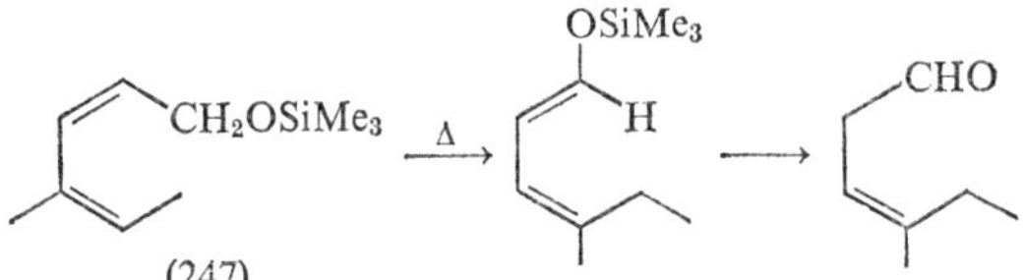

A synthesis which is applicable to highly hindered olefins is the conversion of (249) to (250) with extrusion of X and Y.[187] Thus (251) is converted into the

[183] E. J. Corey and H. Yamamoto, *J. Amer. Chem. Soc.*, 1970, **92**, 6636.
[184] M. Schlosser and K. F. Christmann, *Synthesis*, 1969, 38; E. J. Corey, J. I. Shulman, and H. Yamamoto, *Tetrahedron Letters*, 1970, 447.
[185] R. J. Anderson, C. A. Hendrick, and J. B. Siddall, *J. Amer. Chem. Soc.*, 1970, **92**, 735.
[186] E. J. Corey and D. K. Herron, *Tetrahedron Letters*, 1971, 1641.
[187] D. H. R. Barton and B. J. Willis, *Chem. Comm.*, 1970, 1225; D. H. R. Barton, E. H. Smith, and B. J. Willis, *ibid.*, p. 1226.

highly hindered (252) in 80% yield on heating in the presence of tris(diethylamino)phosphine. Similarly (253), prepared from cyclohexanone, hydrazine,

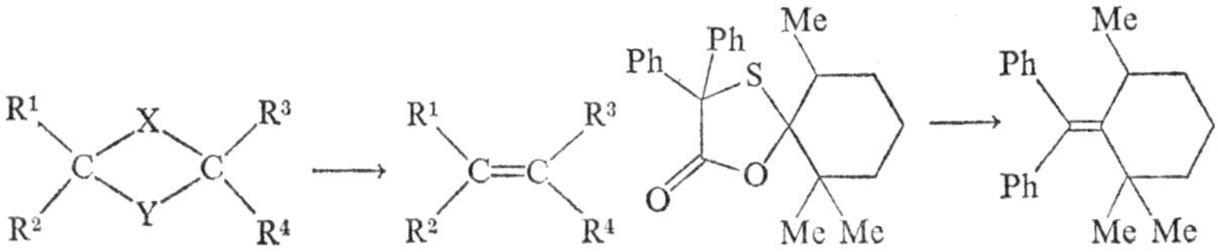

(249) (250) (251) (252)

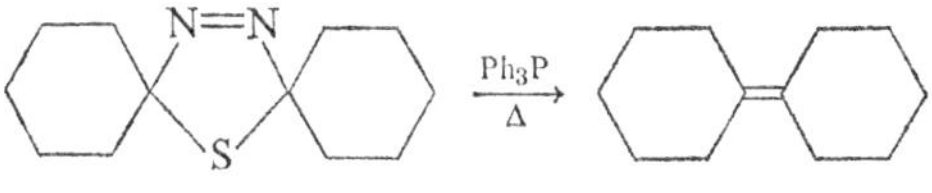

(253)

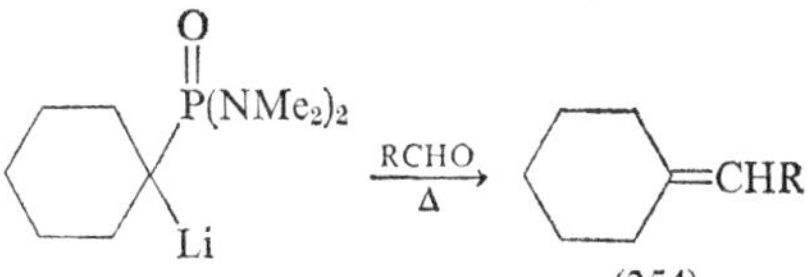

(255) (254)

and H_2S, followed by lead tetra-acetate oxidation, gave bi(cyclohexylidene) (77%).[187] Cyclohexylidene derivatives (254) are also obtained by reaction of lithium cyclohexylphosphonic acid bis(dimethylamide) (255) with aldehydes. The reaction with ketones is not successful.[188]

Hindered olefins are obtained by the sequence shown in Scheme 16. The

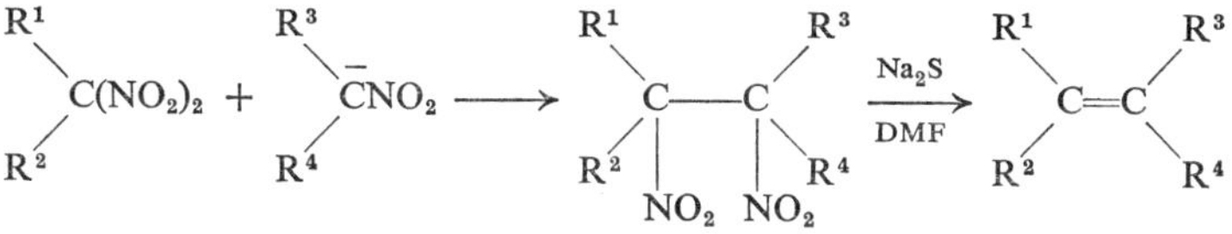

(256)

Scheme 16

required vicinal dinitro-compounds (256) are prepared by reaction of α-dinitro-compounds with a nitroparaffin salt, and treatment with Na_2S gives olefins in excellent yields.[189]

Treatment of the epoxides (257) with lithium diphenylphosphide gives stereospecific ring-opening and quaternization of the crude product yields the betaines (258). The latter generally fragment under the conditions of

[188] J. B. Jones and P. W. Marr, *Canad. J. Chem.*, 1971, **49**, 1300.

[189] N. Kornblum, S. D. Boyd, H. W. Pinnick, and R. G. Smith, *J. Amer. Chem. Soc.*, 1971, **93**, 4316.

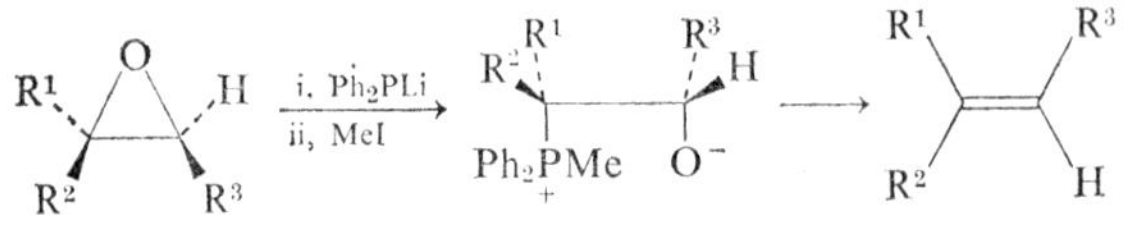

$$\text{(257)} \quad \underset{R^2}{\overset{R^1}{\triangle}}\overset{O}{\underset{R^3}{\triangle}}H \xrightarrow[\text{ii, MeI}]{\text{i. Ph}_2\text{PLi}} \underset{\underset{Ph_2PMe}{+}}{\overset{R^1}{\underset{O^-}{\triangle}}}\overset{R^3}{\underset{}{\triangle}}H \longrightarrow \underset{R^2}{\overset{R^1}{\diagdown}}C=C\overset{R^3}{\diagup}H$$

(257)   (258)

quaternization and olefins are formed with a stereochemistry inverted relative to the starting olefin from which the epoxide was derived. Thus *cis*-octene oxide may be converted to *trans*-octene in 95% yield.[190]

A synthesis analogous to the Wittig reaction but using silicon compounds gives olefins in yields >50% (Scheme 17).[191]

$$\text{Me}_3\text{SiCH}_2\text{Cl} \longrightarrow \text{Me}_3\text{SiCH}_2\text{MgBr} \xrightarrow[\text{ii, H}_2\text{O}]{\text{i, R}^1\text{R}^2\text{C=O}}$$

$$\underset{\text{OH}}{\text{R}^1\text{R}^2\text{CCH}_2\text{SiMe}_3} \xrightarrow[\text{Na Salt}]{\Delta} \text{R}^1\text{R}^2\text{C}=\text{CH}_2$$

Scheme 17

Conversion of epoxides into the corresponding alkenes may be achieved with magnesium amalgam and magnesium bromide[192] or using a Zn–Cu couple,[193] methods which may have advantages over that using the Cr^{II}–ethylenediamine complex[194] in certain cases.

Aliphatic 1,1-dichloro-olefins have been obtained in good yields by treating aldehydes with tris(dimethylamino)phosphine and carbon tetrachloride in THF at low temperatures.[195]

Mild reagents for dehydration of alcohols include the methyl(carboxysulphamoyl)triethylammonium hydroxide inner salt (258a). The trialkylammonium salts (259), formed from secondary or tertiary alcohols, ionize, despite the charge associated with the leaving group, to provide tight ion pairs which undergo fast stereospecific proton transfer, giving olefins in high yield. The stereospecific conversion of (260) to (261a) and (262) to (261b)

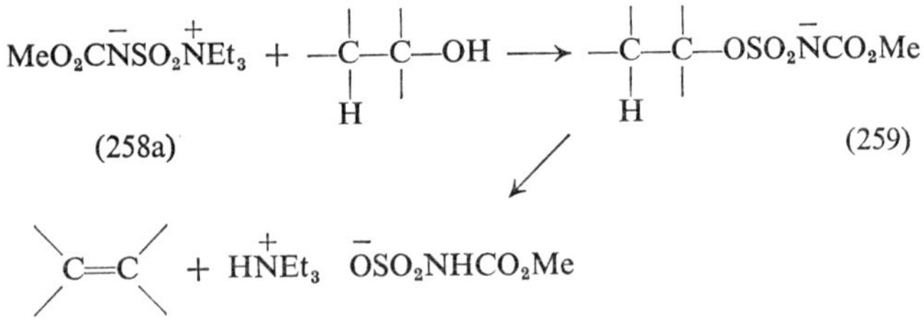

$$\overset{-}{\text{MeO}_2\text{C}}\text{NSO}_2\overset{+}{\text{NEt}_3} + \underset{\text{H}}{-\overset{|}{\underset{|}{\text{C}}}-\overset{|}{\underset{|}{\text{C}}}-\text{OH}} \longrightarrow \underset{\text{H}}{-\overset{|}{\underset{|}{\text{C}}}-\overset{|}{\underset{|}{\text{C}}}-\text{OSO}_2\overset{-}{\text{N}}\text{CO}_2\text{Me}}$$

(258a)   (259)

$$\diagdown\text{C}=\text{C}\diagup + \text{H}\overset{+}{\text{NEt}_3} \quad \overset{-}{\text{O}}\text{SO}_2\text{NHCO}_2\text{Me}$$

[190] E. Vedejs and P. L. Fuchs, *J. Amer. Chem. Soc.*, 1971, **93**, 4070.
[191] T.-H. Chan, E. Chang, and E. Vinokur, *Tetrahedron Letters*, 1970, 1137.
[192] F. Bertini, P. Grasselli, G. Zubiani, and G. Cainelli, *Chem. Comm.*, 1970, 144.
[193] S. M. Kapchan and M. Maruyama, *J. Org. Chem.*, 1971, **36**, 1187.
[194] J. K. Kochi, D. M. Singleton, and L. J. Andrews, *Tetrahedron*, 1968, **24**, 3503.
[195] G. Lavielle, J.-C. Combret, and J. Villieras, *Bull. Soc. chim. France*, 1971, 2047.

Acetylenes, Allenes, and Olefins

(260) $R^1 = H, R^2 = D$
(262) $R^1 = D, R^2 = H$

(261a) $R = D$
(261b) $R = H$

indicates that proton (deuteron) transfer takes place faster than rotational interconversion of the ion pairs.[196]

Another dehydrating reagent applicable to secondary and tertiary alcohols, which are otherwise prone to undergo carbonium ion rearrangements, is the stable crystalline sulphurane (263). An intermediate alkoxysulphonium ion (264) is a postulated intermediate, followed by rapid abstraction of β-proton. For secondary alcohols the mechanism has considerable E_2 character, as shown by a preference for *trans* diaxial elimination, whereas in tertiary alcohols the mechanism is nearer E_1, with a carbonium ion intermediate.[197]

$$Me_3COH + Ph_2S(OR_F)_2 \rightleftharpoons Ph_2S-OCMe_3 + R_FOH$$
$$(263) \qquad \qquad \underset{OR_F}{|}$$
$$(264)$$

$$\downarrow$$

$$Ph_2SO + Me_2C=CH_2 + R_FOH \qquad R_F = -\underset{CF_3}{\overset{CF_3}{|}}C-Ph$$

Homologation of olefins by four carbon atoms is achieved by reaction of the derived borane (265) with buta-1,3-diene monoxide in the presence of catalytic amounts of oxygen.[198] Trialkylboranes are also converted into the 4,4-dialkyl-*cis*-but-2-ene-1,4-diols (266) in good yields by addition to α-lithiofuran and oxidation of the intermediate boroxarocyclohexenes (267) with hydrogen peroxide.[199]

$$R_3B + CH_2=CHCH-CH_2 \xrightarrow{O_2} RCH_2CH=CHCH_2OH + Et_2BOH$$
$$(265) \qquad \underset{O}{\diagdown\diagup} \qquad (89\% \; trans)$$

$R_3B + Li\text{-furan} \longrightarrow$ (267) $\xrightarrow{H_2O_2}$ (266)

---

[196] G. M. Burgess, H. R. Penton, and E. A. Taylor, *J. Amer. Chem. Soc.*, 1970, **92**, 5224.
[197] J. C. Martin and R. J. Arhart, *J. Amer. Chem. Soc.*, 1971, **93**, 4327.
[198] A. Suzuki, N. Miyaura, M. Itoh, H. C. Brown, G. W. Holland, and Ei. Negishi, *J. Amer. Chem. Soc.*, 1971, **93**, 2792.
[199] A. Suzuki, N. Miyaura, and M. Itoh, *Tetrahedron*, 1971, **27**, 2775.

Transformation of vicinal dicarboxylic acids or anhydrides into olefins is a valuable synthetic transformation. An additional method uses various transition metals, exemplified by conversion of (268) into (269) in 53% yield. Even better yields are obtained with thioanhydrides, but mixtures are obtained when abstractable β-hydrogens are present.[200] The transformation of

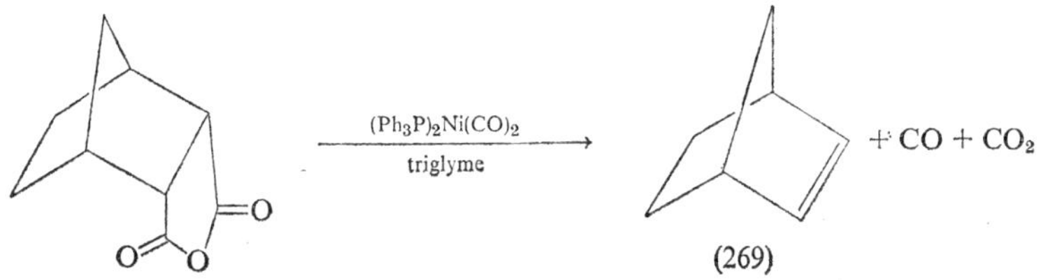

1,2-diols into alkenes is likewise useful synthetically. A new method requires heating the diol, *e.g.* (270), with NN-dimethylformamide dimethylacetal and treatment of the dioxolan (271) with acetic anhydride. Although the reaction appears to be stereospecific, the alkene may undergo acid-catalysed isomerization.[201]

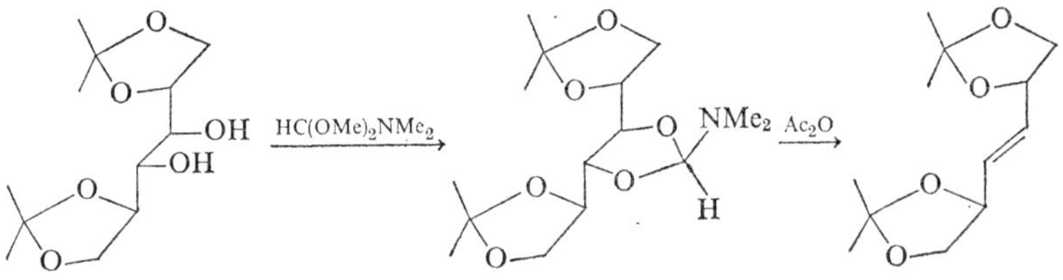

Dehydration of 1,1-di-t-butylethanol with thionyl chloride–pyridine gives (272) with little rearrangement, whereas the major products from the more substituted alcohols are the rearranged olfins [*e.g.* (273) → (274)]. This is rationalized on the basis of the most stable intermediate carbonium ion conformations (275) and (276), formed by ionization of the chlorosulphite. In the first case the preferred conformation (275) favours elimination of the proton which is coplanar with the unoccupied orbital of the electron-deficient carbon, whereas in (276) the lack of such an available proton results in methyl shift from one of the t-butyl groups.[202]

Among particular olefins recently synthesized is tetra-acetylethylene (277), potentially useful for heterocyclic synthesis.[203]

[200] B. M. Trost and F. Chen, *Tetrahedron Letters*, 1971, 2603.
[201] F. W. Eastwood, K. J. Harrington, J. S. Josan, and J. L. Pura, *Tetrahedron Letters*, 1970, 5223.
[202] J. S. Lomas, D. S. Sagatys, and J. E. Dubois, *Tetrahedron Letters*, 1971, 599.
[203] G. Adembri, F. DeSio, R. Nesi, and M. Scotton, *J. Chem. Soc.* (*C*), 1970, 1536.

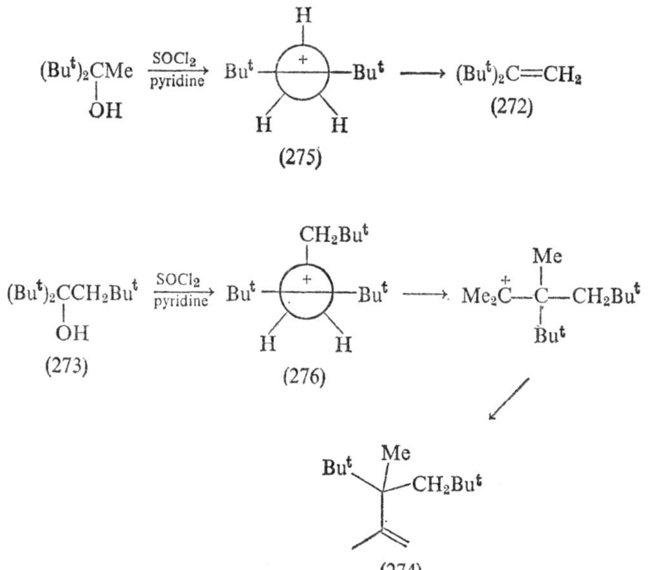

$(Bu^t)_2CMe$ --$\underset{pyridine}{SOCl_2}$→ (275) → $(Bu^t)_2C=CH_2$
|
OH
(272)

$(Bu^t)_2CCH_2Bu^t$ --$\underset{pyridine}{SOCl_2}$→ (276) → $Me_2\overset{+}{C}-\underset{Bu^t}{\overset{Me}{C}}-CH_2Bu^t$
|
OH
(273)

↙

$Bu^t\underset{}{\overset{Me}{\diagdown}}{-}CH_2Bu^t$
(274)

A survey of the methods for synthesizing tetrathioethylenes includes the new method of pyrolysis of orthothio-oxalates, e.g. (278).[204]

$(CH_3CO)_2CH-CH(COCH_3)_2$ → $(CH_3CO)_2\overset{Tl}{C}-\overset{Tl}{C}(COCH_3)_2$ $\xrightarrow{I_2}$

$(CH_3CO)_2C=C(COCH_3)_2$
(277)

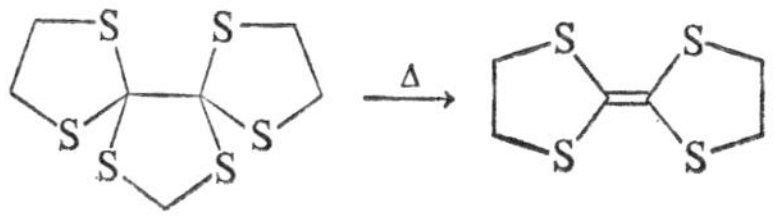

(278) $\xrightarrow{\Delta}$

A synthesis of acyclic 1,5-dienes is exemplified by the route shown in Scheme 18. In all cases cleavage of the more substituted of the two equivalently orientated β,γ-cyclohexane ring bonds occurred in the boronate decomposition (279).[205] A further route to these synthetically valuable 1,5-dienes is

---

[204] D. L. Coffen, J. Q. Chambers, D. R. Williams, P. E. Garrett, and N. D. Canfield, J. Amer. Chem. Soc., 1971, 93, 2258.
[205] J. A. Marshall and J. H. Babler, Tetrahedron Letters, 1970, 3861.

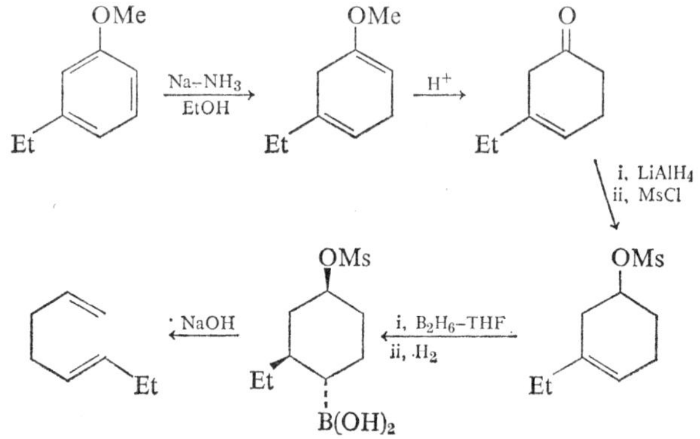

Scheme 18

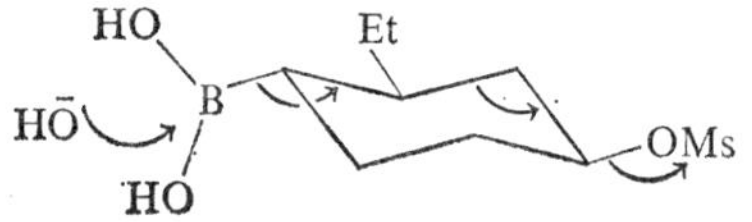

(279)

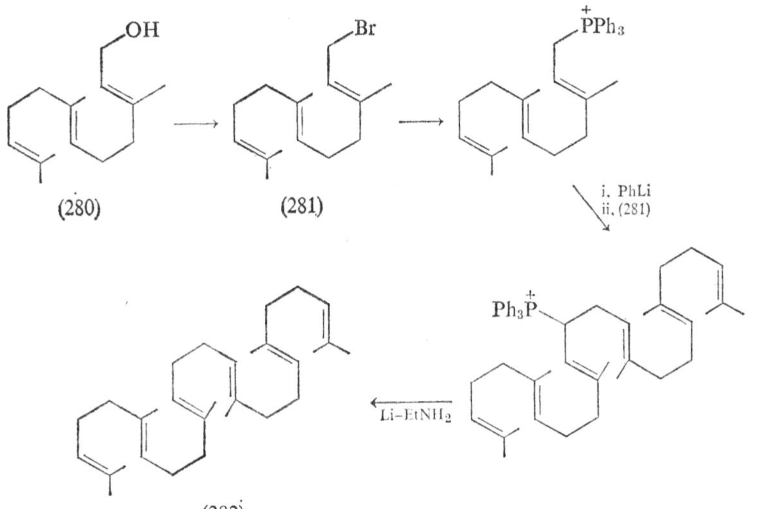

(280) (281) (282)

illustrated in the conversion of *trans,trans*-farnesol (280) into all-*trans*-squalene (282).[206]

Elimination pathways as routes to olefins have also been reviewed[171,172] and studied.[207]

[206] E. H. Axelrod, G. M. Milne, and E. E. van Tamelen, *J. Amer. Chem. Soc.*, 1970, **92**, 2139.

[207] N. Ono, *Bull. Chem. Soc. Japan*, 1971, **44**, 1369; R. D. Bach and D. Andrzejewski, *J. Amer. Chem. Soc.*, 1971, **93**, 7118; D. H. Hunter and D. J. Shearing, *ibid.*, p. 2348; D. J. Lloyd, D. M. Muir, and A. J. Parker, *Tetrahedron Letters*, 1971, 3015.

Acetylenes, Allenes, and Olefins

Cycloaddition Reactions.—Steric selectivity in the addition of carbenes to olefins has been reviewed.[176] The relative reactivity of olefins in the homogeneous reaction with zinc carbenoid generated from diethylzinc and methylene di-iodide shows the order tetramethylethylene > trimethylethylene > cyclohexene > hept-1-ene, in agreement with the electrophilic character of the carbenoid. Steric effects appear to be less important, in contrast to the closely related Simmons–Smith reaction where tetramethylethylene has a very low reactivity. The Hammett ρ-value for addition to substituted styrenes (correlation with σ) has the value −1.61.[208] In addition of the phenylcarbenoid of zinc to olefins, a larger *syn* selectivity is observed than with the phenylcarbenoid of lithium; electron-donating substituents on the phenyl group and ether solvents appear to enhance the selectivity (Scheme 19).[209] The origin of

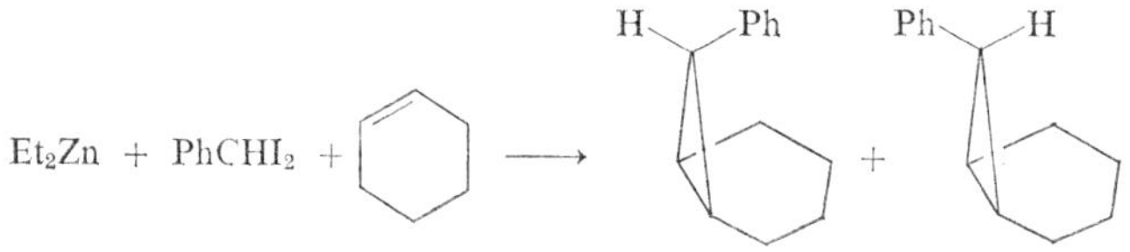

Et$_2$Zn + PhCHI$_2$ + [cyclohexene] → [syn-phenylnorcarane] + [anti-phenylnorcarane]

syn   anti
ratio 17   1   in ether

Scheme 19

this *syn* selectivity is not clear: it does not appear to be the result of London dispersion forces since addition of fluorocarbenoid with olefins gives *syn*-fluorocyclopropane (283) stereoselectively, in spite of its smaller polarizability.[210]

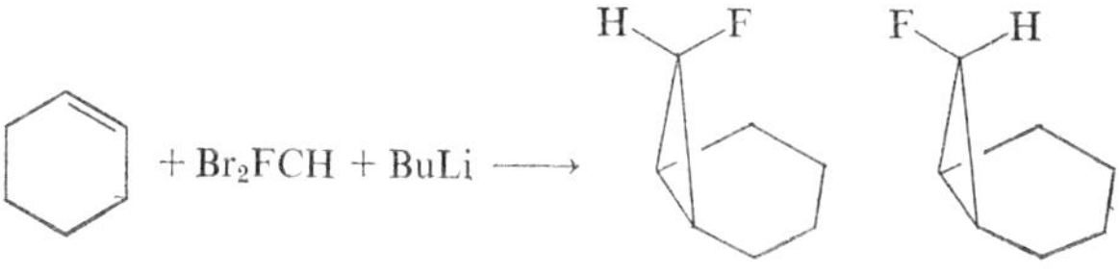

[cyclohexene] + Br$_2$FCH + BuLi ⟶ [syn-fluoronorcarane] + [anti-fluoronorcarane]

(283)

A rationalization of the *cis*:*trans* ratios of olefins produced in carbenic decomposition of diazo-compounds R^{1}CN$_2$CH$_2$R^2 is based on competing electrostatic and steric effects in the intermediate singlet carbene.[211] Heteroatom-containing substituents in the 3-position of cyclohexene can direct the addition of electrophilic species from a *cis* direction and rates of addition are sometimes accelerated.[212] The addition of dichlorocarbene to (284), however,

---

[208] J. Nishimura, J. Furukawa, N. Kawabata, and M. Kitayama, *Tetrahedron*, 1971, **27**, 1799.
[209] J. Nishimura, J. Furukawa, N. Kawabata, and H. Koyama, *Bull. Chem. Soc. Japan*, 1971, **44**, 1127.
[210] M. Schlosser and G. Heinz, *Chem. Ber.*, 1971, **104**, 1934.
[211] Y. Yamamoto and I. Moritani, *Tetrahedron*, 1970, **26**, 1235.
[212] C. D. Poulter, E. C. Friedrich, and S. Winstein, *J. Amer. Chem. Soc.*, 1969, **91**, 6892.

proceeds stereoselectively to give the cyclopropane (285) via apparent *trans* attack.[213] Similarly, the dichlorocarbene addition product of olefin (286) has been shown by an X-ray determination to have structure (287).[214]

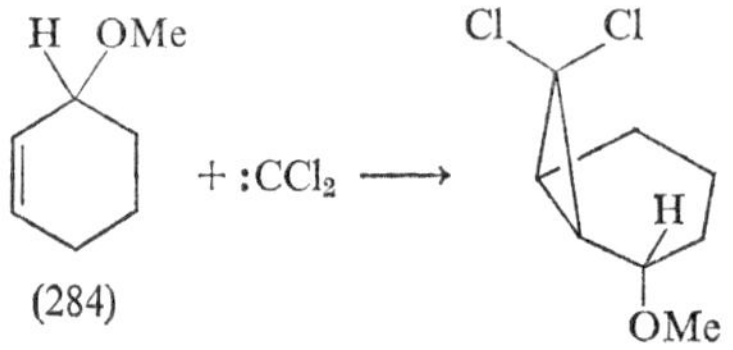

Cope rearrangement of hexa-1,5-dienes generally proceeds by a concerted mechanism with concurrent rupture of the 3,4-bond and formation of the 1,6-bond. The intervention of a biradical mechanism in certain cases is also known.[215] Diene (288) was synthesized to investigate the possible incursion

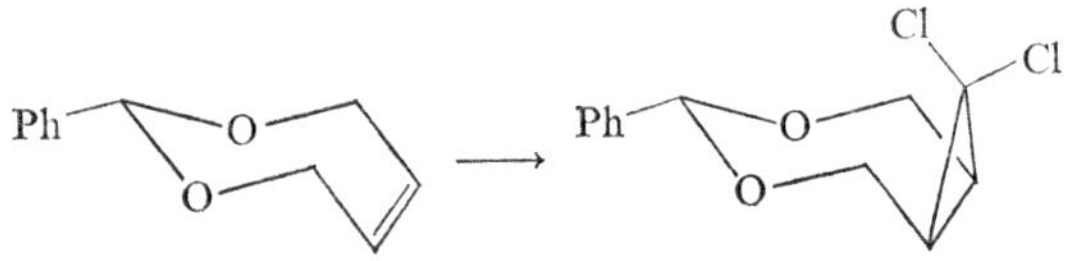

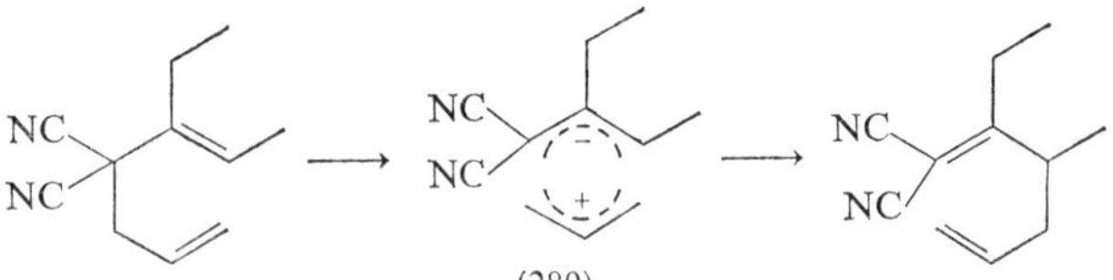

of a third type of mechanism: the formation and recombination of an ion pair (289).[216] The rate of isomerization was followed by observation of the $\alpha\beta$-unsaturated dinitrile absorption band at 240 nm as a function of solvent polarity. There was an increase in rate of only 17 in changing from cyclohexane to ethanol–water, in agreement with secondary deuterium isotope effects[217] in suggesting that the mechanism is concerted.

Thermolysis of 1,5-dien-3-ols (290) leads either to Cope rearrangement (a) or to 1,5-hydrogen shift reaction products (b). These pathways are both observed in rearrangement of 1,2-divinylcycloalkane-1,2-diols; the former leads to ring enlargement by four carbon atoms [(291) → (292)].[218]

[213] M. A. Tobias and B. E. Johnston, *Tetrahedron Letters*, 1970, 2703.
[214] G. R. Clark, B. Fraser-Reid, and G. J. Palenik, *Chem. Comm.*, 1970, 1641.
[215] G. S. Hammond and C. D. Deboer, *J. Amer. Chem. Soc.*, 1964, **86**, 899.
[216] D. G. Wigfield and S. Feiner, *Canad. J. Chem.*, 1970, **48**, 855.
[217] K. Humski, T. Strelkov, S. Borcic, and D. E. Sunko, *Canad. J. Chem.*, 1970, **48**, 855.
[218] P. Leriverend and J.-M. Conia, *Bull. Soc. chim. France*, 1970, 1040; C. Brown and J.-M. Conia, *ibid.*, p. 1050; P. Leriverend and J.-M. Conia, *ibid.*, p. 1060.

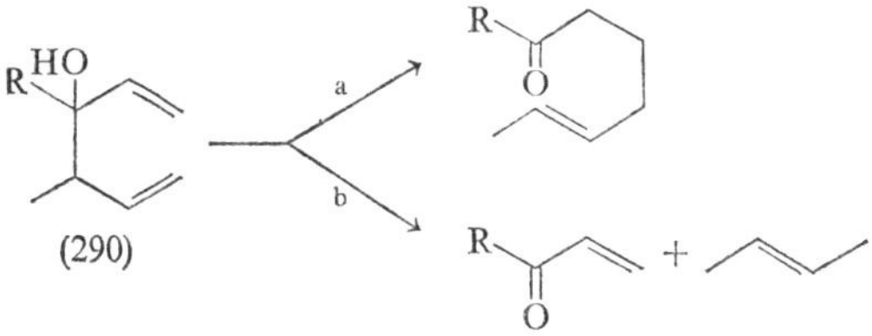

Keten–olefin [2 + 2] cycloadditions have received considerable attention recently.[219] The isomer with the larger keten substituent (L) in the *endo* position (293) is found in addition of keten to cyclopentadiene. This is predicted from the orthogonal keten–olefin transition state $[2_s + 2_a]$ of Woodward and Hoffman. This model also accounts for the reduced reactivity towards ketens of *trans*-butene compared with *cis*-butene.[220] Methylbromo- and methylchloro-keten undergo cycloaddition with a wide variety of olefins with only small variation in the *endo:exo* ratio, *i.e.* the ratio of (295):(294)

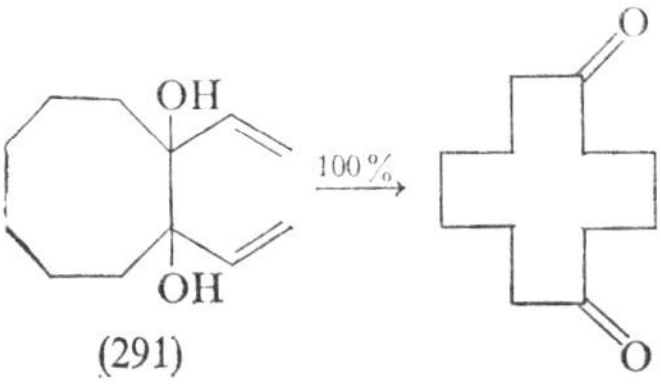

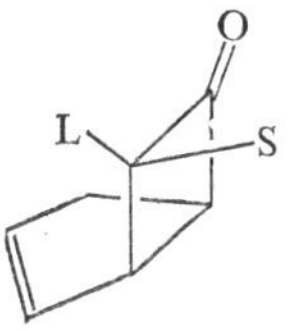

varies only from 5.0 to 4.5 between dihydropyran, cyclo-octene, 1,3-cyclo-octadiene, and cyclohexene. Thus the reaction is controlled by the size of the larger substituent in the keten molecule.[219] Norbornene and norbornadiene are reported to be poor ketenophiles although they show high dipolarophilic activity, and addition of dichloroketen to dicyclopentadiene gives only (296) or (297).[221]

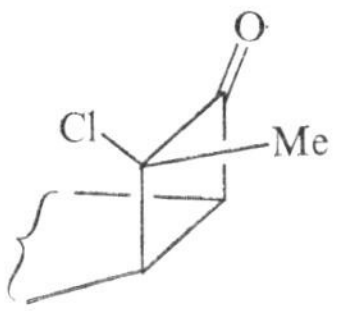

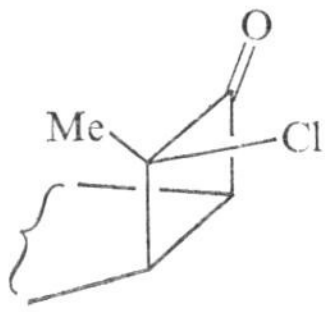

The concerted suprafacial thermal addition of two olefins is a symmetry-forbidden process and thermal 1,2-cycloadditions of simple olefins have been

---

[219] W. T. Brady and R. Roe, *J. Amer. Chem. Soc.*, 1971, **93**, 1662, and refs. 1—8 therein.
[220] N. S. Isaacs and P. F. Stanbury, *Chem. Comm.*, 1970, 1061.
[221] L. Ghosez, R. Montaigue, A. Roussel, H. Vanlierde, and P. Mollet, *Tetrahedron*, 1971, **27**, 615.

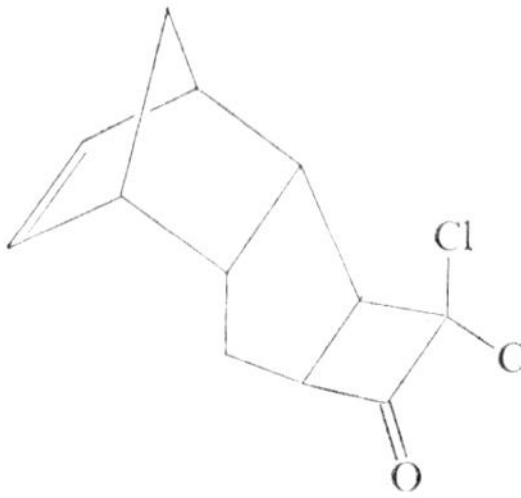

(296) 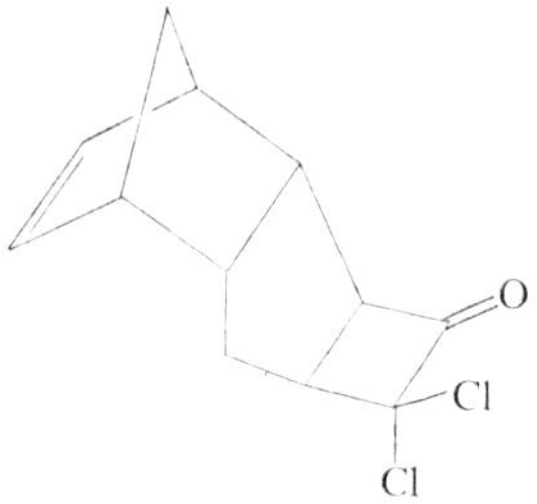   (297)

shown to involve biradical intermediates. The major dimeric product (298), isolated from thermolysis of (299), suggests that the predominant portion of the reaction occurs in a stereochemical sense compatible with orbital symmetry theory $[_\pi 2_s + _\pi 2_a]$.[222] However, the isolation of two other dimers

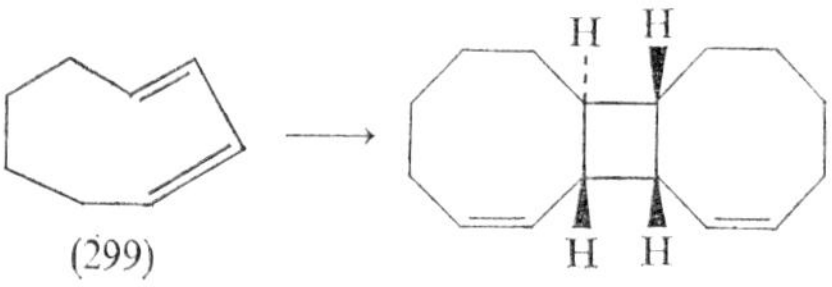

(299) → (298)

differing only in stereochemistry at the ring junctions, and the lack of variation with temperature in the ratios of these dimers obtained, favours a stepwise mechanism with a biradical intermediate.

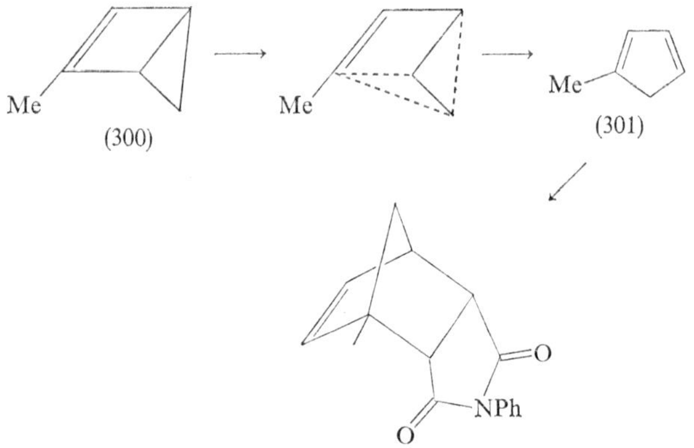

(300) → → (301)

(302)

[222] A. Padwa, W. Koehn, J. Masaracchia, C. L. Osborn, and D. J. Trecker, *J. Amer. Chem. Soc.*, 1971, **93**, 3633.

Acetylenes, Allenes, and Olefins

2-Methylbicyclo[2,1,0]pent-2-ene (300) rearranges thermally to 1-methylcyclopentadiene (301), showing the process as one ascribable to a $[_\sigma 2_s + {}_\sigma 2_a]$ concerted cycloreaction involving C-1—C-2 and C-4—C-5 bonds and not a biradical two-step process. The cyclopentadiene (301) was trapped as it was formed (43 °C) as its N-phenylmaleimide adduct (302), a process which completes favourably with the 1,5-hydrogen shift in the diene itself. It is pointed out that this result undermines the usual either/or concerted/non-concerted question of mechanism in that an unusual or suppressed concerted mechanism may be brought to the fore when the more obvious concerted mechanism is prohibited.[223]

Under the influence of a nickel catalyst, methylenecyclopropane (303) reacts with methyl acrylate to give the methylenecyclopentane (304) in 82% yield, formally by a $[_\pi 2 + {}_\sigma 2]$ cycloaddition.[224] The symmetrical intermediate (305) may be excluded since (306) and (307) are obtained from (308) and (309), respectively. A symmetrical intermediate would give the same adduct or mixture of adducts.

(303) $R^1 = R^2 = H$
(308) $R^1 = Me, R^2 = H$ /
(309) $R^1 = H, R^2 = Me$

(304) $R^1 = R^2 = H$
(306) $R^1 = Me, R^2 = H$
(307) $R^1 = H, R^2 = Me$

(305)

Diels–Alder addition of 2-chloroacrylonitrile and dienes gives the expected adducts, *e.g.* (310), which may be converted *via* azide, isocyanate, and hydrolysis to the ketone (311) in good yield. This constitutes a method for 1,4-addition of the methylenecarbonyl unit (—CH_2CO—) to dienes.[225]

Among the many 1,3-dipolar additions of olefins recently reported are the intramolecular nitrone–olefin cycloadditions,[226] *e.g.* (312) → (313), and the cycloaddition with olefinic dipolarophiles of carbonyl ylides (314)[227] and azomethine ylides (315),[228] obtained by conrotatory ring-opening of the corresponding three-membered rings Attempted cycloaddition of the olefin (316) with benzenesulphonyl isothiocyanate gave the dipolar product (317), whose structure was supported by addition to diphenylketen, giving (318).[229]

---

[223] J. E. Baldwin and A. H. Andrist, *Chem. Comm.*, 1970, 1561.
[224] R. Noyori, T. Odagi, and H. Takaya, *J. Amer. Chem. Soc.*, 1970, **92**, 5780.
[225] E. J. Corey, T. Ravindranathan, and S. Terashima, *J. Amer. Chem. Soc.*, 1971, **93**, 4326.
[226] W. Oppolzer and K. Keller, *Tetrahedron Letters*, 1970, 1117; W. Oppolzer, and H. P. Weber, *ibid.*, p. 1121.
[227] H. Hamberger and R. Huisgen, *Chem. Comm.*, 1971, 1190; A. Dahmen, H. Hamberger, R. Huisgen, and V. Markowski, *ibid.*, p. 1192.
[228] J. H. Hall and R. Huisgen, *Chem. Comm.*, 1971, 1187; J. H. Hall, R. Huisgen, C. H. Ross, and W. Scheer, *ibid.*, p. 1188.
[229] R. Gompper and B. Wetzel, *Tetrahedron Letters*, 1971, 529.

Electrophilic Addition.—A detailed kinetic and product analysis of the reaction of 1,2-dimethylcyclohexene with HCl in acetic acid has been made at <10% conversion, when secondary reactions are unimportant.[230] Addition of HCl takes place by two competing mechanisms: a carbonium chloride ion pair (AdE2) leading to (319) as the major product, and a termolecular (AdE3) mechanism with a transition state resembling (320), the latter dominating at high Cl$^-$ concentrations and giving the *trans* addition product (321). Products from attack of the acetic acid solvent are also obtained. Cyclohexene reacts similarly, though slower by both mechanisms.[231] Addition of HCl to 1-phenyl-4-t-butylcyclohexene has yielded two crystalline geometrical *cis*- and *trans*-isomers. The stereochemistry of addition was determined by the use of deuteriated cyclohexene (322) and the isolation of (323) as the major product implies a *syn* addition process since this is the less-stable

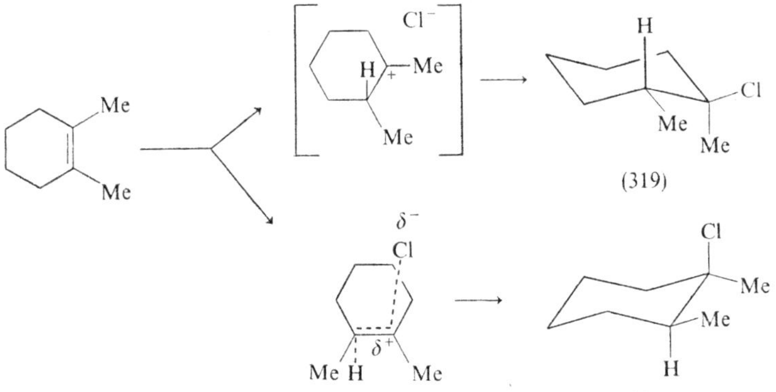

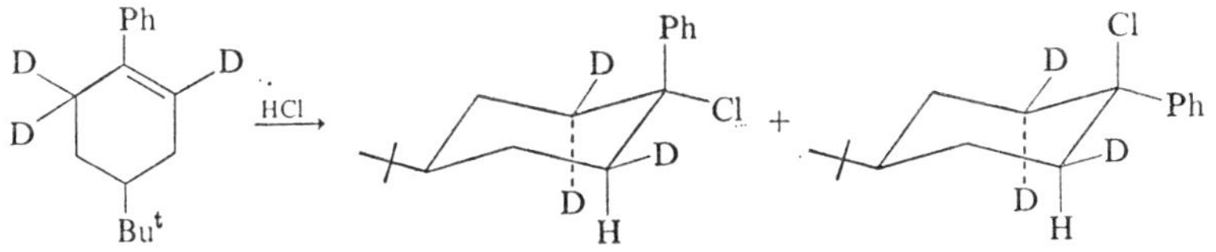

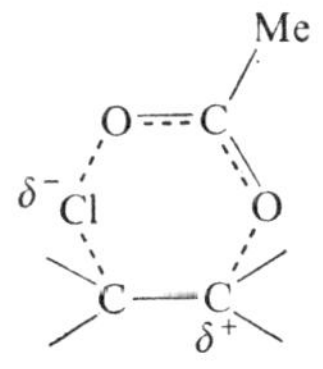

---

[230] R. C. Fahey and C. A. McPherson, *J. Amer. Chem. Soc.*, 1971, **93**, 2445.
[231] R. C. Fahey, M. W. Monahan, and C. A. McPherson, *J. Amer. Chem. Soc.*, 1970, **92**, 2810.

isomer. However, the ratio of isomers produced is dependent on HCl concentration and temperature.[232] A cyclic transition state (324) leading to *cis* or *trans* addition has been proposed for the reaction of chlorine acetate with olefins.[233]

Addition of antimony pentachloride to cyclohexene gives 1,2-dichlorocyclohexanes in a *cis:trans* ratio of 5, whereas chlorination with other metal chlorides gives mainly *trans* addition, as in the case using elemental chlorine.[234] Iodonium nitrate, prepared *in situ* by addition of ICl to AgNO$_3$ in chloroform–pyridine, adds to alkenes at room temperature to form (*i*) iodonitrate esters (325) (*trans* addition), (*ii*) iodoalkane pyridinium nitrates (326), or (*iii*) alkene pyridinium iodides (327) depending on the substrate.[235] Compounds (326) and (327) may be formed by nucleophilic displacement of

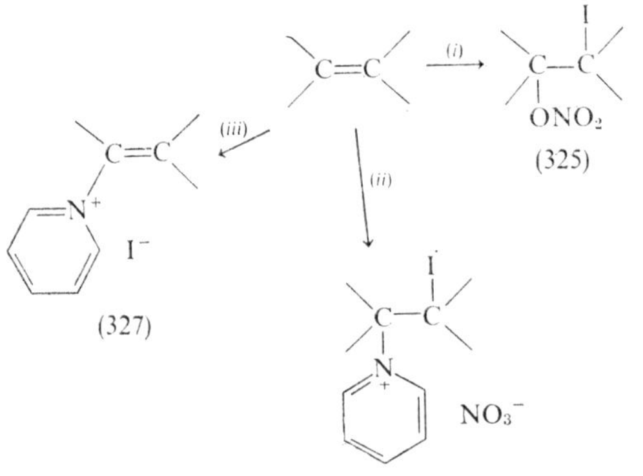

nitrate ion from the initial iodonitrate esters. Iodine azide adds to olefins stereo- and regio-specifically.[177] That *trans* addition obtains in this reaction is shown by reaction with *cis*-deuteriostyrene (328), where the product is (329), elimination with potassium t-butoxide in ether giving only (330). Electrophilic addition of bromine azide is also generally stereospecific but not, however, in the case of (328), indicating that a benzylic cation is involved.[236] Iodine isocyanate also adds stereospecifically *via* an iodonium ion.

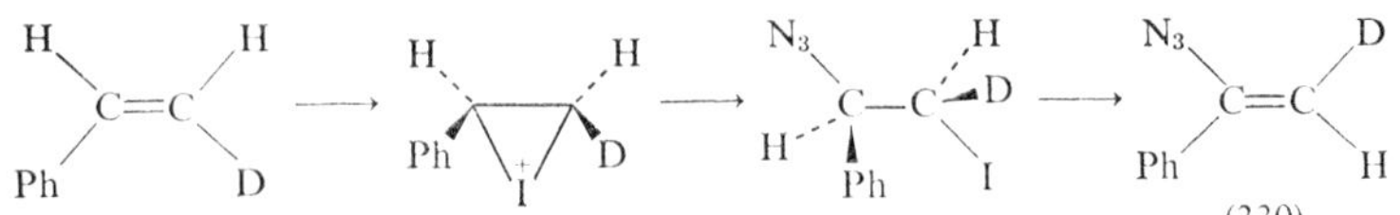

[232] K. D. Berlin, R. O. Lyerla, D. E. Gibbs, and J. P. Devlin, *Chem. Comm.*, 1970, 1246.
[233] P. B. D. de la Mare, C. J. O'Connor, M. J. Rosser, and M. A. Wilson, *Chem. Comm.*, 1970, 731.
[234] S. Uemura, O. Sasaki, and M. Okano, *Chem. Comm.*, 1971, 1064.
[235] U. E. Diner and J. W. Lown, *Chem. Comm.*, 1970, 333.
[236] A. Hassner, F. P. Boerwinkel, and A. B. Levy, *J. Amer. Chem. Soc.*, 1970, **93**, 4879.

Acetylenes, Allenes, and Olefins

The iodocarbamates obtained are easily hydrogenolysed by zinc and acid to provide a synthesis of the carbamates (331) from olefins.[237]

$$\begin{array}{c}\diagdown\\ \diagup\end{array}C{=}C\begin{array}{c}\diagup\\ \diagdown\end{array} \xrightarrow{\text{INCO}} \begin{array}{c}\diagdown\\ \diagup\end{array}\overset{I}{\underset{OCN}{C{-}C}}\begin{array}{c}\diagup\\ \diagdown\end{array} \xrightarrow[\text{ii, Zn-H}^+]{\text{i, EtOH}} \begin{array}{c}\diagdown\\ \diagup\end{array}CH{-}\overset{|}{\underset{|}{C}}{-}NHCO_2R$$

(331)

Anti-Markovnikov addition of HBr to terminal olefins is effected by bromination of the derived organoboranes in the presence of sodium methoxide.[238] Alternatively,[239] alk-1-enes are converted to primary alkyl bromides by the sequence:

$$RCH{=}CH_2 \xrightarrow{R_2BH} RCH_2CH_2BR_2 \xrightarrow[\text{ii, Br}_2]{\text{i, HgO-OH}^-} RCH_2CH_2Br$$

Iodohydrins have not previously been unambiguously prepared from iodine, olefin, and water, but can easily be obtained by iodination of the olefin in the presence of iodic acid, which re-oxidizes the iodide formed in the reaction back to iodine. Alternatively, oxygen in the presence of nitrous acid may act as the oxidant. The corresponding epoxides are readily formed with base.[240]

$$Ph_2C{=}CHMe + Et_3AlCl \longrightarrow \begin{array}{c}\text{Me}\\|\\Ph_2\overset{+}{C}\cdots\overset{CH}{}\cdots\overset{-}{Al(Et)Cl}\\H\diagdown\diagup CH_2\\CH_2\end{array}$$

$$\begin{array}{c}\text{Me}\\|\\Ph_2CH{-}CHOH\end{array} \xleftarrow{H_2O_2} \begin{array}{c}\text{Me}\\|\\Ph_2CH{-}CHAl(Et)Cl\end{array} + C_2H_4$$

(332)

The well-known hydroalumination of olefins with dialkylaluminium hydrides can be extended to the addition of diethylaluminium chloride. With diphenylpropene the intermediate (332) is formed, together with ethylene, and yields 1,1-diphenylpropan-2-ol in 90% yield[241] on oxidation with H_2O_2.

Chlorination of styrenes in DMF gives an isolable immonium chloride (333),[242] which is converted into 1,2-dichloro-1-phenylethane on heating,

---

[237] A. Hassner, R. P. Hoblitt, C. Heathcock, J. E. Kropp, and M. Lorber, *J. Amer. Chem. Soc.*, 1970, **92**, 1326.
[238] H. C. Brown and C. F. Lane, *J. Amer. Chem. Soc.*, 1970, **92**, 6660; see also C. F. Lane and H. C. Brown, *J. Organometallic Chem.*, 1971, **26**, C51.
[239] J. J. Tufariello and M. M. Hovey, *Chem. Comm.*, 1970, 372.
[240] J. W. Cornforth, *J. Chem. Soc. (C)*, 1970, 846.
[241] A. Alberola, *Tetrahedron Letters*, 1970, 3471.
[242] A. E. Roocker and P. de Radzitzky, *Bull. Soc. chim. belges*, 1970, **79**, 531.

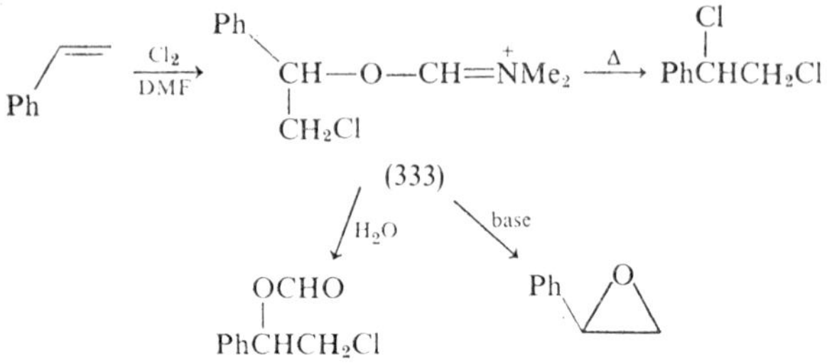

$$\text{Ph}\diagup\!\!=\xrightarrow{\text{Cl}_2,\,\text{DMF}}\underset{\overset{|}{\text{CH}_2\text{Cl}}}{\text{Ph}\text{CH}-\text{O}-\text{CH}=\overset{+}{\text{N}}\text{Me}_2}\xrightarrow{\Delta}\underset{\overset{|}{\text{Cl}}}{\text{PhCHCH}_2\text{Cl}}$$

(333)

↙ H₂O   ↘ base

OCHO
|
PhCHCH₂Cl

(334)

Ph△O (styrene oxide)

into styrene oxide on treatment with base, or into the formate (334) on gentle hydrolysis.

Reaction of olefins with mercuric acetate in aqueous THF followed by *in situ* reduction of the mercurial intermediate is a convenient method for Markovnikov hydration of the double bond (Scheme 20).[243] Using chiral

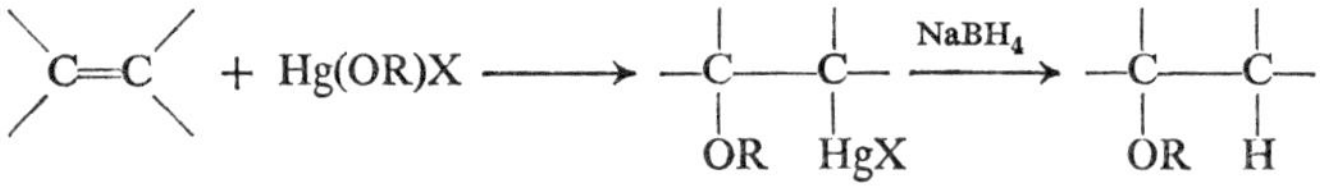

$$\diagdown\!\!\text{C}=\text{C}\!\!\diagup + \text{Hg(OR)X} \longrightarrow -\underset{\underset{\text{OR}}{|}}{\text{C}}-\underset{\underset{\text{HgX}}{|}}{\text{C}}- \xrightarrow{\text{NaBH}_4} -\underset{\underset{\text{OR}}{|}}{\text{C}}-\underset{\underset{\text{H}}{|}}{\text{C}}-$$

Scheme 20

mercury(II) carboxylates in the oxymercuration step leads to the synthesis of optically active alcohols.[244]

Electrophilic addition of mercuric salts to olefins has been thought for a long time to proceed through mercurinium ions.[245] The ion (335) has now been observed by n.m.r. on adding cyclohexene in sulphur dioxide to a solution of mercuric trifluoroacetate in a slight excess of $FSO_3H-SbF_5-SO_2$. Alternatively the σ-route can be employed—the ionization of 2-acetoxymercuricyclohexan-1-ol (336) in $FSO_3H-SbF_5-SO_2$ solution.[246] The mercurinium ions

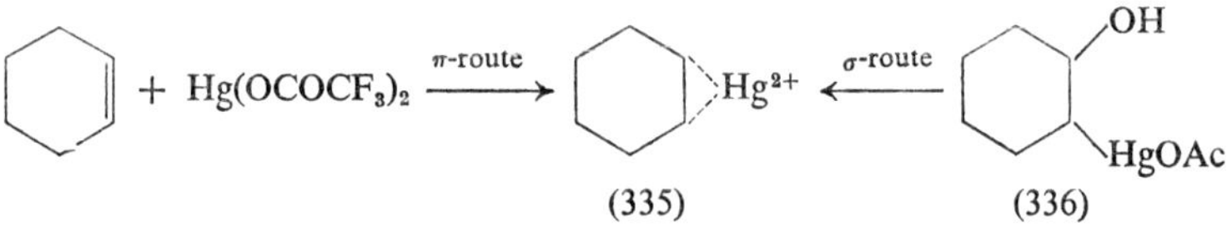

⬡ + Hg(OCOCF₃)₂ —π-route→ ⬡·Hg²⁺ ←σ-route— ⬡(OH)(HgOAc)

(335)   (336)

of ethylene and norbornylene have similarly been observed.[247] Oxymercuration of 4-t-butylcyclohexene and 1-methyl-4-t-butylcyclohexene gives exclusively the *trans*-diaxial products (337) and (338). Bromination and methoxybromination give *trans*-diaxial products *via* bromonium ion intermediates, whereas hydrobromination proceeds by an Ad*E*3 mechanism and gives mixtures of axial and equatorial bromides. By analogy, therefore,

[243] H. C. Brown and P. J. Geohegan, *J. Org. Chem.*, 1970, **35**, 1844.
[244] R. M. Carlson and A. H. Funk, *Tetrahedron Letters*, 1971, 3661.
[245] W. Kitching, *Organometallic Chem. Rev.*, 1968, 3, 61.
[246] G. A. Olah and P. R. Clifford, *J. Amer. Chem. Soc.*, 1971, **93**, 2320.
[247] G. A. Olah and P. R. Clifford, *J. Amer. Chem. Soc.*, 1971, **93**, 1261.

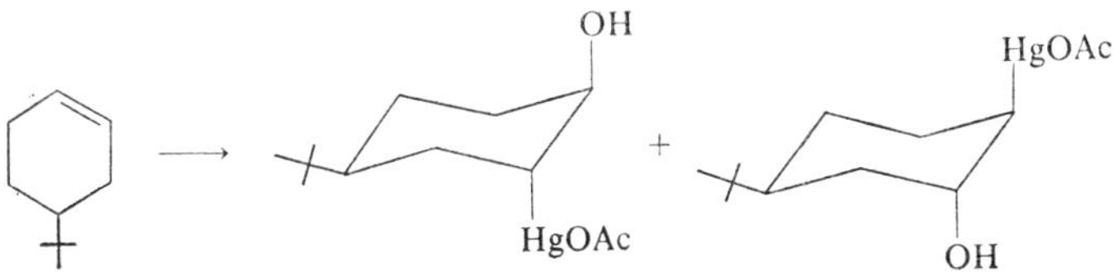

Acetylenes, Allenes, and Olefins 71

mercurinium ions are intermediates in the formation of (337) and (338).[248] The location of olefinic links in long-chain esters may be accomplished by methoxymercuration followed by $NaBH_4$ reduction and gas-chromatography–mass spectrometry of the resulting methoxy-esters.[249]

Nucleophilic Addition.—Addition of alkyl-lithium reagents to unstrained unconjugated double bonds is observed when complexation of the alkyl-lithium is possible by an ether, thioether, or amine within the olefin. Thus (339) is converted into (340) by isopropyl-lithium in 95% yield. A transition

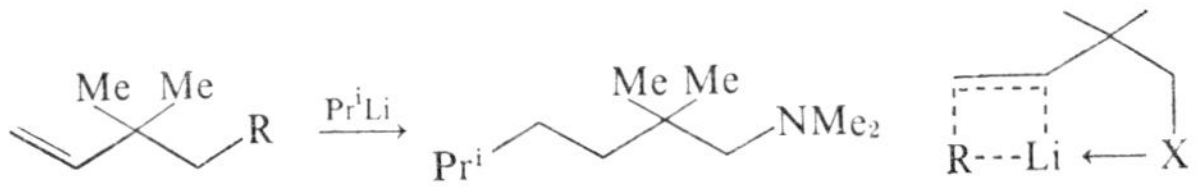

(339) R = NMe₂
(342) R = H

(340)

(341)

state such as (341) has been suggested, but the stereochemistry of addition to the double bond remains to be proved. More action is observed with (342).[250]

Organomagnesium halides containing a β double bond react with olefins by addition without catalysis:

$$ClMgCH_2\overset{R^1}{C}=CH_2 + H_2C=CH-R^2 \longrightarrow ClMgCH_2\overset{R^2}{C}HCH_2\overset{R^1}{C}=CH_2$$

Crotylmagnesium halide reacts with ethylene almost exclusively in the form (343), but with oct-1-ene both forms react.[251]

$$ClMgCH_2CH=CHMe \rightleftharpoons ClMg\overset{Me}{C}HCH=CH_2$$
(343)

$$\downarrow C_2H_4$$

$$MeCH_2\overset{}{C}HCH=CH_2 \xleftarrow{H_2O} ClMgCH_2CH_2\overset{}{C}HCH=CH_2$$
$$\overset{|}{Me} \qquad\qquad\qquad\qquad\qquad \overset{|}{Me}$$

---

[248] D. J. Pasto and J. A. Gontarz, *J. Amer. Chem. Soc.*, 1971, **93**, 6902.
[249] P. Abley, F. J. McQuillin, D. E. Minnikin, K. Kusamran, K. Maskens, and N. Polgar, *Chem. Comm.*, 1970, 348.
[250] A. H. Veefkind, J. V. D. Schaaf, F. Bickelhaupt, and G. W. Klumpp, *Chem. Comm.*, 1971, 722.
[251] H. Lehmkuhl and D. Reinehr, *J. Organometallic Chem.*, 1970, **25**, C47.

Reduction with sodium in diethyl ether of di-t-butyl ketone gives (344). The postulated mechanism of this interesting transformation is addition of the dianion (345) to ethylene, generated *in situ* from diethyl ether; this was

$$Bu^t_2C=O \xrightarrow{Na-ether} Bu^t\underset{OH}{\underset{|}{C}}(CH_2)_2\underset{OH}{\underset{|}{C}}Bu^t_2$$
$$(344)$$

$$Bu^t_2\bar{C}-\bar{O} + CH_2=CH_2 \longrightarrow Bu^t\underset{O^-}{\underset{|}{C}}CH_2\bar{C}H_2$$
$$(345)$$

with $Bu^t_2C=O$ feeding back into the cycle.

supported by the isolation of (344) from di-t-butyl ketone and potassium in benzene when ethylene was bubbled through the solution.[252]

Radical Addition.—Ceric acetate reacts with styrene in acetic acid at 110 °C to give the lactones (346) in 60—70% yield.[253] With lead tetra-acetate (LTA),

$$PhCH=CH_2 \xrightarrow[RCH_2CO_2H]{Ce(OAc)_2} \underset{(346)\ R\ =\ H\ or\ Me}{\begin{array}{c}PhCH-CH_2\\ \diagdown\ \diagup R\\ O\\ \|\\ O\ H\end{array}} \qquad \underset{(347)}{\begin{array}{c}PhCHCH_2Me\\ |\\ OCOMe\end{array}}$$

whose action is known to be decarboxylative, the acetate (347) is obtained. Thermal decomposition of ceric carboxylates, in which exchange with other carboxylic acids present is assumed, is believed to involve carboxyalkyl radicals $R\dot{C}HCO_2H$, but a decarboxylative pathway obtains in photochemical decomposition where the product is (347), the same as that obtained with LTA.

Certain electrophilic species such as Cl_2, Br_2, and BrN_3 can add to olefins by ionic or molecule-induced radical mechanisms depending on the conditions. Under radical-inhibiting conditions, methyl hypobromite adds to hex-1-ene to yield primarily (348), but (349) is obtained under radical-stimulating conditions.[254] Acetyl hypobromite added to hex-1-ene exclusively

$$\underset{(348)}{\begin{array}{c}Me(CH_2)_3CHCH_2Br\\ |\\ OMe\end{array}} \qquad \underset{(349)}{\begin{array}{c}Me(CH_2)_3CHCH_2OMe\\ |\\ Br\end{array}}$$

by an ionic mechanism. The high-temperature pyrolysis of cycloalkenes (350; $n = 0$—2) causes isomerization to α,ω-dienes (351) and/or ring-contracted vinylcycloalkanes (352) probably *via* biradical intermediates (353).[255]

---

[252] P. D. Bartlett, T. T. Tidwell, and W. P. Weber, *Tetrahedron Letters*, 1970, 2919.
[253] E. I. Heiba and R. M. Dessau, *J. Amer. Chem. Soc.*, 1971, **93**, 995.
[254] V. L. Heasley, C. L. Frye, G. E. Heasley, K. A. Martin, D. A. Redfield, and P. S. Wilday, *Tetrahedron Letters*, 1970, 1573.
[255] J. K. Crandall and R. J. Watkins, *J. Org. Chem.*, 1971, **36**, 913.

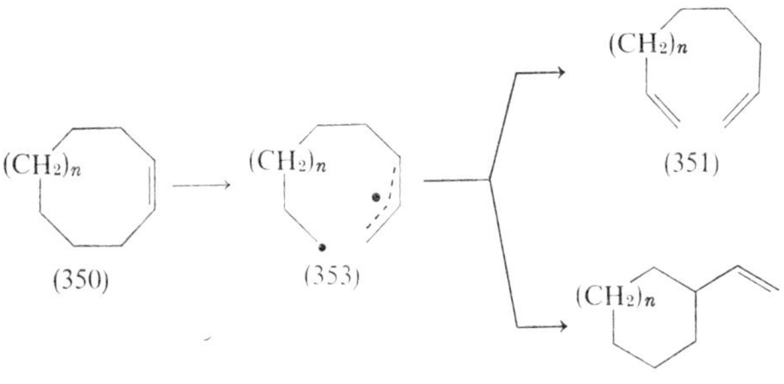

Acetylenes, Allenes, and Olefins

[Structures (350) → (353) → (351) and (352)]

The kinetics of addition of iodine to simple olefins using ^{131}I-labelled iodine support the view that the stereospecific addition is the result either of attack of an iodine atom on a charge-transfer complex between an olefin and an iodine molecule or attack of an iodine molecule on a charge-transfer complex between an olefin and an iodine atom. The products of the reaction, the vicinal di-iodides, were reasonably stable in the dark if care was taken to remove all iodine.[256]

Oxidation and Reduction.—Oxidative rearrangement of olefins with thallium(III) nitrate in methanol provides a simple synthesis of aldehydes and ketones in high yields. Cleavage of the intermediate thallium compound proceeds *via* a transition state with high carbonium ion character, which leads either to carbonyl compounds (354) by Wagner–Meerwein rearrangement or to glycol methyl ethers (355).[257]

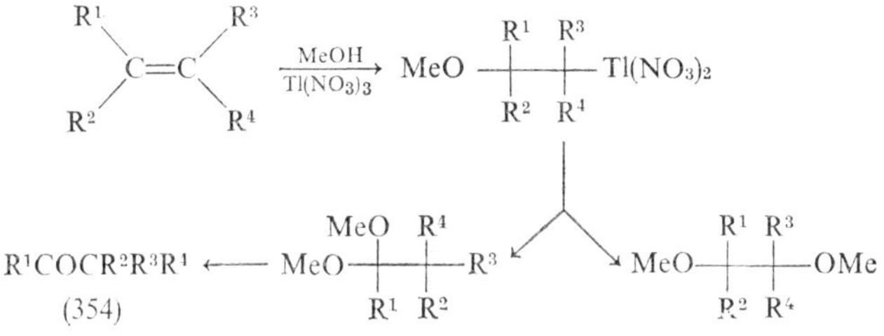

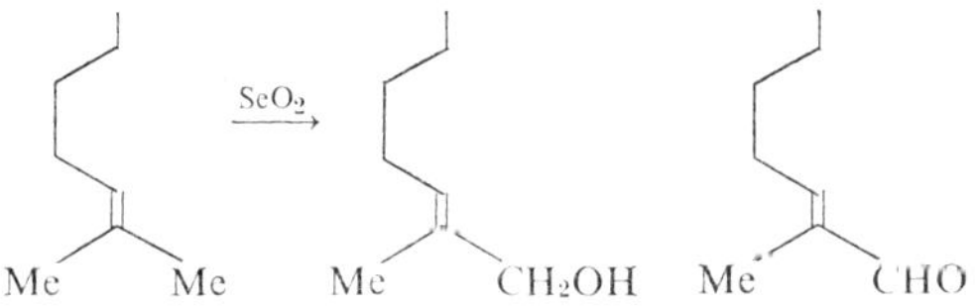

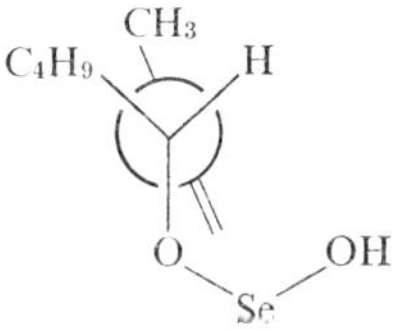

[Scheme showing R¹R²C=CR³R⁴ → MeO-CR¹R²-CR³R⁴-Tl(NO₃)₂ → (354) R¹COCR²R³R¹ and (355) MeO-CR¹R²-CR³R⁴-OMe]

[Structures (356), (357), (358), (359) with SeO₂ reaction sequence]

[256] R. L. Ayres, C. J. Michejda, and E. P. Rack, *J. Amer. Chem. Soc.*, 1971, **93**, 1389.
[257] A. McKillop, J. D. Hunt, E. C. Taylor, and F. Kienzle, *Tetrahedron Letters*, 1970, 5275.

Allylic oxidation by selenium dioxide in ethanol of *gem*-dimethylolefins has revealed some unexpected stereospecificity. Thus oxidation of (356) gives exclusively the *trans*-alcohol (357), or the *trans*-aldehyde (358) after a longer heating period. To accommodate these stereochemical findings, an S_N2' solvolysis of the first-formed allylic selenite ester (359) is assumed for a trisubstituted olefin.[258] Similarly, the oxidation of a series of alkylated cyclohexenes by SeO_2 in wet dioxan has been shown to proceed stereoselectively.[259] The usual pathway for selenium dioxide oxidation in acetic acid was drastically changed when a catalytic amount of concentrated sulphuric acid was added: using cyclohexene the major product was 1,2-diacetoxycyclohexane in a yield of 32% and no 3-acetoxycyclohex-1-ene, the product in the absence of H_2SO_4, was obtained.[260]

Epoxidation of olefins under mild conditions is carried out by using an initially formed complex between H_2O_2 and an isocyanate in non-polar solvents. Using phenyl isocyanate the by-products are 1,3-diphenylurea and carbon dioxide.[261]

Potassium permanganate in acetic anhydride converts acyclic and larger cyclic rings into α-diketones. Thus cyclododecene is converted into the α-diketone (360) in 48% yield, with (361) and (362) as easily removable by-products.[262]

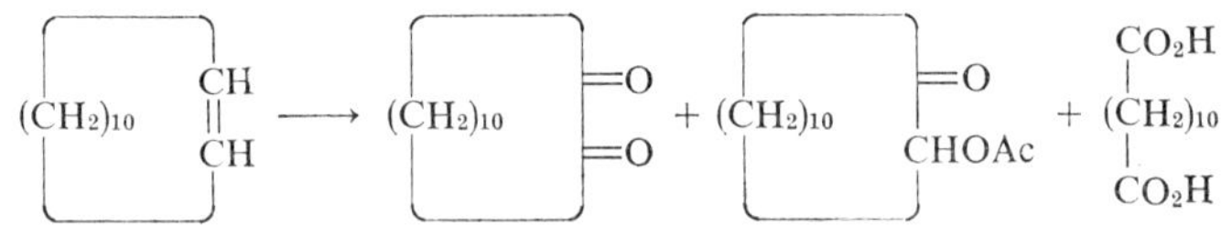

Electron-rich alkenes, *e.g.* (363), are oxidized rapidly and in good yields by carbon tetrabromide:

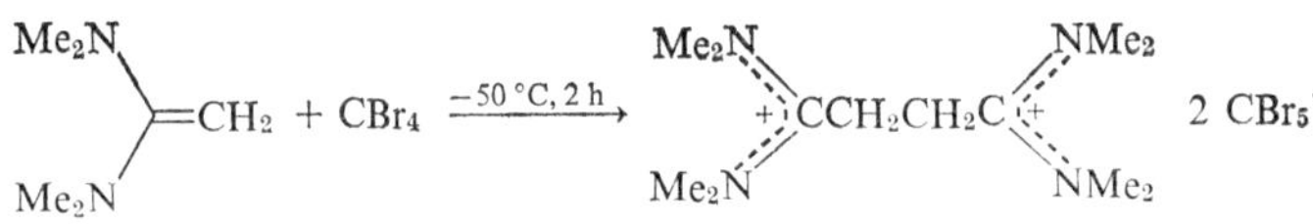

Keten diethyl acetal, however, is not oxidized under these conditions.[263] Electrochemical, oxidative 1,2-addition of azide into non-activated double

---

[258] U. T. Bhalerao and H. Rapoport, *J. Amer. Chem. Soc.*, 1971, **93**, 4835.
[259] E. N. Trachtenberg and J. R. Carver, *J. Org. Chem.*, 1970, **35**, 1646; E. N. Trachtenberg, C. H. Nelson, and J. R. Carver, *ibid.*, p. 1653.
[260] K. A. Javaid, N. Sonoda, and S. Tsutsumi, *Bull. Chem. Soc. Japan*, 1970, **43**, 3475.
[261] N. Matsumura, N. Sonoda, and S. Tsutsumi, *Tetrahedron Letters*, 1970, 2029.
[262] K. B. Sharpless, R. F. Lauer, O. Repič, A. Y. Teranishi, and D. R. Williams, *J. Amer. Chem. Soc.*, 1971, **93**, 3303.
[263] F. Effenberger and O. Gerlach, *Tetrahedron Letters*, 1970, 1669.

Acetylenes, Allenes, and Olefins

bonds has been accomplished: electron-rich olefins dimerize and give 1,4-diazides in superior yields, e.g. stryrene gives 1,4-diazido-2,3-diphenyl-butanes in ca. 50% yield.[264]

The mechanism of ozonolysis appears to grow more complex with the passage of time. Ozonolysis of ethylidenecyclohexane (364) in the presence

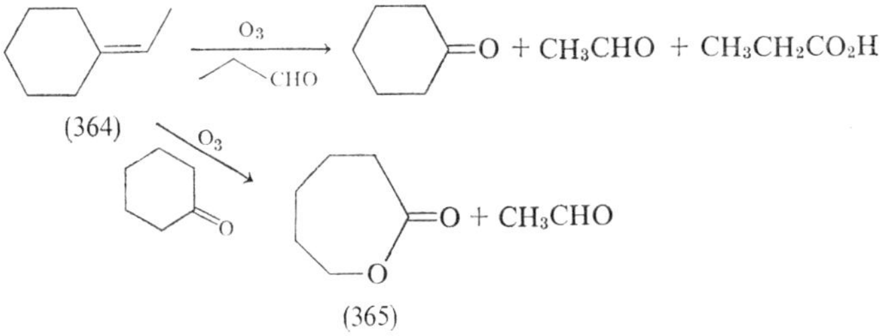

of excess cyclohexanone or propionaldehyde has been found to lead to Baeyer–Villiger reaction products—6-hexanolide (365) and propionic acid—with complete suppression of the normal ozonide formation.[265] It has been suggested that these products arise by attack of the Staudinger molozonide (366) upon the aldehyde (ketone) (Scheme 21). The molozonide is considered

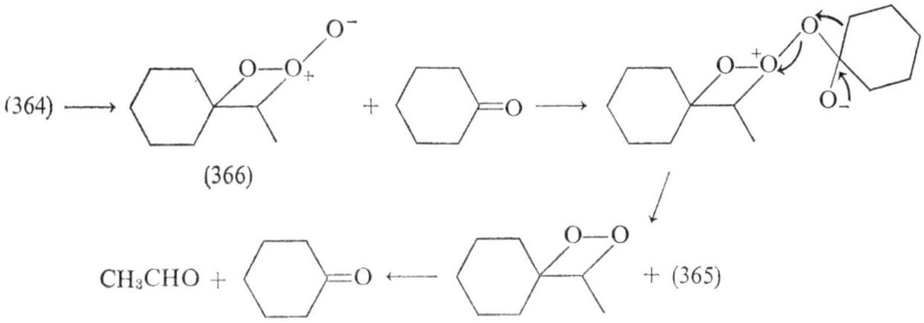

to be either the first addition product of an olefin and ozone or the rearrangement product of a peroxy-epoxide (367). It may itself rearrange either to the trioxolan (368), or to the Criegee zwitterion (369), or to the ozonide (370).

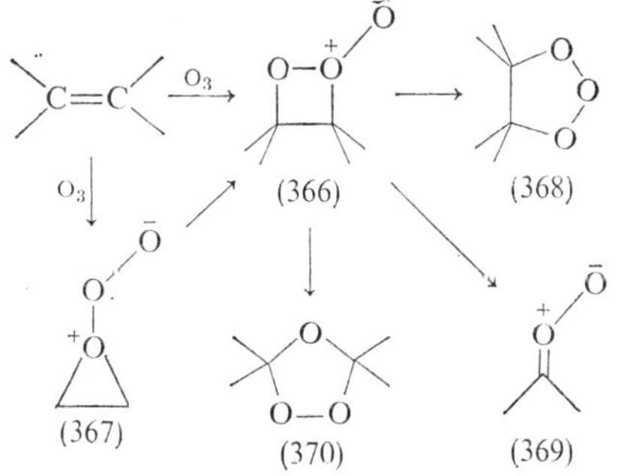

[264] H. Schäfer, *Angew. Chem. Internat. Edn.*, 1970, **9**, 158.
[265] P. R. Story, J. A. Alford, J. R. Burgess, and W. C. Ray, *J. Amer. Chem. Soc.*, 1971, **93**, 3042; P. R. Story, J. A. Alford, W. C. Ray, and J. R. Burgess, *ibid.*, p. 3044.

A new polymer-supported rhodium(I) catalyst has been obtained by chloromethylation of polystyrene beads containing 1.8% cross-linking of divinylbenzene, converting to the diphenylphosphine, and then equilibrating with tris(triphenylphosphine)chlororhodium(I) (Scheme 22). This catalyst reduced

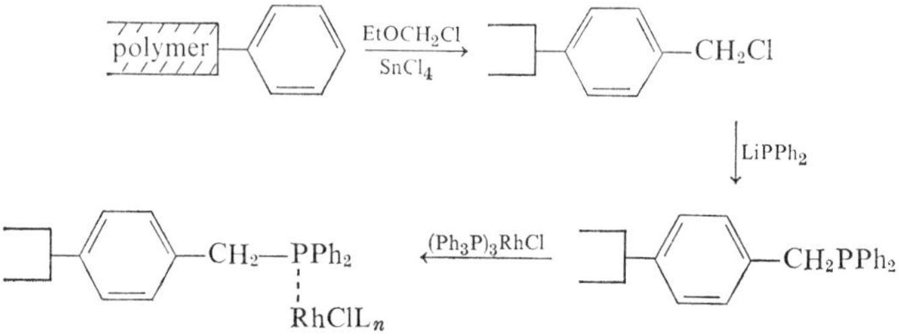

Scheme 22

olefins in benzene at a rate which depended on the molecular size of the olefin, attributable to restriction of the size of the solvent channels by the random cross-linking in the polymer.[266]

Reduction of tetra-alkyl-substituted olefins is possible using sodium in hexamethylphosphoramide (HMPA) containing t-butanol, and offers a promising method of introducing deuterium (using t-butan[^{2}H]ol) without isomerization and scrambling common to catalytic reductions.[267]

Electroreduction of olefinic ketones of the type (371) gave the cyclic alcohol (372) in 66% yield. The exclusive formation of the *cis*-isomer is in contrast to the *trans*-isomer formed from the Grignard reaction of methylmagnesium iodide with the cyclopentanone.[268]

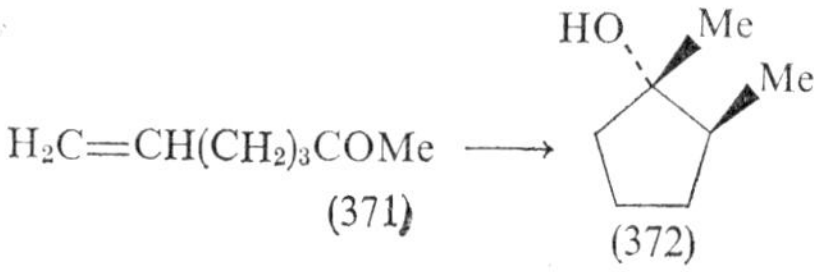

$H_2C=CH(CH_2)_3COMe$
(371)

(372)

Other Reactions and Properties of Olefins.—The addition products of 4-chlorobenzenesulphenyl chloride with a number of cyclic and acyclic olefins, *e.g.* (373), exchange 4-chlorobenzenesulphenyl chloride with oct-1-ene when

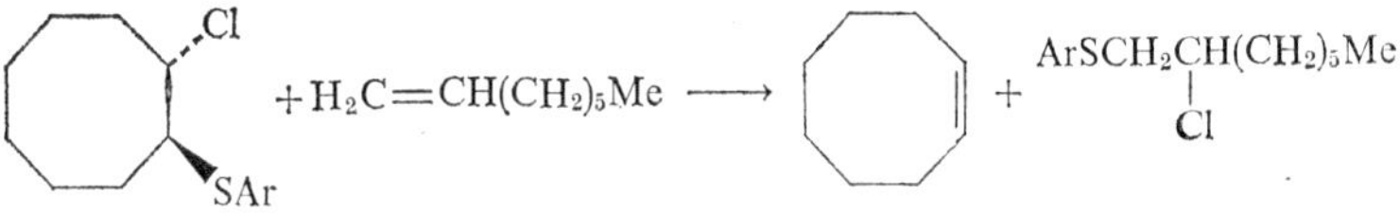

(373) Ar = 4-ClC$_6$H$_4$

[266] R. H. Grubbs and L. C. Krou, *J. Amer. Chem. Soc.*, 1971, **93**, 3062.
[267] G. M. Whitesides and W. J. Ehmann, *J. Org. Chem.*, 1970, **35**, 3565.
[268] T. Shono and M. Mitani, *J. Amer. Chem. Soc.*, 1971, **93**, 5284.

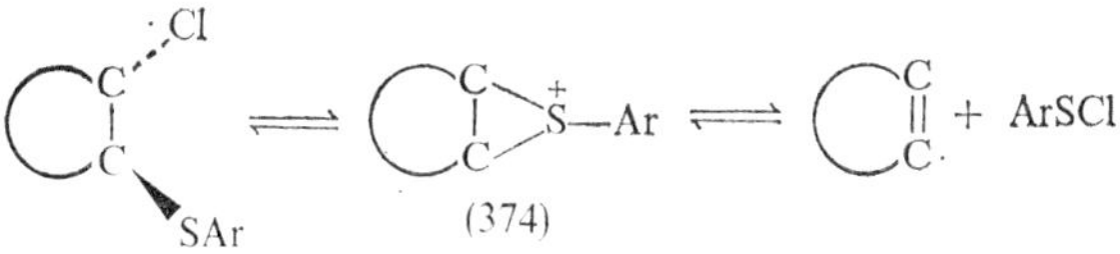

heated in tetrachloroethane.[269] An episulphonium ion intermediate (374) can undergo attack at sulphur to generate the arylsulphenyl chloride.

Reaction of N-bromoacetamide (NBA) with olefins leads to the unknown 2-bromo-N-bromoacetimidates (375).[270] The first stage is a radical reaction of NBA with itself to give (376), followed by ionic addition of the latter to the double bond.

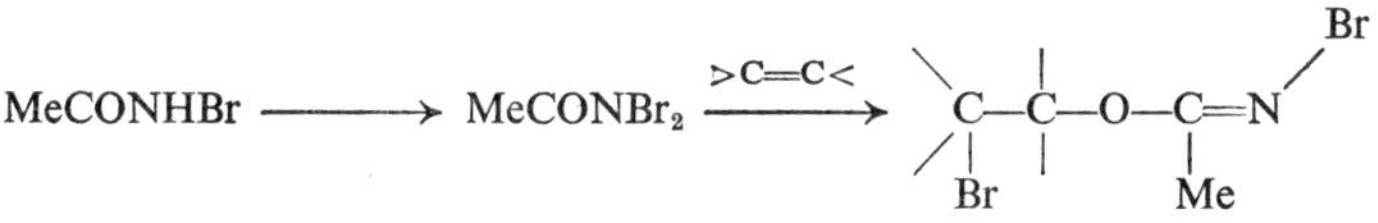

$\alpha\beta$-Unsaturated aldehydes and ketones react with trimethylchlorosilane, magnesium, and HMPA to give 1,4-addition of two trimethylsilyl groups, which on hydrolysis lead to aldehydes and ketones (377), C-silylated in the β-position.[271] Tetrakis(triphenylphosphine)platinum(o) is found to be an effective catalyst for selective hydrosilylation; only terminal double bonds are hydrosilylated [(378) → (379)] and no internal double-bond isomerization takes place.[272]

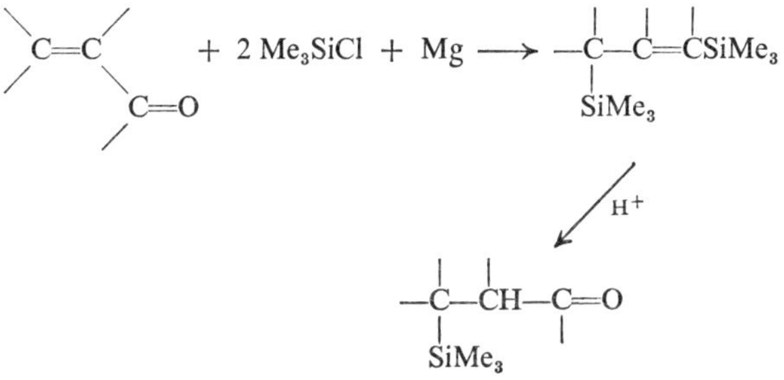

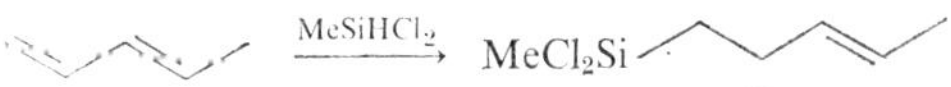

[269] G. H. Schmid and P. H. Fitzgerald, *J. Amer. Chem. Soc.*, 1971, **93**, 2547.
[270] S. Wolfe and D. V. C. Awang, *Canad. J. Chem.*, 1971, **49**, 1384.
[271] R. Calas, J. Dunogues, and M. Bolourtchian, *J. Organometallic Chem* 1971, **26**, 195.
[272] W. Fink, *Helv. Chim. Acta*, 1971, **54**, 1304.

The cycloalkene trifluoromethanesulphonates (triflates) have been used to generate the bent vinyl cations whose preferred geometry is calculated to be linear. In agreement, the solvolysis rates of cyclic triflates decrease with decreasing ring size. The products are exclusively cycloalkanones when water is present, e.g. (380) → (381).[273]

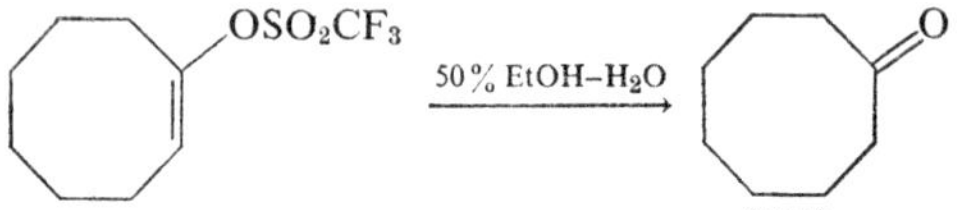

An X-ray crystallographic study of an acetate (382) derived from katonic acid shows it to have a twisted double bond with angles as shown (383). This severe twisting helps to explain the u.v. spectrum of the derived C_{29} alcohol which has λ_{max} 229 nm (ε 5180) in ethanol.[274]

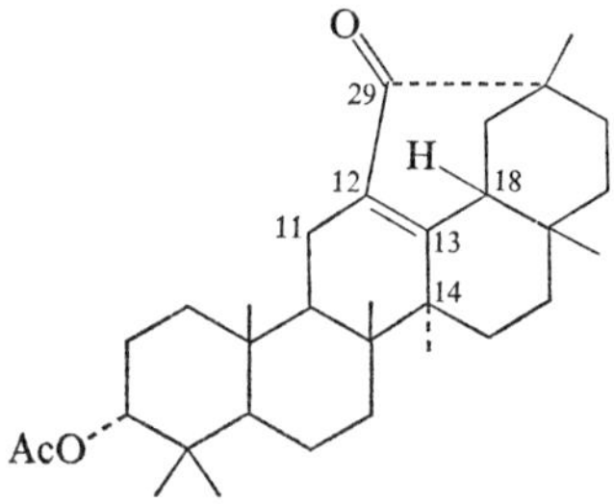

The Cotton effects of many olefins have been correlated with Mills' and Brewster's rules,[275] using the recently derived olefin octant rule.[276] The latter fails, however, in the case of methylenesteroids,[277] and it may be that some twisting is necessary in the olefin for satisfactory application of the rule.

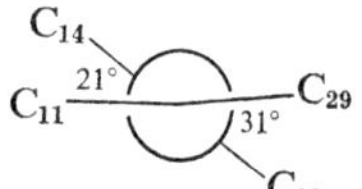

[273] W. D. Pfeifer, C. A. Bahn, P. v. R. Schleyer, S. Bocher, C. E. Harding, K. Hummel, M. Hanack, and P. S. Stang, *J. Amer. Chem. Soc.*, 1971, **93**, 1513.
[274] W. E. Thiessen, H. A. Levy, W. G. Dauben, G. H. Beasley, and D. A. Cox, *J. Amer. Chem. Soc.*, 1971, **93**, 4312.
[275] A. I. Scott and A. D. Wrixon, *Tetrahedron*, 1971, **27**, 4787.
[276] A. I. Scott and A. D. Wrixon, *Chem. Comm.*, 1969, 1182; A. I. Scott and A. D. Wrixon, *Tetrahedron*, 1970, **26**, 3695.
[277] M. Fetizon and I. Hanna, *Chem. Comm.*, 1970, 462.

2
Functional Groups other than Acetylenes, Allenes, and Olefins

BY E. W. COLVIN

1 Alkanes

An up-to-date account of the chemistry of the alkanes has been published.[1] The neopentyl substituent has been reviewed.[2] A general method[3] for the synthesis of hydrocarbons interspersed with *gem*-dimethyl units has been described; the key reaction is the conjugate addition of bifunctional Grignard reagents to diethyl isopropylidenemalonate (Scheme 1). Aluminium alkyls

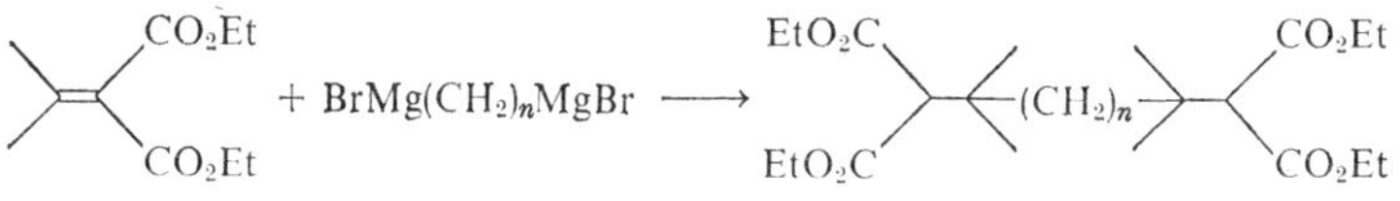

Scheme 1

react with tertiary halides to give quaternary carbon atoms; allylic and benzylic halides react in a similar manner.[4] In a study of the cross-coupling reaction between alkyl halides and anionic metal alkyls, it has been found that lithium dialkyl copper reagents[5] are by far superior to the corresponding manganese, iron, or cobalt analogues.

Pursuing their studies on the activation of sulphur compounds with copper(II) chloride, Japanese authors[6] report a useful method of reductive fission of sulphides and its application to carbon–carbon bond formation, as outlined in Scheme 2; this provides, in effect, a non-hydrogenative method of desulphurization.

The branched alkane fraction from blue-green algae[7] consists of a mixture of 6-, 7-, and 8-methylheptadecane, the total fraction exhibiting optical

[1] F. Asinger and H. II. Vogel, *Houben-Weyl*, 1970, B and V/1a, 1.
[2] M. Montagné, *Bull. Soc. chim. France*, 1970, 347.
[3] R. M. Schisla and W. C. Hammann, *J. Org. Chem.*, 1970, **35**, 3224.
[4] J. P. Kennedy, *J. Org. Chem.*, 1970, **35**, 532.
[5] E. J. Corey and G. H. Posner, *Tetrahedron Letters*, 1970, 315.
[6] T. Mukaiyama, K. Narasaka, K. Maekawa, and M. Furusato, *Bull. Chem. Soc. Japan*, 1971, **44**, 2285.
[7] J. Han and M. Calvin, *Chem. Comm.*, 1970, 1490.

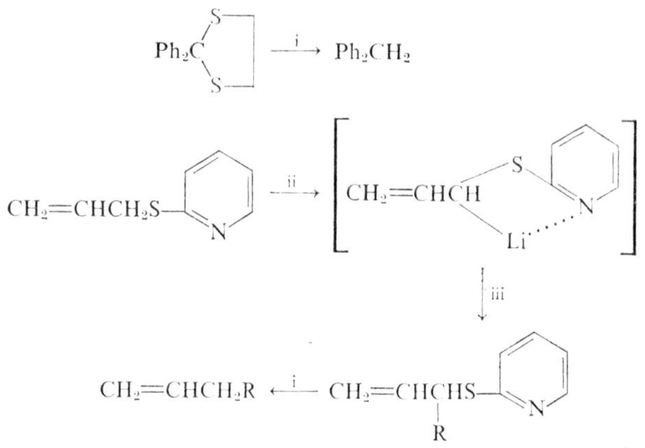

Scheme 2

Reagents: i, LiAlH$_4$–CuCl$_2$; ii, PhLi; iii, RBr

activity. The preparation of dimeric hydrocarbons by reaction of organoboranes with neutral hydrogen peroxide has been described[8] in detail.

Saturated single bonds have been found to undergo electrophilic substitution. In a study of the σ-donor ability of single bonds by two-electron, three-centre bond formation, Olah[9] has noted that carbon–hydrogen bonds show substantial general reactivity in electrophilic reactions, such as isomerization, hydrogen–deuterium exchange, protolysis, alkylation, nitration, and halogenation; this promises a new area of chemistry where alkanes and cycloalkanes are used as substrates in a variety of electrophilic reactions.

Inspired by the highly selective substitution of unactivated positions of substrates in enzyme-mediated reactions, Breslow[10] has described the selective oxidation of unactivated methylene positions by reagent–substrate orientation. While initial experiments[11] attained orientation by chemical bond attachment, it has been found that loose complexation can achieve similar results. In theory, a stabilizing interaction of only 3 kcal mol^{-1} is sufficient to give a 100:1 preference to that transition state over other arrangements lacking such stabilization. In the event, hydrogen bonding in a carboxylic acid mixed dimer was explored. Photolysis of an equimolar solution of n-hexadecanol hemisuccinate and benzophenone-4-carboxylic acid in carbon tetrachloride gave results indicative of selective attack by the excited benzophenone carbonyl oxygen at the C-11 position of the alcohol, a position predicted by inspection of molecular models of the mixed complex (1). Application of this technique to a more rigid, steroidal substrate, 3α,5α-androstanyl hemisuccinate (2) gave, after suitable processing, 16-keto-3α,5α-androstanol (3) in 21% yield as the only keto-steroid produced; once again,

---

[8] D. B. Bigley and D. W. Payling, *J. Chem. Soc. (B)*, 1970, 1811.
[9] G. A. Olah, Y. Halpern, J. Shan, and Y. K. Mo, *J. Amer. Chem. Soc.*, 1971, **93**, 1251; G. A. Olah and J. A. Olah, *ibid.*, p. 1256; G. A. Olah and H. C. Lin, *ibid.*, p. 1259.
[10] R. Breslow and P. C. Scholl, *J. Amer. Chem. Soc.*, 1971, **93**, 2331.
[11] R. Breslow and S. W. Baldwin, *J. Amer. Chem. Soc.*, 1970, **92**, 732.

Functional Groups other than Acetylenes, Allenes, and Olefins

molecular models indicated the susceptibility of such a site to intramolecular attack. The yield of 16-ketone could be raised to 38% when a large excess of benzophenone-4-carboxylic acid was used.

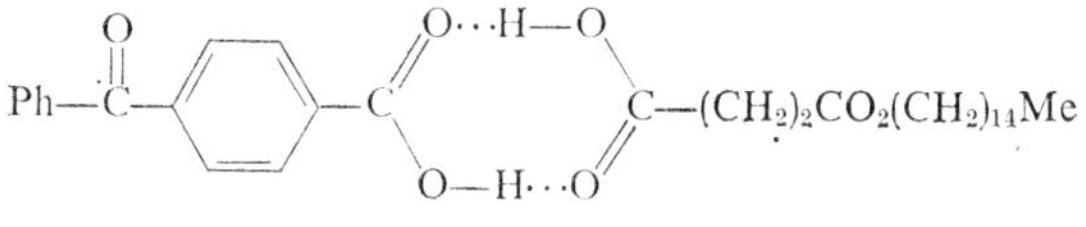

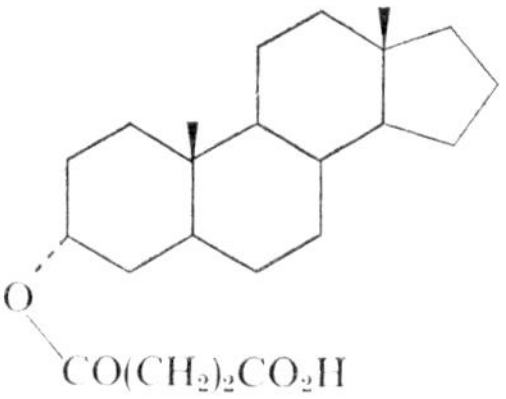

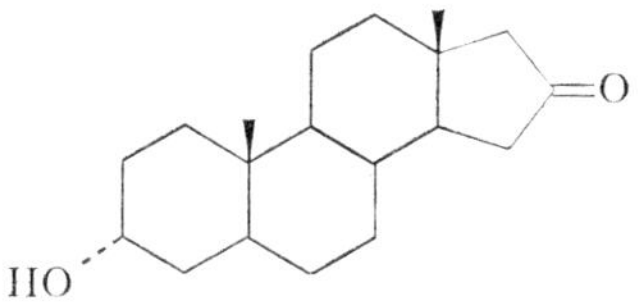

Alkanes are photo-oxidized by nitrobenzene in a process of hydrogen abstraction followed by reaction of the resulting alkyl radical with nitrobenzene; ketones and alcohols are obtained in good yield.[12]

A novel synthesis is reported[13] of $\alpha\beta$-unsaturated ketones by anodic oxidation of alkanes in a fluorosulphonic acid–carboxylic acid electrolyte. The proposed mechanism is shown in Scheme 3, fluorosulphonic acid generating the acylium ion from the carboxylic acid.

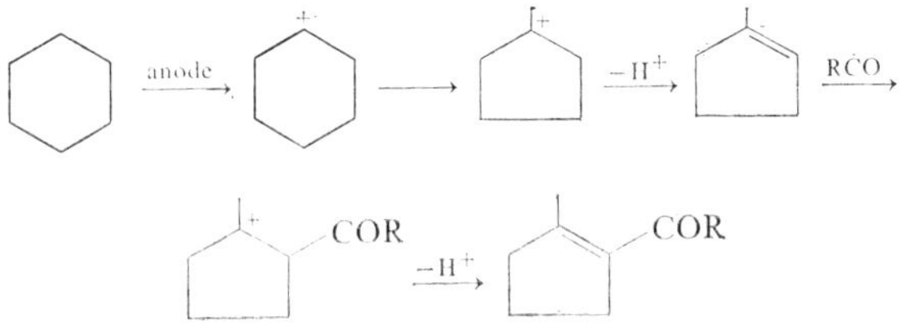

Scheme 3

Atomic Carbon.—Skell and Plonka have extended their studies of the chemistry of atomic carbon, which they generate in a low-intensity carbon arc. The mechanism of the conversion of carbon–hydrogen bonds to carbon–methyl bonds has been investigated thoroughly;[14] the detection of triplet carbenes as intermediates infers that the inserting carbon atoms are in the ^3P state.

Diatomic carbon is proposed[15] as the reacting species in the reaction of

---

[12] J. W. Weller and G. A. Hamilton, *Chem. Comm.*, 1970, 1390.
[13] J. Bertram, M. Fleischman, and D. Pletcher, *Tetrahedron Letters*, 1971, 349.
[14] P. S. Skell, J. H. Plonka, and L. S. Wood, *Chem. Comm.*, 1970, 710.
[15] P. S. Skell, J. H. Plonka, and R. F. Harris, *Chem. Comm.*, 1970, 689; P. S. Skell and J. H. Plonka, *J. Amer. Chem. Soc.*, 1970, **92**, 5620.

carbon with acetone or acetaldehyde to give acetylene; two distinct pathways are implicated: an intermolecular, radical hydrogen-abstraction route, with the ethynyl radical as intermediate, suggestive of a triplet C_2 molecule, and an intramolecular, non-radical abstraction of two hydrogen atoms from the one molecule, with singlet vinylidene as the intermediate, from the singlet C_2 molecule (Scheme 4).

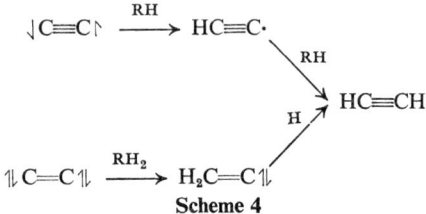

Scheme 4

The generation of carbenes by reaction of metastable singlet carbon atoms with aldehydes and ketones has been described in detail;[16] the distribution of intramolecular rearrangement products is identical with that observed from toluene-*p*-sulphonylhydrazone-derived carbenes. Suitably substituted carbenes generated in this manner, *e.g.* that prepared from (4), can be trapped[17] by olefins, when the cyclopropane (5) produced retains the original olefin geometry, indicative of a singlet species.

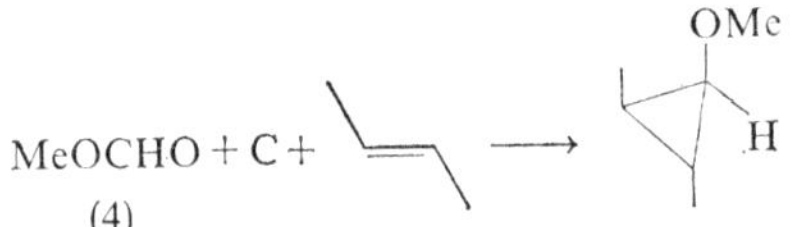

$$\text{MeOCHO} + \text{C} + \diagup\!\!\diagdown \longrightarrow \text{(5)}$$
(4)

The first instance of a carbene with hydroxy-group bonded to the bivalent carbon has been encountered in the deoxygenation of acetic acid to acetaldehyde by atomic carbon; methyl acetate reacts in an analogous manner[18] (Scheme 5).

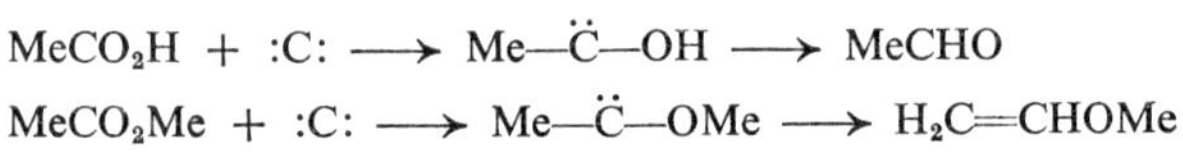

$$\text{MeCO}_2\text{H} + :\text{C}: \longrightarrow \text{Me—}\ddot{\text{C}}\text{—OH} \longrightarrow \text{MeCHO}$$
$$\text{MeCO}_2\text{Me} + :\text{C}: \longrightarrow \text{Me—}\ddot{\text{C}}\text{—OMe} \longrightarrow \text{H}_2\text{C=CHOMe}$$

Scheme 5

Epoxides[19] and oxetans[20] are deoxygenated to give olefins in a non-stereospecific manner; triplet intermediates are probably involved, though not necessarily exclusively. The desulphurization of sulphides[21] is described, the radicals formed undergoing subsequent reaction.

---

[16] P. S. Skell and J. H. Plonka, *J. Amer. Chem. Soc.*, 1970, **92**, 836.
[17] P. S. Skell and J. H. Plonka, *J. Amer. Chem. Soc.*, 1970, **92**, 2160.
[18] J. H. Plonka and P. S. Skell, *Tetrahedron Letters*, 1970, 4557.
[19] J. H. Plonka and P. S. Skell, *Chem. Comm.*, 1970, 1108.
[20] P. S. Skell, K. J. Klabunde, and J. H. Plonka, *Chem. Comm.*, 1970, 1109.
[21] K. J. Klabunde and P. S. Skell, *J. Amer. Chem. Soc.*, 1971, **93**, 3807.

Unsaturated hydrocarbons are converted by atomic carbon into dienes,[22] in addition to the allenes reported previously; a singlet precursor is assigned to this insertion reaction, in which the stereochemistry of the original olefinic site is essentially retained (Scheme 6).

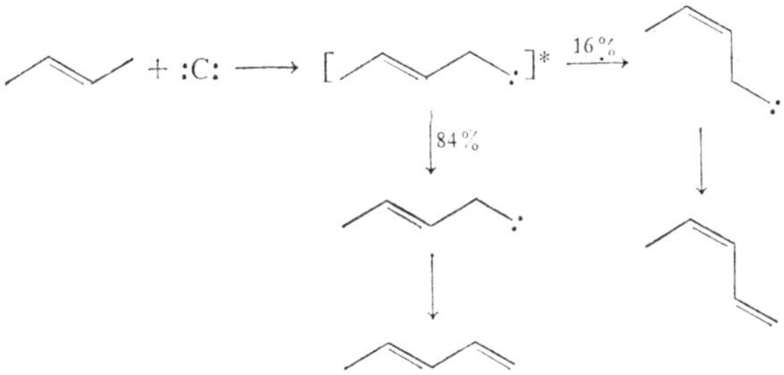

Scheme 6

Carbon atoms have been produced in a chemical reaction; thermolysis of the hydrazone anion (6) afforded benzene and toluene,[23] by rearrangement of the carbene (7).

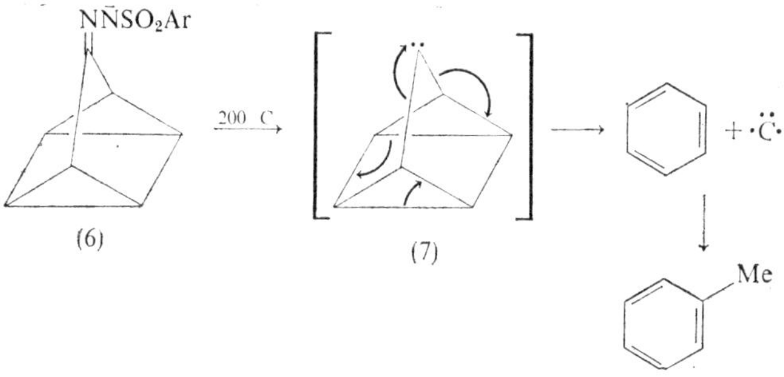

2 Carboxylic Acids

Preparation and Protection.—An ingenious, high-yield method[24] has been published for the chain-lengthening of carboxylic acids by three carbon atoms, *via* the alkylation of oxazol-5-one anions. The method owes its success to the electrophilic C-2 atom in the oxazolone (8), in effect a masked carboxyl carbon, becoming nucleophilic on transition to the anion (9). Acids are readily converted into oxazol-5-ones by treatment of the corresponding acid chloride with *dl*-valine, followed by cyclization with acetic anhydride, as shown in Scheme 7. Reaction of the oxazolone with acrylonitrile in the presence of triethylamine gives the alkylated product (10), which can be readily cleaved to a γ-ketonitrile or a γ-ketoacid.

A similar methodology is apparent in the masking of carboxylic acids as

---

[22] P. S. Skell, J. E. Villaume, J. H. Plonka, and F. A. Fagone, *J. Amer. Chem. Soc.*, 1971, **93**, 2699.
[23] P. B. Shevlin and A. P. Wolf, *Tetrahedron Letters*, 1970, 3987.
[24] P. Gruber and W. Steglich, *Angew. Chem. Internat. Edn.*, 1971, **10**, 655.

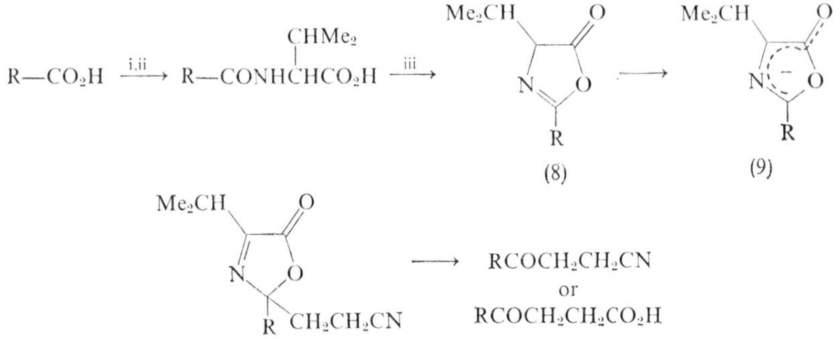

R—CO₂H $\xrightarrow{\text{i,ii}}$ R—CONHCHCO₂H $\xrightarrow{\text{iii}}$

(with CHMe₂ substituent on middle carbon)

(8) (9)

(10) → RCOCH₂CH₂CN
or
RCOCH₂CH₂CO₂H

Reagents: i, SOCl₂; ii, *dl*-valine; iii, Ac₂O

Scheme 7

their 2-oxazoline derivatives,[25] readily prepared from the acids and aminoethanols, 2-amino-2,2-dimethylethanol being the most suitable. The 2-alkyl group of such oxazolines (11) is readily metallated with n-butyl-lithium, and subsequently alkylated with a range of electrophiles. Acid-catalysed ethanolysis affords the alkylated esters; it was unfortunately not possible to reduce the imine function to give masked aldehydes. The 2-oxazoline group is inert to Grignard reagents, and as such has found use as a carboxyl protecting group.[26]

The protection of carboxylic acids as their phenacyl esters has been described;[27] regeneration is effected with great facility at room temperature by zinc in glacial or dilute aqueous acetic acid. Of notable applicability is the formation of γ-hydroxyesters from γ-lactones, an otherwise indirect transformation; an example is the obtention of the ester (13) from the sodium salt of santonin (12). This protection can be extended to phenols.

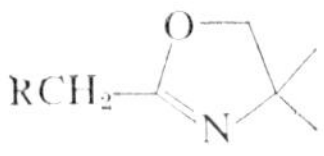

(11)

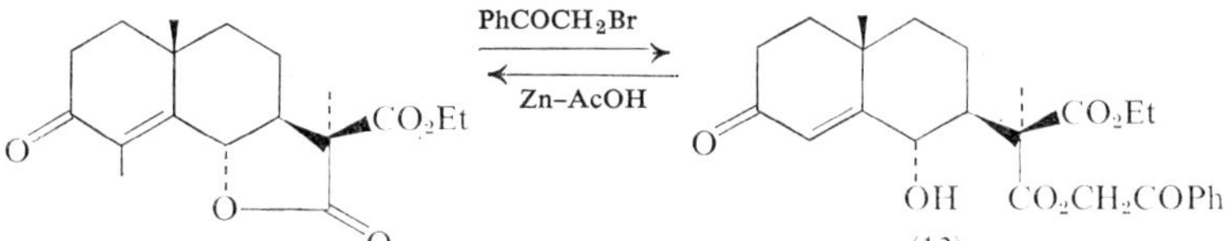

(12) as Na salt $\xrightarrow[\text{Zn-AcOH}]{\text{PhCOCH}_2\text{Br}}$ (13)

[25] A. I. Meyers and D. L. Temple, jun., *J. Amer. Chem. Soc.*, 1970, **92**, 6644.
[26] A. I. Meyers and D. L. Temple, jun., *J. Amer. Chem. Soc.*, 1970, **92**, 6646.
[27] J. B. Hendrickson and C. Kandall, *Tetrahedron Letters.*, 1970, 343.

Methoxy-substituted benzoin esters, such as (14), function[28] as photosensitive protecting groups for acids; regeneration is virtually quantitative, the acid being readily separated from the benzofuran (15): the protected acid has a short excited-state lifetime, minimizing the chance of racemization or photo-oxidation (Scheme 8).

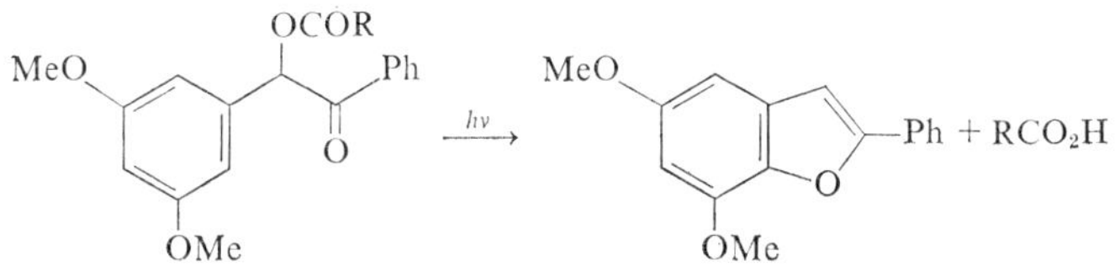

Scheme 8

1,5-Diazabicyclo[4,3,0]non-5-ene smoothly and rapidly eliminates carboxylate anions from 2-toluene-p-sulphonylethyl esters; this alternative method[29] of cleavage obviates the risk of secondary hydrolysis present in the use of aqueous base,[30] and enhances the utility of this ester as a carboxylic acid protecting group.

Absolute Configuration.—Bestmann[31] has described a new method for the determination of the absolute configuration of carboxylic acids. When the optically active acid chloride (16) reacts with two equivalents of the racemic phosphorane (17) to give the phosphonium salt (18), partial kinetic resolution is observed. The second equivalent of now enantiomerically enriched phosphorane then reacts with the salt (18) in a process of transylidation, giving the optically active phosphonium chloride (19), which is precipitated, and the diastereoisomeric acyl ylide (20), which remains dissolved. The absolute configurations of the chlorides (19) have been proven by chemical correlation with the corresponding bromides, whose absolute configurations are known.

A further cautionary note[32] has been sounded on the assignment of absolute configuration to alcohols on the basis of hydride reduction of the corresponding phenyl glyoxylates.

Reactions.—Several pathways are open for the diborane reduction of carboxylic acids. The lower aliphatic acids in THF form triacyloxyboranes (21) which may be further reduced, whereas such intermediates derived from higher aliphatic acids rapidly dismute and may be reduced *via* the carboxylic anhydride (22) and mixed anhydride (23) formed.[33]

---

[28] A. W. Oxford, J. C. Sheehan, and R. Wilson, *J. Amer. Chem. Soc.*, 1971, **93**, 7222.
[29] E. W. Colvin, T. A. Purcell, and R. A. Raphael, *J.C.S. Chem. Comm.*, 1972, 1031.
[30] A. W. Miller and C. J. M. Stirling, *J. Chem Soc.* (*C*), 1968, 2612.
[31] H.-J. Bestmann, H. Scholtz, and E. Kranz, *Angew. Chem. Internat. Edn.*, 1970, **9**, 796.
[32] J. A. Dale and H. S. Mosher, *J. Org. Chem.*, 1970, **35**, 4002.
[33] M. G. Hutchings, T. E. Levitt, A. Pelter, and K. Smith, *Chem. Comm.*, 1970, 347.

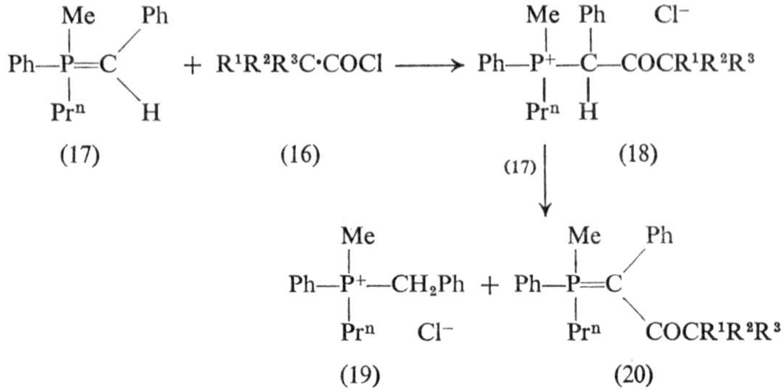

$$\begin{array}{c}\text{Me} \quad \text{Ph} \\ | \quad / \\ \text{Ph--P=C} \\ | \quad \backslash \\ \text{Pr}^n \quad \text{H} \\ (17) \end{array} + \text{R}^1\text{R}^2\text{R}^3\text{C·COCl} \longrightarrow \begin{array}{c} \text{Me} \ \text{Ph} \quad \text{Cl}^- \\ | \quad | \\ \text{Ph--P}^+\text{--C--COCR}^1\text{R}^2\text{R}^3 \\ | \quad | \\ \text{Pr}^n \ \text{H} \\ (18) \end{array}$$

(16) (17) ↓

$$\begin{array}{c} \text{Me} \\ | \\ \text{Ph--P}^+\text{--CH}_2\text{Ph} \\ | \\ \text{Pr}^n \quad \text{Cl}^- \\ (19) \end{array} + \begin{array}{c} \text{Me} \quad \text{Ph} \\ | \quad / \\ \text{Ph--P=C} \\ | \quad \backslash \\ \text{Pr}^n \quad \text{COCR}^1\text{R}^2\text{R}^3 \\ (20) \end{array}$$

N-Bromosuccinimide is recommended as a mild, efficient reagent for the α-bromination of acids; this reaction, previously considered to be a free-radical process, has now been shown[34] to be ionic in nature, and also to be subject to acid catalysis. A general synthesis[35] of per-acids is provided by per-hydrolysis of acyldiethylphosphates under acid catalysis (Scheme 9); the intermediate esters are prepared by acylation of silver diethyl phosphate (24). This method is useful for the preparation of methoxyaromatic and sterically hindered per-acids, *e.g.* trimethylperacetic acid being obtained from pivalic acid in 84% yield. A complementary[36] method involving the intermediacy of acid imidazolides has been published. The electrolysis of some α-ketoacids has been studied; Kolbé processes can be suppressed under certain conditions.[37]

$$(\text{RCO}_2)_3\text{B} \longrightarrow (\text{RCO})_2\text{O} + [(\text{RCO}_2)_2\text{B}]_2\text{O}$$
$$\quad (21) \qquad\qquad (22) \qquad\qquad (23)$$

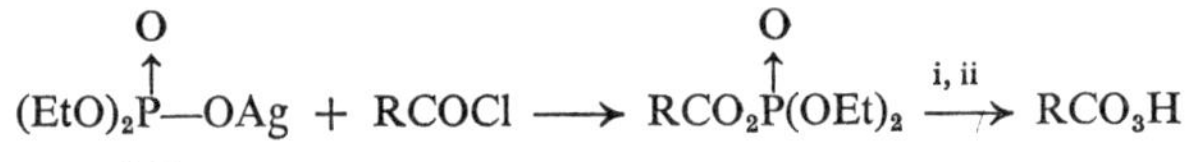

$$\begin{array}{c} \text{O} \\ \uparrow \\ (\text{EtO})_2\text{P--OAg} \end{array} + \text{RCOCl} \longrightarrow \begin{array}{c} \text{O} \\ \uparrow \\ \text{RCO}_2\text{P(OEt)}_2 \end{array} \xrightarrow{\text{i, ii}} \text{RCO}_3\text{H}$$
(24)

Reagents: i, H_2O_2; ii, H_3O^+

Scheme 9

The mould metabolite zymonic acid (25) is now suspected[38] to be a chemical artifact; it is obtained in appreciable amounts by treatment of a dilute aqueous solution of pyruvic acid with calcium carbonate.

A study[39] has been made on the scope and mechanism of the preparation of cinnamic acid derivatives by reaction of saturated aliphatic acids with

---

[34] J. G. Gleason and D. N. Harpp, *Tetrahedron Letters.*, 1970, 3431.
[35] D. A. Konen and L. S. Silbert, *J. Org. Chem.*, 1971, **36**, 2162.
[36] U. Folli and D. Iarossi, *Boll. Sci. Fac. Chim. Ind. Bologna*, 1968, **26**, 61.
[37] B. Wladislaw and J. P. Zimmerman, *J. Chem. Soc. (B)*, 1970, 290.
[38] J. L. Bloomer, M. A. Gross, F. E. Kappler, and G. N. Panday, *Chem. Comm.*, 1970, 1030.
[39] K. Kimura, I. Minato, S. Nishimura, Y. Odaira, and T. Sakakibara, *J. Org. Chem.*, 1970, **35**, 3884.

aromatic compounds in the presence of palladium(II) chloride; oxidative formation of an intramolecular π-complex of a palladium acrylate is suggested.

A rhodium(I) complex with the optically active phosphine (26) is an efficient catalyst for asymmetric hydrogenation;[40] atropic acid is reduced to (S)-hydratropic acid in an optical yield of 63%; this method has been extended to the preparation of optically active amino-acids.

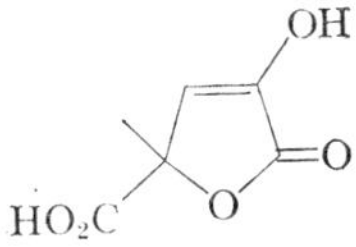

(25) 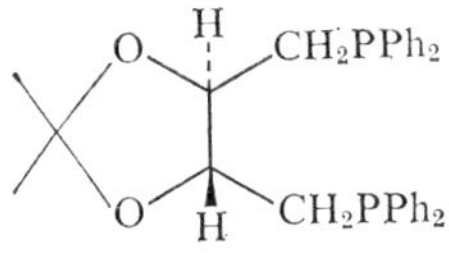 (26)

The Simonini reaction is now considered[41] to be a variant of the Hunsdiecker reaction; if a sufficiently large excess of halogen is not present, the course of the Hunsdiecker is deflected to give coupled ester as Simonini product (Scheme 10). The synthetic limitations of the Hunsdiecker reaction are discussed.

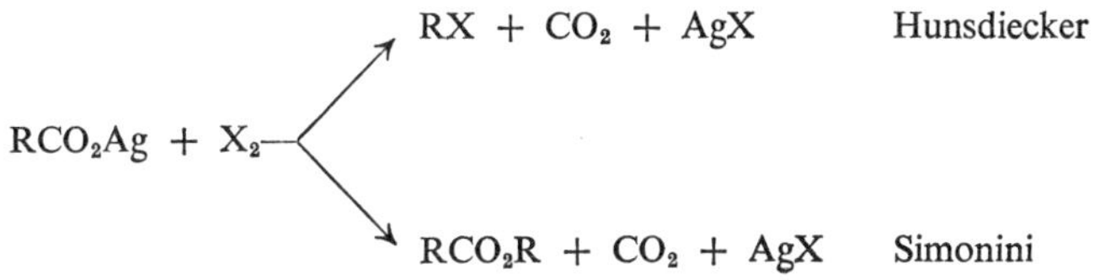

$$RCO_2Ag + X_2 \longrightarrow \begin{cases} RX + CO_2 + AgX & \text{Hunsdiecker} \\ RCO_2R + CO_2 + AgX & \text{Simonini} \end{cases}$$

$$RX + RCO_2Ag \longrightarrow RCO_2R + AgX$$

Scheme 10

In a continuing search for reagents to effect terminal chlorination of long alkyl chains, an American group[42] reports on several significant advances. Adsorption of fatty acids on a carbon tetrachloride–alumina interface causes alignment in a close-packed, rigid array in the axis perpendicular to the interface, affording protection to the alkyl chains but exposing the terminal methyl groups; free-radical chlorination in such circumstances raises the yield of terminal chlorination to 33%, as compared with that of 17% for the homogeneous counterpart.[42a] A parallel to the Hofmann–Löffler chlorination of amines has been observed in the free-radical chlorination of long-chain acids in strongly acidic media.[42b] Oxygen cation-radicals are invoked in this process,

---

[40] T. P. Dang and H. B. Kagan, *Chem. Comm.*, 1971, 481.
[41] N. J. Bunce and N. G. Murray, *Tetrahedron*, 1971, 27, 5323.
[42] (a) N. C. Deno, R. Fishbein, and C. Pierson, *J. Amer. Chem. Soc.*, 1970, 92, 1451; (b) N. C. Deno, R. Fishbein, and J. C. Wyckoff, *ibid.*, p. 5274; (c) W. E. Billups, N. C. Deno, and R. Fishbein, *ibid.*, 1971, 93, 438; (d) N. C. Deno, R. Fishbein, and J. C. Wyckoff, *ibid.*, p. 2065.

88   *Aliphatic, Alicyclic, and Saturated Heterocyclic Chemistry*

which is in effect a McLafferty rearrangement in solution (Scheme 11). In addition to selective chlorination at C-4, comparable amounts of terminal chlorination in hexanoic and octanoic acids were found.

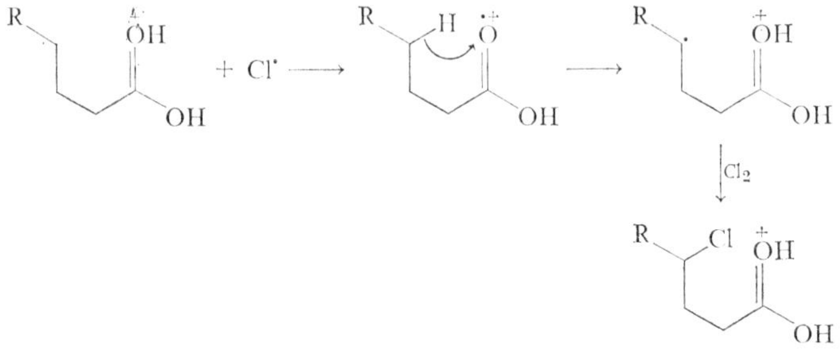

Scheme 11

Nitrogen cation-radical chlorinations of a variety of substrates have been studied in an extension of the intermolecular version of the Hofmann–Löffler reaction; selectivity at the $\omega - 1$ position of C_6 and C_8 substrates is often better than 90%, with almost exclusive monochlorination.[42c] The use of sterically hindered nitrogen cation-radicals leads to a reversal of the normal selectivity of hydrogen-atom abstraction to one of primary > secondary > tertiary.[42d]

3 Carboxylic Acid Esters

Preparation.—Synthesis of acid-labile esters can be achieved by treatment of the corresponding lithium alkoxide, from the alcohol and n-butyl-lithium, with an acid chloride;[43] t-butyl pivalate can be obtained in 64% yield by this procedure. Hindered acids are esterified by triethyloxonium fluoroborate[44] in the presence of ethyl di-isopropylamine. The use of boron trifluoride etherate–alcohol[45] is recommended as a mild, efficient esterification method, as is the employment of alkyl t-butyl ethers,[46] which react with carboxylic acids under acid catalysis to give the alkyl ester, isobutene, and water, no free alcohol being ever present.

The activation of alcohols for ester preparation is described;[47] reaction of diethylazodicarboxylate with the alkyl NN'-tetraethylphosphorodiamidite (27) gives the zwitterion (28), which undergoes rapid attack by a carboxylic acid to give the desired ester, as shown in Scheme 12; the alcohol carbon, if

---

[43] E. M. Kaiser and R. A. Woodruff, *J. Org. Chem.*, 1970, **35**, 1198; *Org. Syntheses*, 1971, **51**, 96.
[44] D. J. Raber and P. Gariano, *Tetrahedron Letters*, 1971, 4741.
[45] J. L. Marshall, K. C. Erickson, and T. K. Folsom, *Tetrahedron Letters*, 1970, 4011.
[46] V. A. Derevitskaya, E. M. Klimov, and N. K. Kochetkov, *Tetrahedron Letters*, 1970, 4269.
[47] O. Mitsunobu and M. Eguchi, *Bull. Chem. Soc. Japan*, 1971, **44**, 3427.

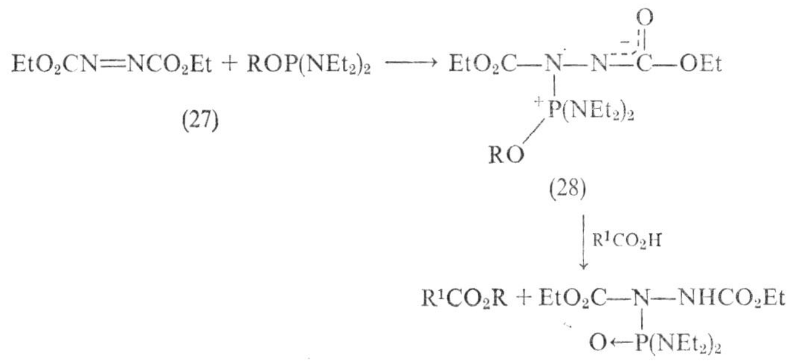

$EtO_2CN=NCO_2Et + ROP(NEt_2)_2 \longrightarrow EtO_2C-N-N=C-OEt$

(27)

$$\begin{array}{c} \overset{+}{P}(NEt_2)_2 \\ RO \end{array}$$

(28)

$\downarrow R^1CO_2H$

$R^1CO_2R + EtO_2C-N-NHCO_2Et$

$\qquad\qquad\qquad O \leftarrow P(NEt_2)_2$

Scheme 12

chiral, suffers inversion of configuration. This procedure is also suitable for the synthesis of phosphate esters.

The full paper[48a] has been published on the preparation of esters by ester exchange, using molecular sieves[48b] as selective alcohol traps. A method of esterification using a 'catalytic dehydrator' has been reported,[49] where the combination of an acid ion-exchange resin and a drying agent is employed.

Iodine triacylates (29) are readily obtained by ozonization of solutions of iodine in aliphatic carboxylic acid anhydrides; thermolysis with mercuric oxide affords the symmetrical esters (30) in good yield.[50] Mercuric iodate can perform the same transformation by direct reaction with the anhydride.

$$(RCO_2)_3I \longrightarrow RCO_2R + RI + CO_2$$

(29) (30)

The adduct (31) formed from dialkyl sulphates and DMF is a potent esterifying agent for a range of carboxylic acids.[51] Carbon dioxide-catalysed ester exchange has been described.[52] Carboxylic acids react readily with diphenyl disulphide in the presence of triphenylphosphine to afford the corresponding phenyl thioesters.[53]

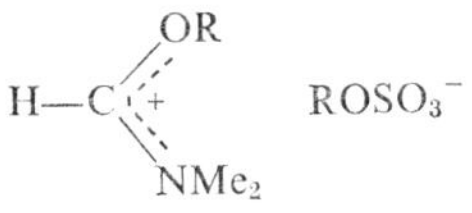

$$H-C\begin{array}{c}OR\\ +\\ NMe_2\end{array} \quad ROSO_3^-$$

(31)

---

[48] (a) H. van Bekkum, J. W. M. De Graaf, J. A. Hagendoorn, D. P. Roelofsen, and H. M. Verschoor, *Rec. Trav. chim.*, 1970, **89**, 193; (b) H. van Bekkum, J. A. Hagendoorn, and D. P. Roelofsen, *Chem. and Ind.*, 1966, 1622.
[49] V. I. Stenberg and G. F. Vesley, *J. Org. Chem.*, 1971, **36**, 2548.
[50] G. B. Bachman, G. F. Kite, S. Tuccarbasu, and G. M. Tullman, *J. Org. Chem.*, 1970, **35**, 3167.
[51] B. Funke and W. Kantlehner, *Chem. Ber.*, 1971, **104**, 3711.
[52] Y. Otsuji, N. Matsumura, and E. Imoto, *Bull. Chem. Soc. Japan*, 1971, **44**, 852.
[53] T. Endo, S. Ikenaga, and T. Mukaiyama, *Bull. Chem. Soc. Japan*, 1970, **43**, 2362.

Carbonate esters are obtained by selenium-mediated reaction of alcohols with carbon monoxide in the presence of the alcohol alkoxide.[54] The synthesis of allylic acetates by acetate-ion displacement on allylic chlorides is markedly catalysed by palladium(II) chloride; an addition–elimination mechanism is invoked[55] to explain the overall stereochemistry of an S_N2' process.

A generally applicable method for the preparation of orthocarboxylic acid enol esters (32) is reported;[56] interchange of the corresponding triethyl orthoesters with β-chloroalkanols and subsequent dehydrohalogenation gives the tris enol esters. Orthoformic acid tris isopropenyl ester, prepared in this manner, gives triacetoxymethane on ozonolysis.

A Swiss group has reported on a new fragmentation mode of bicyclic enol ethers, when macrocyclic lactones are obtained;[57] vicinal α-hydroperoxytetrahydropyranyl ethers (33) annelated to cyclododecane, prepared *via* the corresponding cyclododecanone enol ethers, are smoothly converted homolytically into a mixture of macrocyclic lactones. This fragmentation has been developed to provide an economically feasible synthesis of exaltolide (34).

Improved procedures[58a] have been presented for the preparation of di- and tri-cyclohexylidene peroxides, (35) and (36), valuable macrocyclic lactone precursors.[58b]

Five- and six-membered lactones are readily available by reduction of the

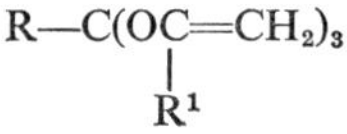

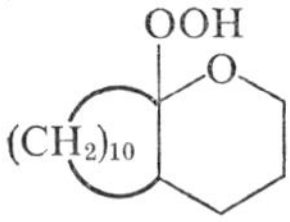

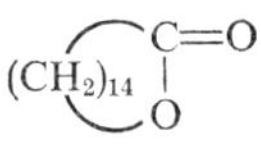

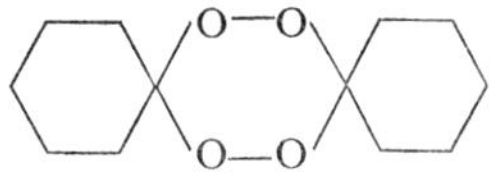

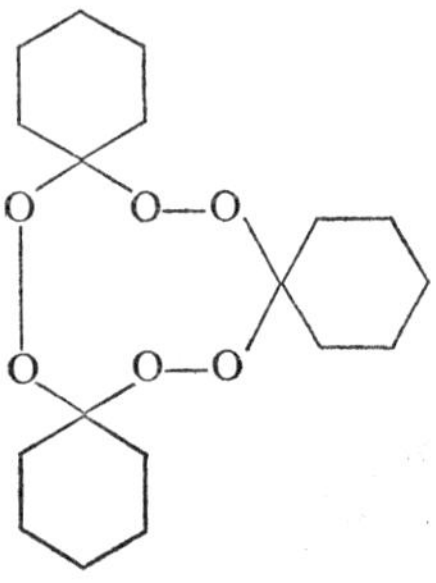

[54] K. Kondo, N. Sonoda, and S. Tsutsumi, *Tetrahedron Letters*, 1971, 4885.
[55] D. G. Brady, *Chem. Comm.*, 1970, 434.
[56] E. Reske and H. Stetter, *Chem. Ber.*, 1970, **103**, 639.
[57] J. Becker and C. Ohloff, *Helv. Chim. Acta*, 1971, **54**, 2889.
[58] (*a*) P. R. Storey, B. Lee, C. E. Bishop, D. D. Denson and P. Busch, *J. Org. Chem.*, 1970, **35**, 3059; (*b*) P. R. Storey, D. D. Denson, C. E. Bishop, B. C. Clarke, jun., and J.-C. Farine, *J. Amer. Chem. Soc.*, 1968, **90**, 817.

corresponding cyclic anhydrides with sodium borohydride in THF or DMF at room temperature.[59]

A synthesis of β-lactones, performed originally on a conformationally rigid molecule, has been found to be of general applicability.[60] Treatment of an open-chain βγ-unsaturated carboxylic acid salt (37) with bromine gives the bromolactone (38) as kinetic product (Scheme 13); this reaction is successful even with but-3-enoic acid. Iodolactonization may also be kinetically controlled, the difference in product ring size being due to the different electrophilicities of the two halogens.

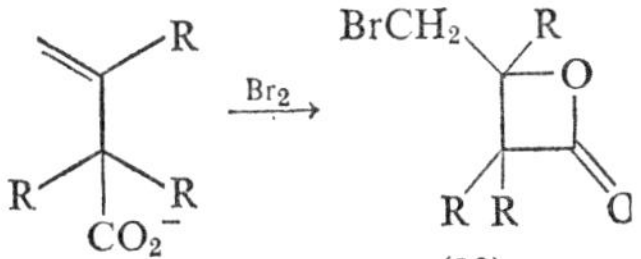

Scheme 13

But-2-enolides are obtained by thermolysis of γ-bromo-αβ-unsaturated acid methyl esters in the presence of iron powder, which seems to catalyse the lactonization by facilitating both the elimination of methyl bromide and rotation about the double bond; a mixture of cis- and trans-γ-bromosenecioate (39) gives 3-methylbut-2-enolide (40).[61] The requisite γ-bromoesters are obtained by Wohl–Ziegler bromination of the esters with N-bromosuccinimide.[62] This overall transformation can also be effected on the corresponding acids, though less smoothly.

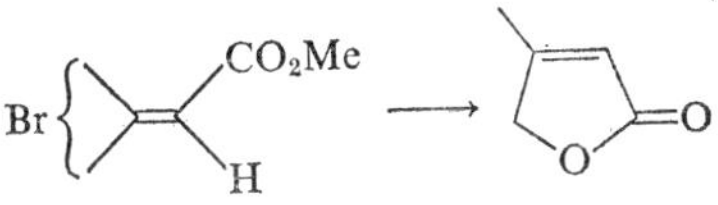

A number of methods describing improved syntheses of α-methylenebutyrolactones have been published. Magnesium methyl carbonate has been found[63] to carboxylate butyrolactones (though not six-membered lactones) in high yield; treatment of the acid (41) with aqueous formaldehyde and diethylamine, followed by separate treatment of the crude product with sodium acetate in acetic acid, gives the desired α-methylenebutyrolactone (42). An independently developed, complementary route[64] involves thermal fragmentation of the salt (43), iodide ion functioning as a nucleophile for

---

[59] D. M. Bailey and R. E. Johnson, *J. Org. Chem.*, 1970, **35**, 3574.
[60] W. E. Barnett and J. C. McKenna, *Tetrahedron Letters*, 1971, 2595; *Chem. Comm.*, 1971, 551.
[61] A. S. Dreiding, A. Löffler, F. Norris, K. L. Svanholt, and W. Taub, *Helv. Chim. Acta*, 1970, **53**, 403.
[62] A. S. Dreiding, A. Löffler, R. J. Pratt, and H. P. Rüesch, *Helv. Chim. Acta*, 1970, **53**, 383.
[63] J. Martin, P. C. Watts, and F. Johnson, *Chem. Comm.*, 1970, 27.
[64] E. S. Behare and R. B. Miller, *Chem. Comm.*, 1970, 402.

cleavage of the methyl ester to the acid anion. These methods failed when applied to the malonate ester (44), an intermediate in a synthesis of the mould metabolite versimide (45), but the difficulty was circumvented[65] by potassium iodide-induced demethoxycarbonylative β-elimination on the sulphone (46), giving the desired α-methylene ester in high yield.

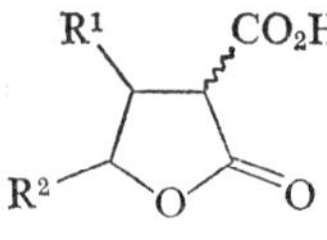

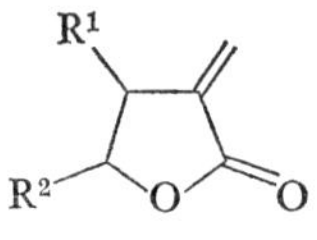

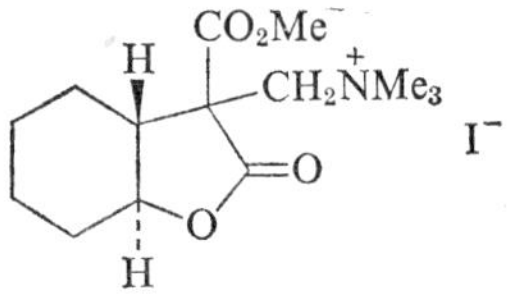

A potentially more simple procedure[66] is the Reformatsky condensation of the bromoester (47) with ketones; α-methylenebutyrolactones are produced, in one step, in fair to good yield. An efficient three-step synthesis consisting of ketone enamine formation, Michael addition to ethyl β-nitroacrylate, and borohydride reduction has been reported;[67] the last process affords a mixture of the *cis*-butyrolactone and the *trans*-hydroxyester as shown in Scheme 14.

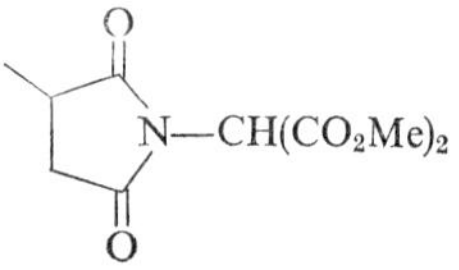

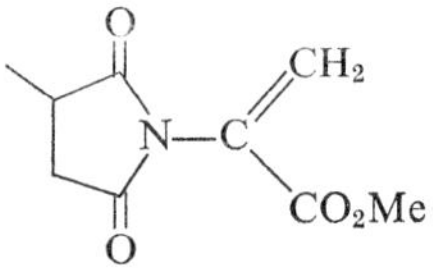

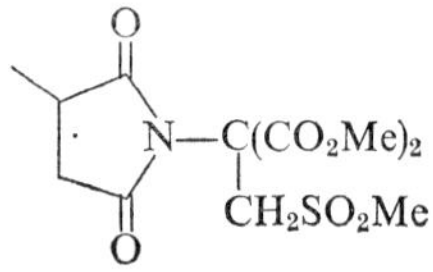

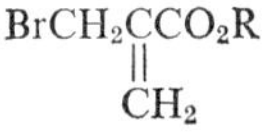

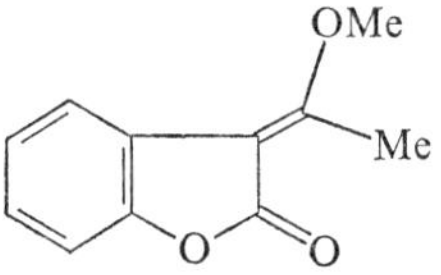

The thermal reaction of 1,1-dimethoxyethylene with butyrolactones is reported[68] to give substituted α-methylenebutyrolactones, *e.g.* (48), in good yield. The methylene group of α-methylenebutyrolactones can be protected by a reversible thiol addition reaction.[69]

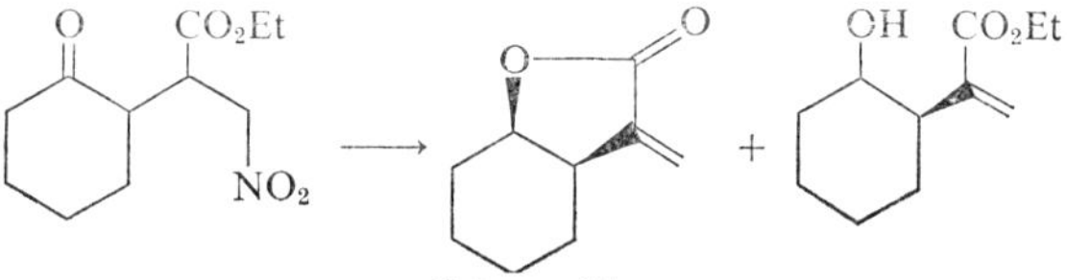

Scheme 14

[65] P. R. Atkins and I. T. Kay, *Chem. Comm.*, 1971, 430.
[66] E. Öhler, K. Reininger, and U. Schmidt, *Angew. Chem. Internat. Edn.*, 1970, **9**, 457.
[67] J. W. Patterson and J. E. McMurry, *Chem. Comm.*, 1971, 488.
[68] O. L. Chapman, C. L. McIntosh, and J. C. Clardy, *Chem. Comm.*, 1971, 384.
[69] S. M. Kupchan, T. J. Giacobbe, and I. S. Krull, *Tetrahedron Letters*, 1970, 2859.

Functional Groups other than Acetylenes, Allenes, and Olefins 93

A facile method for the introduction of an alkoxyalkyl group has been reported.[70] Treatment of the organozinc derivatives of α-bromoesters with acetals and acid chlorides gives the corresponding β-ethers, as shown in Scheme 15; acylium ion attack on the acetal to give the intermediate alkoxycarbonium ion is postulated.

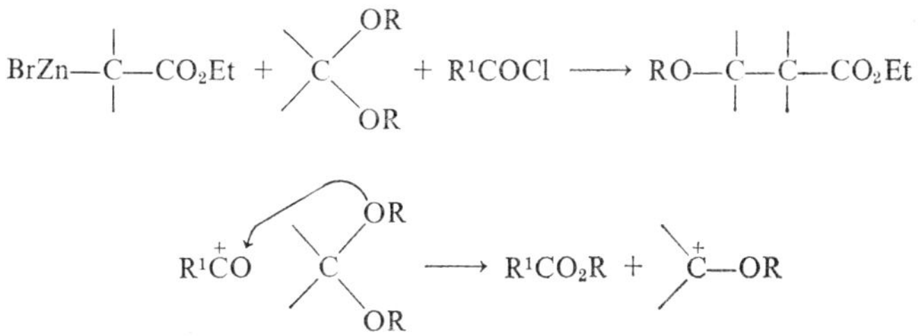

$$BrZn-\underset{|}{\overset{|}{C}}-CO_2Et + \underset{OR}{\overset{OR}{C}} + R^1COCl \longrightarrow RO-\underset{|}{\overset{|}{C}}-\underset{|}{\overset{|}{C}}-CO_2Et$$

$$R^1\overset{+}{C}O \quad \underset{OR}{\overset{OR}{C}} \longrightarrow R^1CO_2R + \underset{}{\overset{+}{C}}-OR$$

Scheme 15

Reactions.—Base-induced cyclization of dimethyl esters of αα'-dibromoalkanedioic acids (49) is reported to give the corresponding dimethyl cycloalk-1-ene-1,2-dicarboxylates (50); good yields are obtained for C_6—C_9 diacids.[71] Evidence is presented supporting the transitory existence of dimethyl cycloprop-1-ene-1,2-dicarboxylate in the reaction of dimethyl αα'-dibromoglutarate with potassium t-butoxide. The use of fatty acid isopropenyl esters as acylating agents has been described.[72] A synthesis of isopropenyl formate is reported;[73] this ester shows promise as a formylating agent for weakly basic amines under neutral conditions.

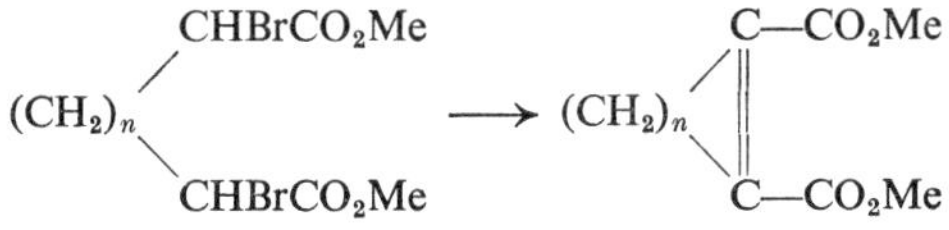

$$(CH_2)_n \underset{CHBrCO_2Me}{\overset{CHBrCO_2Me}{\diagup}} \longrightarrow (CH_2)_n \underset{C-CO_2Me}{\overset{C-CO_2Me}{\diagup}}$$
$$(49) \qquad\qquad (50)$$

Ambident anion alkylation of an ester enolate has been observed;[74] alkylation of methyl O-(tetrahydropyranyl)mandelate at room temperature gave the expected C-alkylated product, whereas at 80 °C transesterification *via* O-alkylation was observed, as outlined in Scheme 16.

The role of the acidity of an alcohol in the facilitation of the reduction of its esters has been studied.[75] A simple and inexpensive method for the production of αα-dideuteriocarboxylic acids and 1,1- or 2,2-dideuterioalcohols has been reported, and is shown in Scheme 17. This method avoids the use of expensive reagents, and can be performed on a relatively large scale.[76]

---

[70] J. M. Cure and M. Gaudemar, *Bull. Soc. chim. France*, 1970, 2962; M. Gaudemar and F. Gaudemar-Bardone, *Bull. Soc. chim. France*, 1970, 2968.
[71] R. N. McDonald and R. R. Reitz, *Chem. Comm.*, 1971, 90.
[72] G. G. Moore and E. S. Rothman, *J. Org. Chem.*, 1970, **35**, 2331.
[73] J. E. W. van Melick, J. W. Scheeren, and R. J. F. Nivard, *Tetrahedron Letters*, 1971, 2083.
[74] K. L. Shepard and J. I. Stevens, *Chem. Comm.*, 1971, 951.
[75] L. A. Cohen and S. Takahashi, *J. Org. Chem.*, 1970, **35**, 1505.
[76] A. N. Yeo, *Chem. Comm.*, 1971, 609.

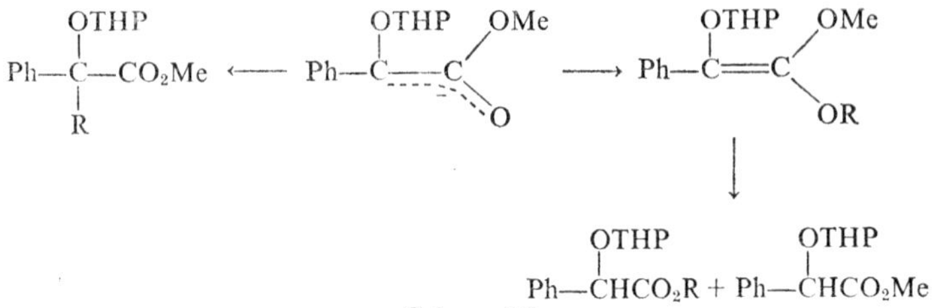

Scheme 16

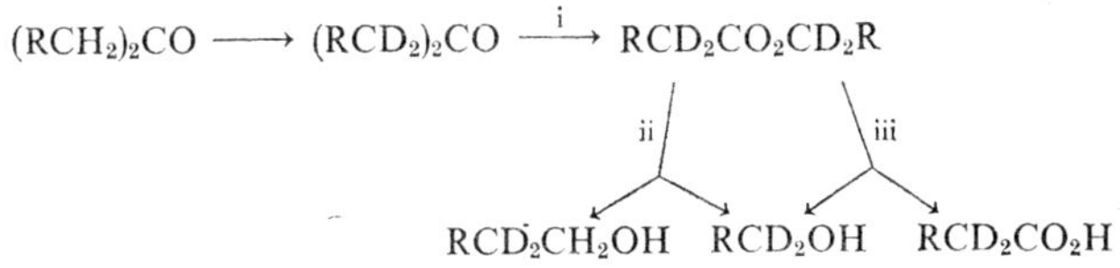

$(RCH_2)_2CO \longrightarrow (RCD_2)_2CO \xrightarrow{i} RCD_2CO_2CD_2R$

ii ↙   ↘ iii

$RCD_2CH_2OH \quad RCD_2OH \quad RCD_2CO_2H$

Reagents: i, $(CF_3CO)_2O-H_2O_2$; ii, $LiAlH_4$; iii, OH^-

Scheme 17

Darzens condensation of an aromatic aldehyde with α-bromo-γ-valerolactone has been used[77] in the preparation of ketides. The acid-catalysed decarboxylation of α-phenylglycidic acids is reported to give abnormal products;[78] a non-concerted mechanism involving oxiran ring-opening is invoked. α-Halogenoglycidic esters can be obtained by reaction of tris(dimethylamino)phosphine with trichloroacetic acid derivatives in the presence of carbonyl compounds (Scheme 18); cyclohexanone affords the isomeric halogenated pyruvic acid derivative (51) in quantitative yield.[79]

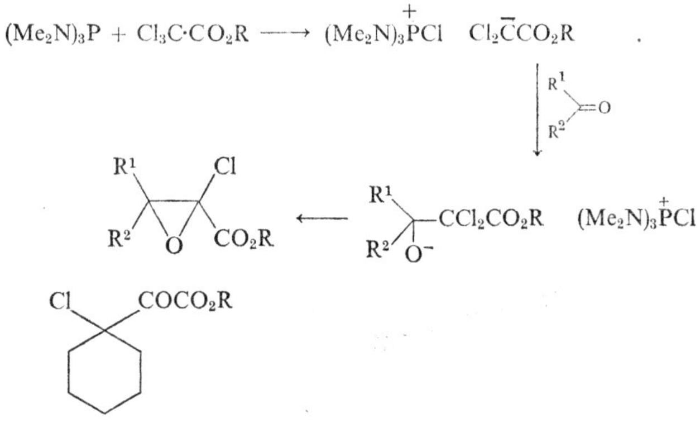

(51)

Scheme 18

[77] J. B. Bremner, M. J. Dimsdale, R. L. Garcea, and J. D. White, *J. Amer. Chem. Soc.*, 1971, **93**, 281.
[78] J. Kagan and S. P. Singh, *J. Org. Chem.*, 1970, **35**, 2203.
[79] J.-C. Combret, G. Lavielle, and J. Villieras, *Bull. Soc. chim. France*, 1971, 898.

Functional Groups other than Acetylenes, Allenes, and Olefins 95

α-**Anions of Acids and Esters.**—The α-anions of the lithium salts of carboxylic acids and the lithium enolates of carboxylic acid esters show promise of impressive synthetic utility. Following the initial report by Creger[80] of the preparation of the α-anion of lithium isobutyrate and the subsequent alkylation of this species to afford αα-dimethyl trisubstituted acetic acids, a number of significant developments have been described. In the majority of cases, the method of choice for generation of the anion has been treatment of the sodium salt of the acid with lithium di-isopropylamide in THF containing hexamethylphosphoramide.[81] The anions produced have been shown to undergo ready alkylation, giving di- or tri-alkyl acetic acids (52);[82] they are valuable precursors for the preparation of the corresponding aldehydes (53),[83] by formylation with ethyl formate and subsequent decarboxylation, and the corresponding nitroparaffins (54) by reaction with n-propyl nitrate; they react smoothly with ketones, affording β-hydroxyacids (55),[84] providing one of a number of recent alternatives to the Reformatsky reaction (*loc. cit*). In the condensation of α-anions with esters, β-ketoacid anions are produced, which, if trapped as the trimethylsilyl esters with trimethylsilyl chloride, provide an efficient route to β-ketoacids[85] by neutral cleavage of the silyl ester with methanol; these acids, in turn, on decarboxylation give a very good entry to highly hindered ketones (56), *e.g.* t-butyl isopropyl ketone is obtained in 70% yield by such a reaction sequence between the α-anion of lithium isobutyrate and methyl pivalate. They are also capable of undergoing conjugate addition to $\alpha\beta$-unsaturated esters, yielding substituted glutaric acid derivatives (57). Bistrimethylsilyl enol ethers (58) are obtained[86] on treatment of α-anions with chlorotrimethylsilane; on thermal dimerization, the esters (59) are formed, giving β-ketoacids on hydrolysis. Aeration of solutions of α-anions in THF affords a very facile general route to α-hydroxyacids (60).[87] These transformations are summarized in Scheme 19. An alternative method of preparation of the α-anion of acetic acid, where the reagent is the ion-radical system produced from lithium and naphthalene, has been reported independently by two groups;[88] the anion so formed reacts in comparable fashion to those described above. A study[89] of the positional and stereochemical isomerization of $\alpha\beta$- and $\beta\gamma$-unsaturated acid dianions has been made.

A similar rapid expansion of interest followed the report by Rathke[90] of the preparation of lithium ethyl acetate, produced quantitatively by

---

[80] P. L. Creger, *J. Amer. Chem. Soc.*, 1967, **89**, 2500; *Org. Syntheses*, 1970, **50**, 58.
[81] P. E. Pfeffer and L. S. Silbert, *J. Org. Chem.*, 1970, **35**, 262.
[82] P. L. Creger, *J. Amer. Chem. Soc.*, 1970, **92**, 1397.
[83] P. E. Pfeffer and L. S. Silbert, *Tetrahedron Letters*, 1970, 699.
[84] G. W. Moersch and A. R. Burkett, *J. Org. Chem.*, 1971, **36**, 1149.
[85] Y.-N. Kuo, J. A. Yahner, and C. Ainsworth, *J. Amer. Chem. Soc.*, 1971, **93**, 6321.
[86] Y.-N. Kuo, F. Chen, C. Ainsworth, and J. J. Bloomfield, *Chem. Comm.*, 1971, 136.
[87] G. W. Moersch and M. L. Zwiesler, *Synthesis*, 1971, 647.
[88] B. Angelo, *Bull. Soc. chim. France*, 1970, 1848; S. Watanabe, K. Suga, T. Fujita, and K. Fujiyoshi, *Israel J. Chem.*, 1970, **8**, 731.
[89] P. E. Pfeffer and L. S. Silbert, *J. Org. Chem.*, 1971, **36**, 3290.
[90] M. W. Rathke, *J. Amer. Chem. Soc.*, 1970, **92**, 3222.

96   Aliphatic, Alicyclic, and Saturated Heterocyclic Chemistry

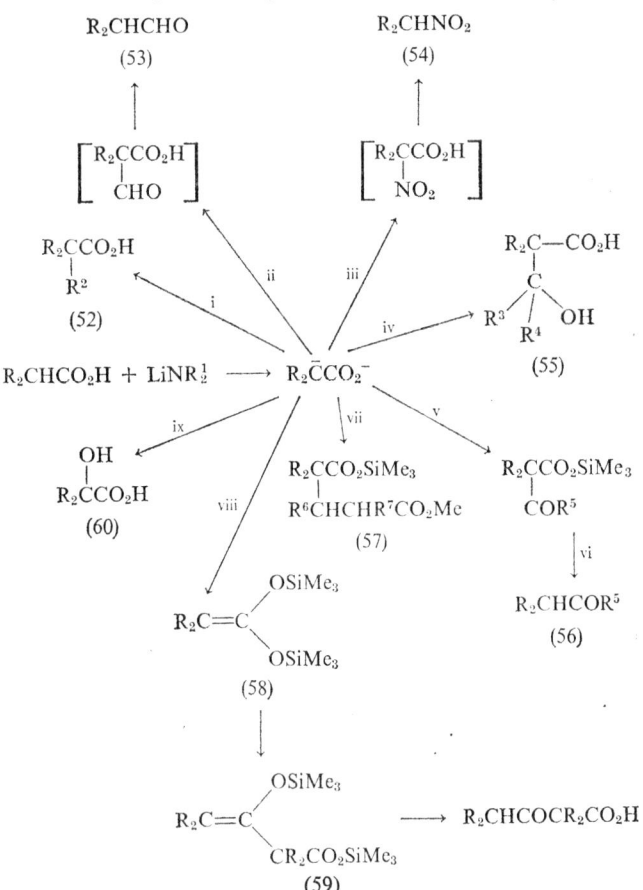

Reagents: i, R^2X; ii, HCO_2Et; iii, Pr^nONO_2; iv, R^3COR^4; v, $R^5CO_2Et-Me_3SiCl$; vi, H_3^+O; vii, $R^6CH=CR^7CO_2Me-Me_3SiCl$; viii, Me_3SiCl; ix, O_2

Scheme 19

treatment of ethyl acetate with lithium bistrimethylsilylamide in THF at −78 °C; at this temperature, the enolate anion reacts rapidly with aldehydes and ketones, giving the corresponding β-hydroxyesters (61) in excellent yield. It was later reported[91] that lithium isopropyl cyclohexylimide (from n-butyl-lithium and isorpopyl cyclohexylamine) could effect the same enolate anion formation on a series of esters at −78 °C, and that the anions so produced were stable at room temperature, as distinct from the earlier method of generation; this provides the first general method of preparation of stable solutions of ester enolates without the normal problem of self-condensation;

[91] M. W. Rathke and A. Lindert, *J. Amer. Chem. Soc.*, 1971, **93**, 2318.

Functional Groups other than Acetylenes, Allenes, and Olefins

these enolates could be alkylated readily at room temperature in THF and DMSO. Interestingly, when solutions of ester enolates generated in this manner were quenched with D_2O, monodeuteriation to a consistent extent of 50—60% was observed, a phenomenon previously noted[92] and ascribed to some degree of association between the enolate anion and the free amine produced on its formation, quenching leading to exclusive proton transfer from the amine to the anion in the complex; this association will provide a degree of steric hindrance, thus explaining the relative stability of the anions with respect to self-condensation. The ester enolates are effective precursors for the preparation of β-ketoesters[93] (62) by reaction with acid chlorides (use of acetyl chloride is not discussed, and indeed the reaction may be unsuccessful), of malonic acid half-esters (63)[94] by carboxylation with carbon dioxide, and of α-iodo- and α-bromo-esters (64)[95] by reaction with iodine

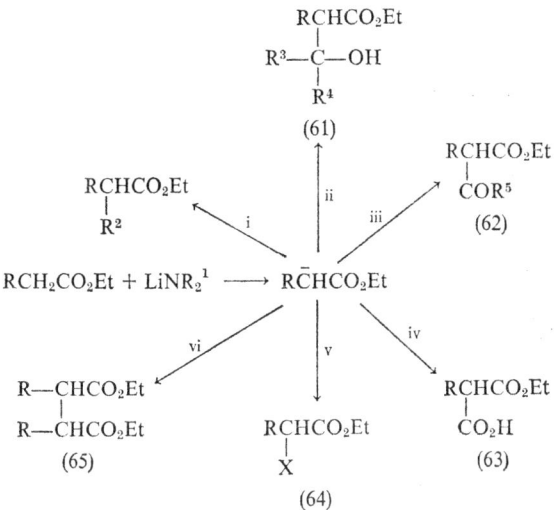

Reagents: i, R^2X; ii, R^3COR^4; iii, R^5COCl; iv, CO_2; v, X_2; vi, Cu^{II} salts

Scheme 20

or bromine. Simple unsubstituted ester enolates couple readily in the presence of cupric salts giving substituted succinic esters (65),[96] but the yields in this reaction drop markedly with α-substitution. These reactions are outlined in Scheme 20.

---

[92] P. L. Creger, *J. Amer. Chem. Soc.*, 1970, **92**, 1396.
[93] M. W. Rathke and J. Deitch, *Tetrahedron Letters*, 1971, 2953.
[94] S. Reiffers, H. Wynberg, and J. Strating, *Tetrahedron Letters*, 1971, 3001.
[95] M. W. Rathke and A. Lindert, *Tetrahedron Letters*, 1971, 3995.
[96] M. W. Rathke and A. Lindert, *J. Amer. Chem. Soc.*, 1971, **93**, 4605.

The preparation of lithium ethyl diazoacetate has been reported;[97] this compound reacts as an enolate anion in analogous fashion to those reported above, e.g. with cyclohexanone the corresponding α-diazo-β-hydroxyester is produced.

Reformatsky Reactions.—A trimethyl borate–THF solvent system is recommended[98] for the Reformatsky reaction, rapid reaction and high yields being attained at ambient temperatures; functioning as a weak acid, trimethyl borate neutralizes the basic alkoxides which, by reaction with the starting materials, are the main cause of the low yields frequently encountered in this reaction (Scheme 21).

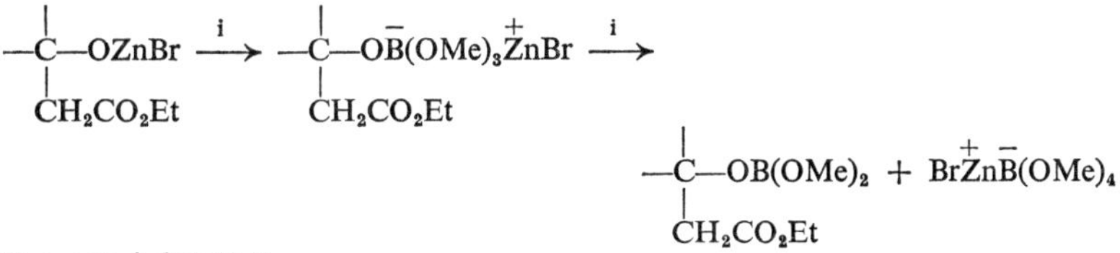

$$\begin{array}{c} | \\ -\text{C}-\text{OZnBr} \\ | \\ \text{CH}_2\text{CO}_2\text{Et} \end{array} \xrightarrow{i} \begin{array}{c} | \\ -\text{C}-\text{O}\bar{\text{B}}(\text{OMe})_3\overset{+}{\text{Zn}}\text{Br} \\ | \\ \text{CH}_2\text{CO}_2\text{Et} \end{array} \xrightarrow{i}$$

$$\begin{array}{c} | \\ -\text{C}-\text{OB}(\text{OMe})_2 + \text{Br}\overset{+}{\text{Zn}}\bar{\text{B}}(\text{OMe})_4 \\ | \\ \text{CH}_2\text{CO}_2\text{Et} \end{array}$$

Reagent: i, (MeO)$_3$B

Scheme 21

The conjugate addition of Reformatsky reagents to αβ-unsaturated substrates has been described.[99] Optically active β-hydroxyesters are obtained by condensation in the presence of (−)-sparteine; optical purities of 35—98% are reported.[100] The use of trimethylsilyl α-bromoesters is advised[101] where isolation of the β-hydroxyacid is required, mild acid treatment of the immediate product effecting ready cleavage (Scheme 22).

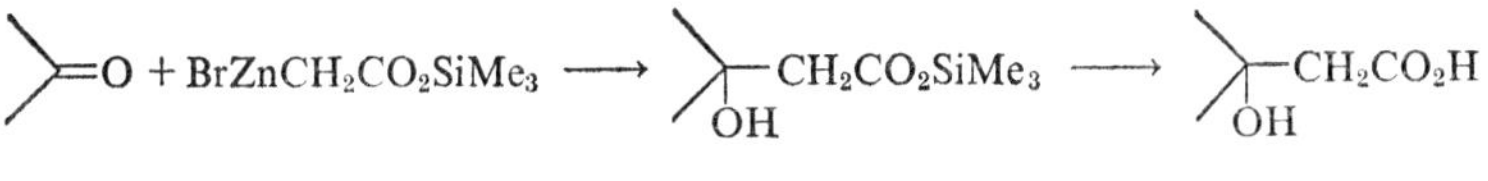

$$\begin{array}{c} \\ \\ \end{array}\!\!=\!\text{O} + \text{BrZnCH}_2\text{CO}_2\text{SiMe}_3 \longrightarrow \begin{array}{c} \\ \text{—CH}_2\text{CO}_2\text{SiMe}_3 \\ \text{OH} \end{array} \longrightarrow \begin{array}{c} \\ \text{—CH}_2\text{CO}_2\text{H} \\ \text{OH} \end{array}$$

Scheme 22

The organozinc reagent from ethyl α-bromoisobutyrate has been found to comprise two entities; one of these, the bromozinc enolate (66), reacted as expected: a dimeric structure (67) has been assigned[102] to the other, inactive component.

[97] U. Schöllkopf and H. Frasnelli, *Angew. Chem. Internat. Edn.*, 1970, **9**, 301.
[98] M. W. Rathke and A. Lindert, *J. Org. Chem.*, 1970, **35**, 3966.
[99] G. Daviaud, M. Massy, and Ph. Miginiac, *Tetrahedron Letters*, 1970, 5169.
[100] M. Guetté, J.-P. Guetté, and J. Capillon, *Tetrahedron Letters*, 1971, 2863.
[101] A. Horeau, *Tetrahedron Letters*, 1971, 3227.
[102] W. R. Vaughan and H. P. Knoess, *J. Org. Chem.*, 1970, **35**, 2394.

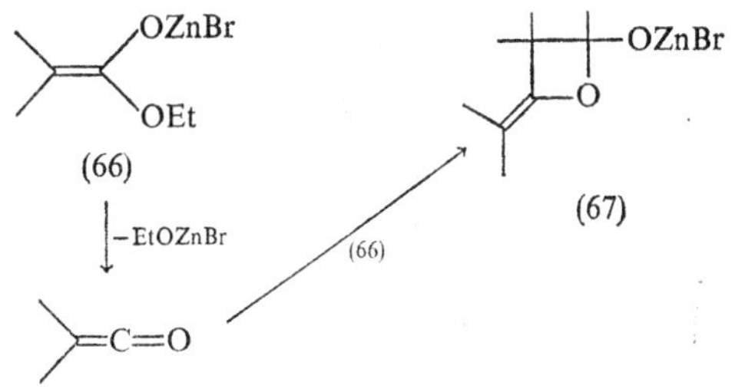

Functional Groups other than Acetylenes, Allenes, and Olefins

[Structures (66) and (67) with OZnBr, OEt groups; arrows showing conversion via −EtOZnBr to ketene >C=O]

Solid-phase Synthesis.—Patchornik and others, exploring an area of great promise, have reported on the use of polymers to attain, at will, conditions of extremely high or low concentration. The directed monoacylation[103] and monoalkylation[104] of esters have been described, as has the use of polymer-supported Wittig ylides.[105] At the other extreme of high concentration, unidirectional 'intrapolymer' mixed ester condensations have been achieved (Scheme 23).[106] A polymer (69) was obtained by treating chloromethylated polystyrene-2% divinylbenzene with a limited amount of enolizable acid (68); this polymer was then exposed to an excess of a second, non-enolizable acid (70), to give a polymer (71) containing a relatively high concentration of non-enolizable ester functions of (70). Formation of the anion of the ester moiety of (68) then exposed it to a very high concentration of the ester moieties of (70) and a very low concentration of itself. Condensation and subsequent isolation gave the single ketone (72), and unreacted starting acids.

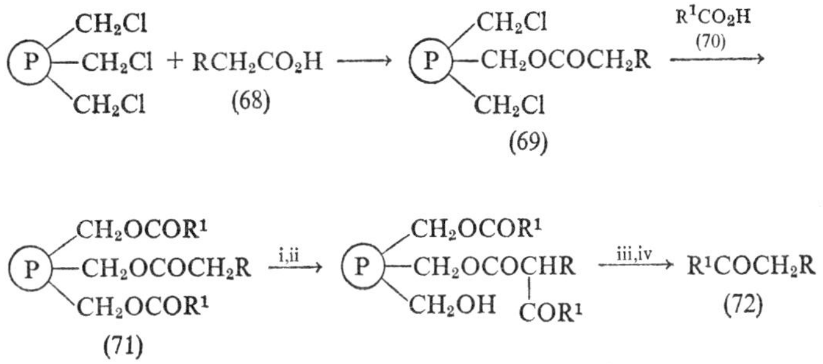

[Scheme showing polymer (P) with CH$_2$Cl groups reacting with RCH$_2$CO$_2$H to give (69), then with R^{1}CO$_2$H (70) to give (71), then reagents i, ii give intermediate, then iii, iv give R^{1}COCH$_2$R (72)]

Reagents: i, Ph$_3$CLi; ii, H$_3$O$^+$; iii, HBr–CF$_3$CO$_2$H; iv, heat

Scheme 23

---

[103] M. A. Kraus and A. Patchornik, *J. Amer. Chem. Soc.*, 1970, **92**, 7587.
[104] M. A. Kraus and A. Patchornik, *Israel J. Chem.*, 1971, **9**, 269; F. Camps, J. Castells, M. J. Ferrando, and J. Font, *Tetrahedron Letters*, 1971, 1713.
[105] F. Camps, J. Castells, J. Font, and F. Vela, *Tetrahedron Letters*, 1971, 1715.
[106] M. A. Kraus and A. Patchornik, *J. Amer. Chem. Soc.*, 1971, **93**, 7325.

A related study of the Dieckmann cyclization of polymer-bound triethylcarbinyl pimelate half-esters (73) led to essentially exclusive formation of the β-ketoester (74), only trace amounts of the isomeric (75) being observed;[107] this result is comparable to that obtained by self-condensation of benzyl triethylcarbinyl pimelate.

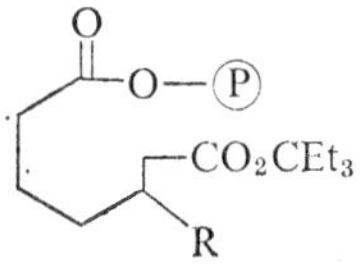

β-Ketoesters.—A synthesis of α-branched β-ketoesters is provided by electrolysis[108] of the phosphonium chloride (76), which is formed on reaction of an acid chloride with an alkoxycarbonylalkylidenephosphorane; if excess ylide is present, transylidation of the salt (76) occurs, leading to formation of the allenic ester (77), as shown in Scheme 24.

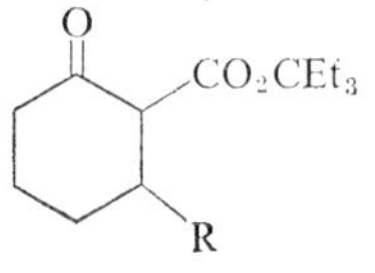

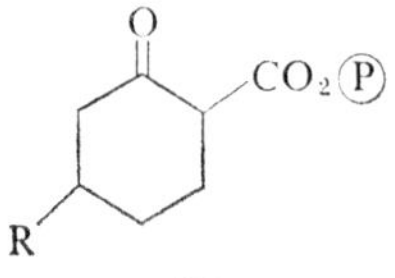

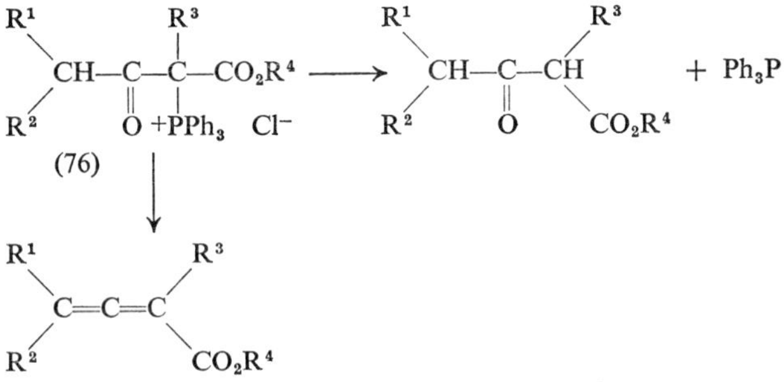

Scheme 24

Phosphorus-based synthons for acetoacetic ester and acetylacetone derivatives are described; the phosphonate (78) has been used in the preparation of the β-ketoester (79), a key intermediate[109] in a synthesis of the fungal sex hormone, trisporic acid B methyl ester (80). Michael addition of an α-methylene ketone to 2,2-diethoxyvinylidenetriphenylphosphorane followed by loss of ethanol gives a valuable reagent (81) for the preparation of 1,3-dioxopent-4-enes[110] by Wittig reaction with aldehydes; the intermediate enol ethers (82) can be isolated if desired.

---

[107] J. I. Crowley and H. Rapaport, *J. Amer. Chem. Soc.*, 1970, **92**, 6363.
[108] H.-J. Bestmann, G. Graf, H. Hartung, S. Kolewa, and E. Vilsmaier, *Chem. Ber.*, 1970, **103**, 2794.
[109] J. A. Edwards, J. Fajkos, J. H. Fried, M. L. Maddox, and V. Schwartz, *Chem. Comm.*, 1971, 292.
[110] H.-J. Bestmann and R. W. Sallfrank, *Angew. Chem. Internat. Edn.*, 1970, **9**, 367.

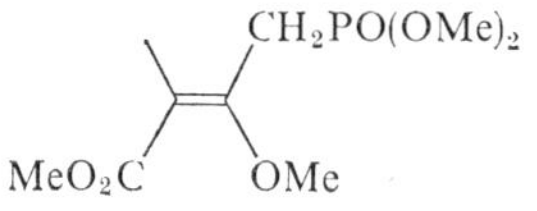

(78) 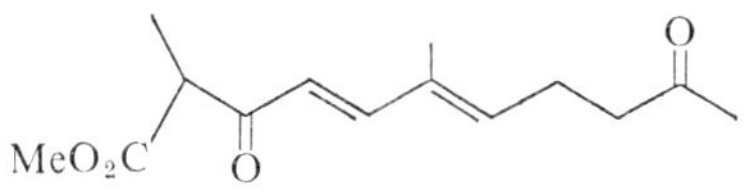 (79) 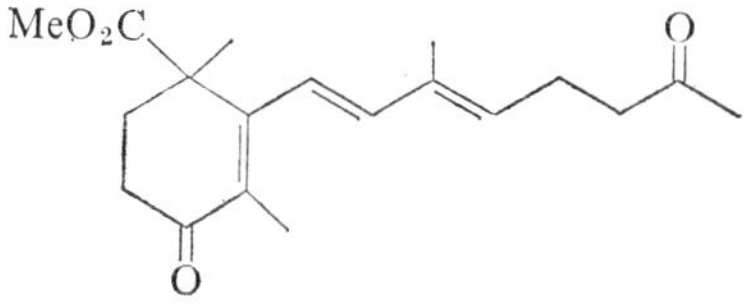 (80) 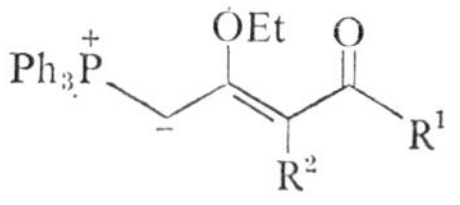 (81) 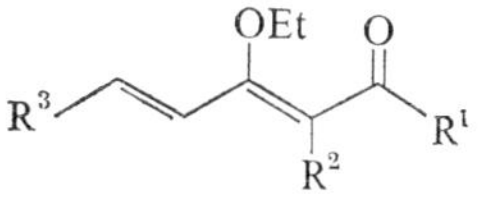 (82)

Zinc salts[111] function as effective catalysts in the conversion of 1-ethoxy-vinyl esters (83) into β-ketoester enol esters (84); the starting esters are readily obtained by mercury-catalysed addition of carboxylic acids to ethoxyacetylene. This rearrangement is most effective with the esters of weak acids, whereas an analogous thermal reaction[112] functions best with esters of strong acids.

A novel alkoxycarbonylmethylene carbon–carbon insertion process is reported to occur in the triethyloxonium fluoroborate-catalysed homologation of ketones with ethyl diazoacetate;[113] β-ketoesters are produced in high yield. Although the analogous reaction with diazoacetone does not occur, the intramolecular counterpart is successful, providing a route to polycyclic β-diketones, e.g. the conversion of (85) into (86).

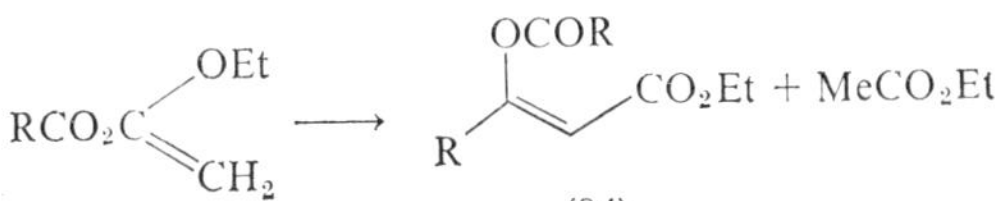

(83) → (84) + $MeCO_2Et$

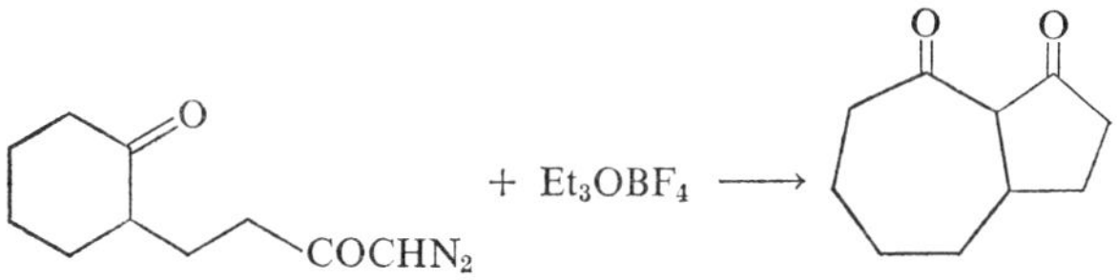

(85) + Et_3OBF_4 → (86)

[111] H. H. Wasserman and S. H. Wentland, *Chem. Comm.*, 1970, 1.
[112] B. Zwanenburg *Rec. Trav. chim.*, 1968, **82**, 593.
[113] W. L. Mock and M. E. Hartman, *J. Amer. Chem. Soc.*, 1970, **92**, 5767.

The dianion[114] derived from a β-ketoester by sequential treatment with sodium hydride and n-butyl-lithium alkylates exclusively at the γ-position; initial sodium enolate formation precludes the formation of carbonyl addition products and subsequent artifacts observed[115] in the use of n-butyl-lithium alone. In addition to providing a convenient route to γ-alkylated acetoacetic esters, such dianions undergo ready aldol condensation,[116] leading, after dehydration, to γδ-unsaturated β-ketoesters (Scheme 25).

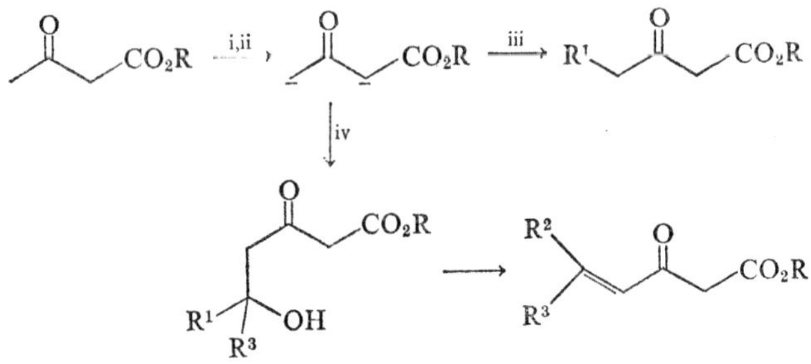

Reagents: i, NaH; ii, Bu^{n}Li; iii, R^{1}X; iv, R^{2}COR^3

Scheme 25

A French group have described[117] the low-temperature dealkoxycarbonylation of β-ketoesters by boric anhydride, B_2O_3. A further example of thermal demethoxycarbonylation of substituted malonate esters has been reported;[118] halogen participation in charge dispersal has been invoked to account for the production of methyl bromoacetate from dimethyl bromomalonate (Scheme 26).

Disubstituted malonic esters are conveniently dealkoxycarbonylated by reduction with sodium in the presence of chlorotrimethylsilane;[119] the reactive keten alkyl trimethylsilyl acetals (87) produced are readily hydrolysed.

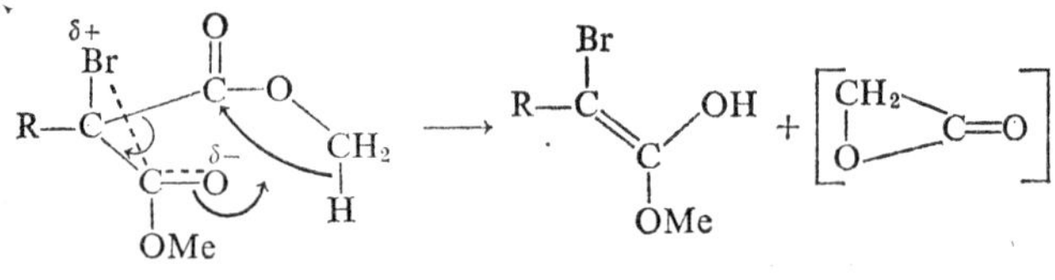

Scheme 26

[114] L. Weiler, *J. Amer. Chem. Soc.*, 1970, **92**, 6702.
[115] G. Brieger and D. G. Spencer, *Tetrahedron Letters*, 1971, 4585.
[116] S. N. Huckin and L. Weiler, *Tetrahedron Letters*, 1971, 4835.
[117] J. M. Lalancette and A. Lachance, *Tetrahedron Letters*, 1970, 3903.
[118] W. Ando, H. Matuyama, T. Migita, and S. Nakaido, *Chem. Comm.*, 1970, 1156.
[119] See ref. 86.

Functional Groups other than Acetylenes, Allenes, and Olefins 103

The 'double reduction' of β-ketoester methoxymethyl enol ethers (88) is reported by an American school[120] as a high-yield method for the hydrogenolysis of the β-carbonyl function; treatment with lithium in ammonia causes reduction of the conjugated double bond, elimination of methoxymethanol, and further reduction to give the saturated ester (89). This procedure is equally effective on the corresponding β-ketoacids.[121]

$$R_2C=C\begin{array}{c}OSiMe_3\\OMe\end{array} \qquad \begin{array}{c}OCH_2OMe\\ \diagup\!\!\diagdown CO_2R\end{array} \qquad \diagup\!\!\diagdown CO_2R$$
$$(87) \qquad\qquad (88) \qquad\qquad (89)$$

General.—A catalyst system for the production of linear acids or esters by carbonylation of terminal olefins has been described;[122] excellent yields and a high degree of linearity enhance the utility of this system. A related process concerning the carbonylation–dimerization of butadiene is noted; a palladium acetonylacetonate–triphenylphosphine complex[123] is catalytically active for the preparation of ethyl nona-3,8-dienoate from butadiene, carbon monoxide, and ethanol. The preparation of but-3-enylsuccinic acid by the catalysed carbonylation of allyl chlorides with double-bond insertion is reported; the intermediacy of but-3-enoic acid is suggested.[124]

An improved reagent for the *O*-alkyl cleavage of methyl esters by nucleophilic displacement has been found by a Stanford group;[125] lithium n-propyl mercaptide (from n-butyl-lithium and n-propyl mercaptan) in hexamethylphosphoramide smoothly and selectively cleaves methyl esters to the corresponding carboxylic acid salts. The use of boron trichloride in dichloromethane is recommended[126] for the cleavage of sterically hindered methyl esters; methyl ethers are unaffected.

$\alpha\beta$-Unsaturated acid chlorides are formed in the condensation of aldehydes and ketones with keten in the presence of boron trichloride;[127] other Friedel–Crafts catalysts usually give β-lactones, from which $\alpha\beta$-unsaturated acids are obtained by cleavage with mineral acid. The reaction between α-iodo- or α-bromo-esters and alkylidenephosphoranes provides a new synthesis of $\alpha\beta$-unsaturated acids;[128] a second equivalent of ylide is required to transylidate the initially formed phosphonium salt (Scheme 27).

---

[120] R. M. Coates and J. E. Shaw, *J. Org. Chem.*, 1970, **35**, 2597, 2601.
[121] K. K. Knutson and J. E. Shaw, *J. Org. Chem.*, 1971, **36**, 1151.
[122] L. J. Kehoe and R. A. Schell, *J. Org. Chem.*, 1970, **35**, 2846.
[123] W. E. Billups, W. E. Walker, and T. C. Shields, *Chem. Comm.*, 1971, 1067.
[124] G. P. Chiusoli and S. Merzoni, *Chem. Comm.*, 1971, 522.
[125] P. A. Bartlett and W. S. Johnson, *Tetrahedron Letters.*, 1970, 4459.
[126] P. S. Manchand, *Chem. Comm.*, 1971, 667.
[127] P. I. Paetzold and S. Kosma, *Chem. Ber.*, 1970, **103**, 2003.
[128] H.-J. Bestmann, H. Dornauer, and K. Rostock, *Chem. Ber.*, 1970, **103**, 685, 2011.

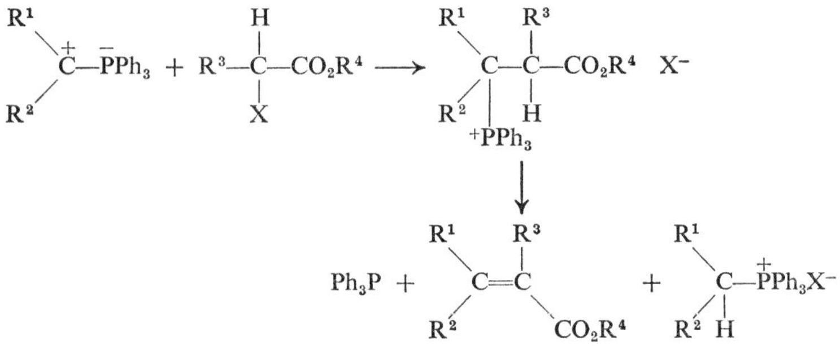

Scheme 27

The preparation of succinic esters by anion coupling with polyhalides has been described,[129] The scope of the Wurtz coupling of α-halogenoesters[130] to give alkyl-substituted succinic esters has been investigated; optimum conditions[131] utilize α-bromoesters and zinc in THF, the presence of copper(II) salts being mandatory to avoid Dieckmann-type condensation to form β-ketoesters.

Organoboranes react with ethyl γ-bromocrotonate (90) to give the βγ-unsaturated esters (91), providing a convenient four-carbon homologation procedure.[132]

$$R_3B + BrCH_2CH=CHCO_2Et \longrightarrow RCH=CHCH_2CO_2Et$$
$$(90) \qquad\qquad\qquad (91)$$

Thallium(III) acetate effects a one-step synthesis of aliphatic α-acyloxy-carboxylic acids (Scheme 28); the transition state (92) is proposed.[133] α-Aminoesters are readily oxidized to the corresponding α-diazoesters[134] by isoamyl nitrite in the presence of acetic acid. A review has been published on the radical addition of carboxylic acids and derivatives to unsaturated linkages.[135]

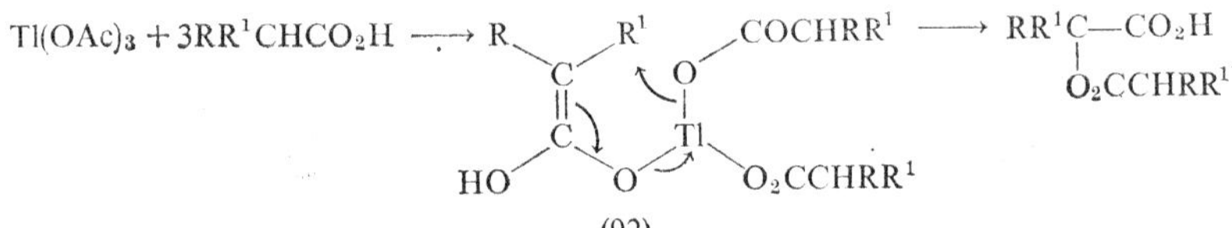

Scheme 28

[129] W. G. Kofron and C. R. Hauser, *J. Org. Chem.*, 1970, **35**, 2085.
[130] B. Kurtev, M. Mladenova, and B. Blagoev, *Compt. rend.*, 1970, **271**, C, 871.
[131] C. Fouquey and J. Jacques, *Synthesis*, 1971, 306.
[132] H. C. Brown and H. Nambu, *J. Amer. Chem. Soc.*, 1970, **92**, 1761.
[133] E. C. Taylor, H. W. Altland, and G. McGillivray, *Tetrahedron Letters*, 1970, 5285.
[134] N. Takamura, T. Mizoguchi, K. Koga, and S. Yamada, *Tetrahedron Letters*, 1971, 4495.
[135] H.-H. Vogel, *Synthesis*, 1970, 99.

A number of papers have been published concerning the t-butoxycarbonyl protecting group. The synthesis of the protecting reagent, t-butyl azidoformate, from t-butylcarbonic diethylphosphoric anhydride and potassium azide has been described in detail.[136] A simple preparation of t-butyl carbazate, a precursor of t-butyl azidoformate, by reaction of t-butylethylcarbonate with hydrazine has been reported:[137] this reaction was formerly considered to be unsuccessful.[138] A route to t-butyl azidoformate by treatment of t-butylchloroformate with tetramethylguanidinium azide is noted.[139] A Japanese group sound a caution on the preparation of t-butyl azidoformate by reaction of t-butyl alcohol with phosgene, followed by addition of sodium azide: care must be taken to expel excess phosgene before addition of azide, to preclude formation of the highly explosive carbazide, $(N_3)_2CO$.[140]

Boron trifluoride etherate in glacial acetic acid or acetic acid–chloroform smoothly cleaves t-butoxycarbonyl groups, leaving N-benzyloxycarbonyl groups untouched.[141]

4 Carboxylic Acid Amides

The study of methods to enhance ester- or amide-bond formation continues to provide much stimulus. A Japanese group, continuing their researches in this vein, have reported on amide[142] and ester[143] bond formation *via* oxidation–reduction condensation by use of 2,2'-dipyridyl disulphide as oxidant; the reduced reagent, 2-mercaptopyridine, isomerizes to the more stable, less reactive thione form. This is shown in Scheme 29.

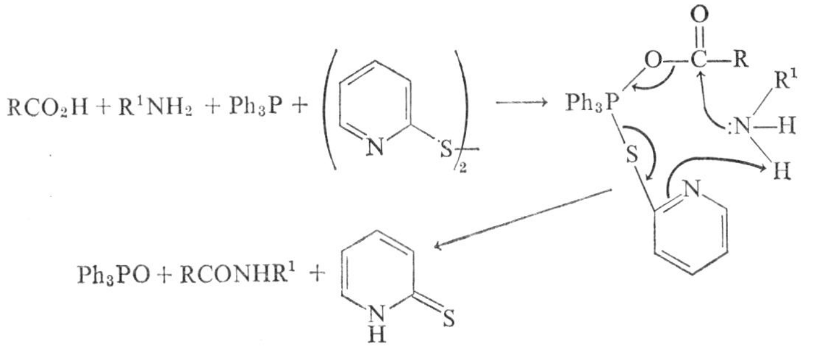

Scheme 29

---

[136] *Org. Syntheses*, 1970, **50**, 9.
[137] M. Muraki and T. Mizoguchi, *Chem. and Pharm. Bull.* (*Japan*), 1970, **18**, 217.
[138] L. Carpino, *J. Amer. Chem. Soc.*, 1960, **82**, 2725.
[139] K. Sakai and J.-P. Anselme, *J. Org. Chem.*, 1971, **36**, 2387.
[140] H. Yajima, H. Kawatani, and Y. Kiso, *Chem. and Pharm. Bull.* (*Japan*), 1970, **18**, 850.
[141] R. G. Hiskey, L. M. Beacham, V. G. Matl, J. N. Smith E. B. Williams, jun., A. M. Thomas, and E. T. Wolters, *J. Org. Chem.*, 1971, **36**, 488.
[142] T. Mukaiyama, R. Matsueda, and M. Suzuki, *Tetrahedron Letters*, 1970, 1901.
[143] T. Mukaiyama, K. Goto, R. Matsueda, and M. Ueki, *Tetrahedron Letters*, 1970, 5293.

In the latter paper,[143] use is made of the water-soluble phosphine (93), which also gives rise to a water-soluble phosphine oxide, as shown in Scheme 30. This general method has been used in peptide synthesis, where it is reported to provide a high-yield, high optical purity mode of condensation.

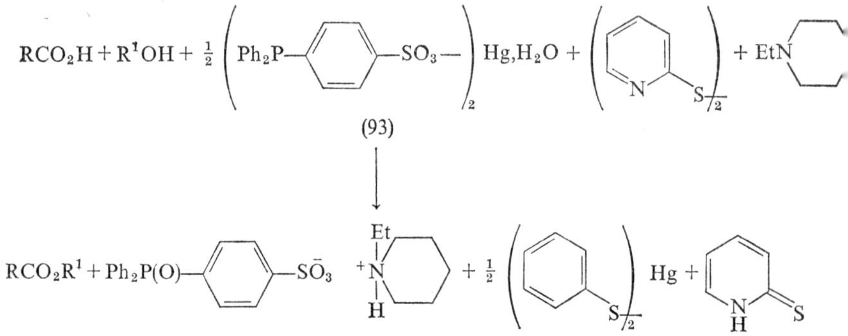

$$RCO_2H + R^1OH + \tfrac{1}{2}\left(Ph_2P\text{-}\underset{(93)}{\bigcirc}\text{-}SO_3^-\right)_2 Hg, H_2O + \left(\bigcirc_N\right)_{S/2}^{} + EtN\bigcirc$$

$$\downarrow$$

$$RCO_2R^1 + Ph_2P(O)\text{-}\bigcirc\text{-}S\bar{O}_3 \quad {}^+\!\!\underset{H}{\overset{Et}{N}}\!\!\bigcirc + \tfrac{1}{2}\left(\bigcirc_{}\right)_{S/2}^{} Hg + \underset{H}{\bigcirc_N}\!\!S$$

Scheme 30

Use of the triphenylphosphine–carbon tetrachloride reagent, with[144] or without[145] the addition of triethylamine, provides an efficient peptide coupling procedure, both methods involving the intermediate phosphonium ester salt (94); the presence of triethylamine obviates the use of two equivalents of reacting amine.

$$Ph_3\overset{+}{P}OCOR \quad X^-$$
(94)

A related technique[146] involves a tris(dimethylamino)phosphine–carbon tetrachloride reagent, where interaction with carboxylic acids gives anhydrides and tris(dimethylamino)phosphonium chloride (95), which can, in turn, activate remaining acid for further reaction, e.g. to complete anhydride formation, giving the overall process shown in Scheme 31. Alternatively, it can be used to effect amide-bond formation, when nucleophilic attack by amine on

$$RCO_2H + (Me_2N)_3P + CCl_4 \longrightarrow$$
$$\tfrac{1}{2}(RCO)_2O + CHCl_3 + \tfrac{1}{2}(Me_2N)_3PO + \tfrac{1}{2}(Me_2N)_3\overset{+}{P}Cl\bar{Cl}$$
(95)

$$2RCO_2H + Et_3N + (Me_2N)_3P + CCl_4 \longrightarrow$$
$$(RCO)_2O + CHCl_3 + Et_3NHCl + (Me_2N)_3PO$$

Scheme 31

[144] T. Wieland and A. Seeliger, *Chem. Ber.*, 1971, **104**, 3992.
[145] L. E. Barstow and V. J. Hruby, *J. Org. Chem.*, 1971, **36**, 1305.
[146] B. Castro and J. R. Dormoy, *Bull. Soc. chim. France*, 1971, 3034.

the initially formed anhydride produces amide and acid anion, which is subsequently activated by chlorophosphonium chloride for further attack by amine, as formulated in Scheme 31a.

$$RCO_2H + 2R^1NH_2 + (Me_2N)_3P + CCl_4 \longrightarrow$$
$$RCONHR^1 + CHCl_3 + R^1NH_3Cl + (Me_2N)_3PO$$

Scheme 31a

The activation of acids by oxidation or dehydration of their derivatives has been studied by a German school.[147] Oxidation of an acid diphenylhydrazide by N-bromosuccinimide yields the azonium ion (96), which functions as a highly activated acid derivative, azobenzene being eliminated on nucleophilic attack. A related activation by oxidation has been performed in the solid phase by production of the polymer-bound azo-compound (97), nitrogen being expelled on amide-bond formation. Dehydrative activation is exemplified by conversion of the ester (98), prepared from the acid and 1,1-diphenylethylene glycol, into the enol ester (99). The generation of azo-compounds by anodic oxidation of hydrazides has been reported.[148] All of these procedures have been utilized successfully in peptide synthesis.

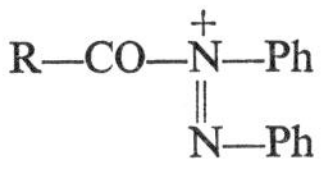

$$R-CO-\overset{+}{\underset{\underset{Ph}{\parallel}}{N}}-Ph \qquad \text{\textcircled{P}}-CH_2-N=N-COR \qquad RCO_2CH_2\underset{Ph}{\overset{Ph}{\underset{|}{\overset{|}{C}}}}-OH$$

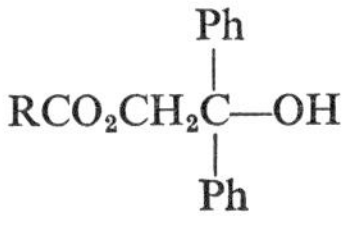

(96)   (97)   (98)

$$RCO_2CH=C\overset{Ph}{\underset{Ph}{\diagdown}}$$

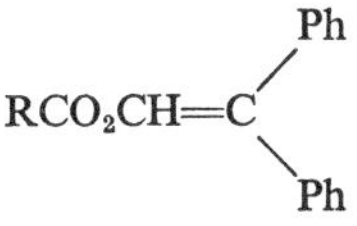

(99)

In the dicyclohexylcarbodi-imide method of peptide-bond formation, the problems of racemization and rearrangement to N-acylureas can be avoided by interception of the intermediate activated ester with 1-hydroxybenzotriazole.[149] The ester so produced functions efficiently in amide-bond formation; similar beneficial properties have already been noted with N-hydroxysuccinimide.

An intriguing report[150] on peptide-bond formation by cupric complexes has been published; the ethyl ester of glycine is converted into the di- tri-, and tetra-peptide ethyl esters in yields of 28, 23, and 3% respectively, based on glycine consumed. Titanium tetrachloride[151] is a satisfactory reagent for the preparation of amides from acids and amines.

[147] T. Wieland, J. Lewalter, and C. Birr, *Annalen*, 1970, **740**, 31.
[148] J. Lewalter and C. Birr, *Annalen*, 1970, **740**, 48.
[149] W. König and R. Geiger, *Chem. Ber.*, 1970, **103**, 788.
[150] S. Yamada, S. Terashima, and M. Wagatsuma, *Tetrahedron Letters*, 1970, 1501.
[151] J. D. Wilson and H. Weingarten, *Canad. J. Chem.*, 1970, **48**, 983.

The γ-alkylation of metallated αβ-ynamines has been proposed by Corey[152] as a convenient method for chain extension, amides being formed on hydrolysis (Scheme 32).

$$MeC\equiv CNR_2 \xrightarrow{i} LiCH_2C\equiv CNR_2 \xrightarrow{ii} R^1CH_2C\equiv CNR_2 \xrightarrow{iii} RCH_2CH_2CONR_2$$

Reagents: i, Bu^{n}Li–TMEDA; ii, R^{1}X; iii, H$_3$O$^+$

Scheme 32

The nucleophilicity of amines in the aminolysis of esters can be greatly improved by use of sodium hydride in DMSO or hexamethylphosphoramide; the amine anion produced reacts smoothly with esters at room temperature to afford the desired amide in excellent yield.[153] A similar use of n-butyl-lithium for amine anion production is recorded.[154]

A facile synthesis of imides by manganese ion-catalysed peroxide oxidation of amides is reported.[155] The direct conversion of cyclic anhydrides into imide carbamates is noted;[156] use is made of alkoxycyanates, thermal rearrangement of the intermediate (100) being proposed (Scheme 33).

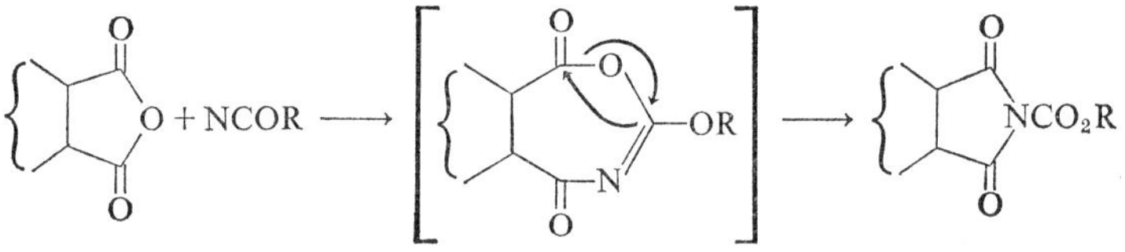

(100)

Scheme 33

A method for the preparation of amides from aldehydes by oxidation of the corresponding cyanohydrin is described;[157] this reaction is subject to the same selectivity as the previously reported method[158] of conversion of aldehydes into esters: the aldehyde must be aromatic or αβ-unsaturated to permit allylic oxidation of the cyanohydrin to the α-oxonitrile, which reacts with amines *in situ* to give the derived amides.

The synthesis of ureas by reaction of ammonia or aliphatic primary and secondary amines with carbon monoxide at atmospheric pressure under selenium catalysis is noted.[159] The intermediate salt can be oxidized to the *sym*-urea, or an exchange can be effected with a second amine, leading to *unsym*-ureas, as shown in Scheme 34. Ureas are readily dehydrated to carbodi-imides by the triphenylphosphine–carbon tetrachloride–triethylamine combination

---

[152] E. J. Corey and D. E. Cane, *J. Org. Chem.*, 1970, **35**, 3405.
[153] B. Singh, *Tetrahedron Letters*, 1971, 321.
[154] K.-W. Yang, J. G. Cannon, and J. G. Rose, *Tetrahedron Letters*, 1970, 1791.
[155] A. Doumax, jun. and D. J. Trecker, *J. Org. Chem.*, 1970, **35**, 2121.
[156] E. Grigat, *Angew. Chem. Internat. Edn.*, 1970, **9**, 68.
[157] N. W. Gilman, *Chem. Comm.*, 1971, 733.
[158] E. J. Corey, N. W. Gilman, and B. E. Ganem, *J. Amer. Chem. Soc.*, 1968, **90**, 5617.
[159] N. Sonoda, T. Yasuhara, K. Kondo, T. Ikeda, and S. Tsutsumi, *J. Amer. Chem. Soc.*, 1971, **93**, 6344.

Functional Groups other than Acetylenes, Allenes, and Olefins 109

$2RNH_2 + Se + CO \longrightarrow$

$$R\overset{+}{N}H_3(RNH\overset{O}{\overset{\|}{C}}-Se)^- \overset{i}{\longrightarrow} (RNH)_2CO + H_2O + Se$$

$\downarrow$ ii

$$R^1\overset{+}{N}H_3(RNH\overset{O}{\overset{\|}{C}}-Se)^- \overset{i}{\longrightarrow} RNHCONHR^1 + H_2O + Se$$

Reagents: i, O_2; ii, R^1NH_2

Scheme 34

of Appel.[160] Formamides are formed[161] in the transition-metal catalysed reaction of amines with carbon dioxide and hydrogen.

Amides can be N-alkylated in good yield with alkyl iodides in the presence of potassium hydroxide in DMSO.[162] A study has been made of the factors affecting the preparation of triamides[163] by diacylation of primary amides with acid chlorides in the presence of pyridine.

As part of a projected synthesis of corrins, a simple synthesis of γ-substituted γ-butyrolactams *via* the conjugate addition of hydrogen cyanide to $\alpha\beta$-unsaturated ketones is described;[164] contrary to a much earlier report,[165] the β-cyanohydrin (101) is not produced, but a mixture of the two butyrolactams (102) and (103) is isolated, (103) being convertible into (102) by reaction with basic cyanide solution. Reaction of one of those, (102), with potassium t-butoxide in t-butyl alcohol gave, *inter alia*, the semi-corrinoid (104). These transformations are outlined in Scheme 35.

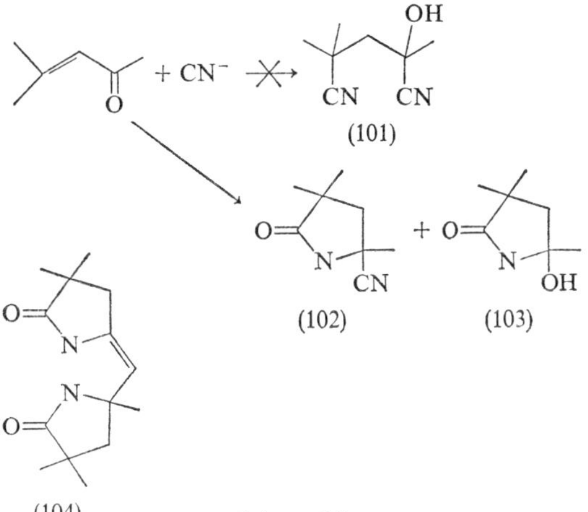

Scheme 35

[160] R. Appel, R. Kleinstück, and K.-D. Ziehn, *Chem. Ber.*, 1971, **104**, 1335.
[161] P. Haynes, L. S. Slaugh, and J. F. Kohnle, *Tetrahedron Letters*, 1970, 365.
[162] G. L. Isele and A. Lüttringhaus, *Synthesis*, 1971, 266.
[163] R. T. LaLonde and C. B. Davis, *J. Org. Chem.*, 1970, **35**, 771.
[164] R. V. Stevens and M. Kaplan, *Chem. Comm.*, 1970, 823.
[165] A. Lapworth, *J. Chem. Soc.*, 1904, 1214.

A convenient alternative to the Beckmann rearrangement is reported by Barton.[166] Alkyl nitrones, readily available from the corresponding ketones, are smoothly transformed into N-alkylamides by treatment with toluene-p-sulphonyl chloride, as shown in Scheme 36. Beckmann rearrangement of the oxime of (105) affords the isomeric lactam (106). This reaction is applicable also to saturated ketones.

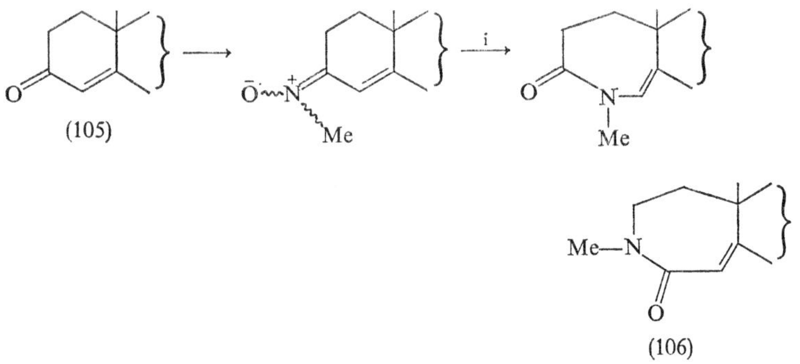

Reagent: i, TosCl–py

Scheme 36

Extending an investigation of the double β-scission reaction of alcohols with nitrogen substitution at the β-carbon (Scheme 37), Japanese authors[167] have reported that the intermediate alkoxyl radical can be generated more effectively photochemically, using hypoiodite, or oxidatively, with lead tetra-acetate, rather than by the earlier method[168] of nitrite ester irradiation.

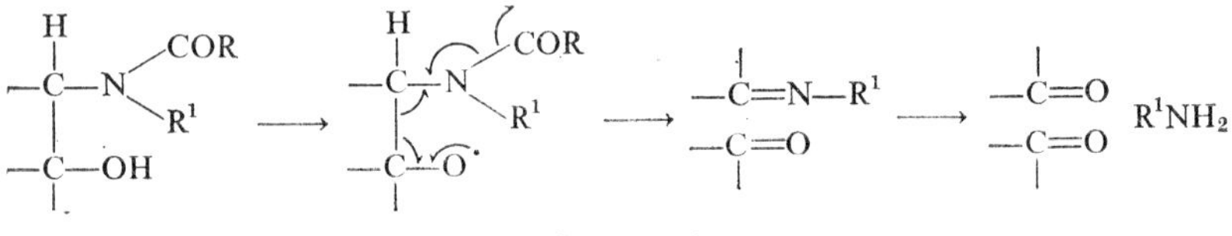

Scheme 37

The reaction of nitronium fluoroborate[169] with olefins in acetonitrile leads, via a Ritter reaction of the intermediate carbonium ion with solvent, to 1-nitro-2-acetamidoalkanes. The inherent instability of DMF towards aqueous base has been noted.[170]

---

[166] D. H. R. Barton, M. J. Day, R. H. Hesse, and M. M. Pechet, *Chem. Comm.*, 1971, 945.
[167] H. Suginome, H. Umeda, and T. Masamune, *Tetrahedron Letters*, 1970, 4571.
[168] H. Suginome, M. Murakami, and T. Masamune, *Bull. Chem. Soc. Japan*, 1968, **41**, 468.
[169] M. L. Scheinbaum and M. Dines, *J. Org. Chem.*, 1971, **36**, 3641.
[170] E. Buncel and E. A. Symons, *Chem. Comm.*, 1970, 164; E. Buncel, S. Kesmarky and E. A. Symons, *ibid.*, 1971, 120.

5 Nitriles

Preparation.—A plethora of methods has been reported for the preparation of nitriles, many of which involve overall dehydration of amides. In this context, much use has been made of phosphines as reagents, amide dehydration being achieved under virtually neutral conditions by heating the substrate in THF with triphenylphosphine and carbon tetrachloride, the nitrile being formed *via* the imino-chloride[171] (107). The yields and generality of this reaction can be markedly enhanced by the presence of triethylamine[172] which, by proton abstraction, induces elimination of triphenylphosphine oxide from the initially formed imidol phosphonium salt (108). Dehydration can also be effected in good yield by heating the amide in hexamethylphosphoramide at elevated temperature[173] or by treatment with polyphosphate ester.[174] All of these methods are obviously mechanistically similar. Amide dehydration at elevated temperature by silazanes, alkoxysilanes, and chlorosilanes allows a choice of mildly basic, neutral, or mildly acidic conditions,[175] respectively. Titanium tetrachloride[176] in the presence of pyridine is a satisfactory reagent for the dehydration of amides under mild conditions.

$$\underset{(107)}{\overset{\overset{\displaystyle Cl}{|}}{R-C=NH}} \quad \underset{(108)}{\overset{\overset{\displaystyle \overset{+}{O}PPh_3}{|}}{R-C=NH}} \quad \underset{(109)}{\overset{\overset{\displaystyle H}{|}}{R-C=N-O-\overset{+}{P}Ph_3}}$$

Methods have been devised to permit dehydration of aldoximes under mild conditions. Treatment with the triphenylphosphine–carbon tetrachloride–triethylamine combination[177] dehydrates aldoximes to nitriles *via* the oximino-phosphonium salt (109). Two procedures utilizing oxime esters have been reported. The first involves pyrolysis of the phenyl oximinocarbonate (110a), prepared from the aldoxime and phenyl chloroformate; the ester smoothly eliminates phenol and carbon dioxide at 100 °C, giving the desired nitrile in high yield under effectively neutral conditions.[178] In an even milder route,[179] the oxime is converted into the corresponding thionoformate with *p*-chlorophenyl- or methyl-chlorothionoformate in the presence of pyridine; the thionoformate ester (110b) decomposes at room temperature to afford

---

[171] E. Yamamoto and S. Sugasawa, *Tetrahedron Letters*, 1970, 4383.
[172] R. A. Appel, R. Kleinstück, and K.-D. Ziehn, *Chem. Ber.*, 1971, **104**, 1030.
[173] R. S. Monson and D. N. Priest, *Canad. J. Chem.*, 1971, **49**, 2897.
[174] Y. Kanadka, T. Kuga, and K. Tanizawa, *Chem. and Pharm. Bull.* (*Japan*), 1970, **18**, 397.
[175] W. E. Dennis, *J. Org. Chem.*, 1970, **35**, 3253.
[176] W. Lehnert, *Tetrahedron Letters*, 1971, 1501.
[177] R. A. Appel, R. Kleinstück, and K.-D. Ziehn, *Chem. Ber.*, 1971, **104**, 2025.
[178] J. M. Prokipcak and P. A. Forte, *Canad. J. Chem.*, 1971, **49**, 1321.
[179] D. L. J. Clive, *Chem. Comm.*, 1970, 1014.

the nitrile in high yield. In both reactions, a cyclic transition state seems plausible (Scheme 38). Oximes are dehydrated to nitriles by titanium tetra-

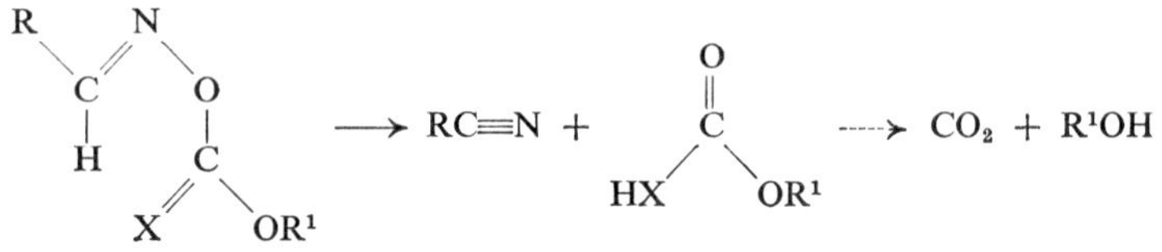

$$\underset{\underset{X}{\overset{H}{\underset{|}{R}}}{\overset{N}{\overset{\diagup}{C}}}\overset{O}{\underset{OR^1}{\overset{\diagdown}{\underset{\diagup}{C}}}}}{} \longrightarrow RC{\equiv}N + \underset{HX \quad OR^1}{\overset{O}{\overset{\|}{C}}} \dashrightarrow CO_2 + R^1OH$$

(110) a; X = S, R^1 = p-chlorophenyl
b; X = O, R^1 = Ph

Scheme 38

chloride[180] in the presence of pyridine. Primary nitro-compounds give the corresponding nitriles on reduction with sulphurated sodium borohydride, $NaBH_2S_3$; secondary nitro-compounds give mixtures of ketones and the simple amines, whereas tertiary nitro-compounds do not react at all.[181]

An ingenious development in nitrile synthesis *via* an acid–nitrile exchange reaction is reported by Klein.[182] The method is based on the equilibrium known to exist between acids and nitriles, the equilibrium being displaced towards the weaker acid (Scheme 39). By use of short-chain dinitriles, such

$$RCO_2H + R^1CN \rightleftharpoons RCONHCOR^1 \rightleftharpoons RCN + R^1CO_2H$$

Scheme 39

as succinodinitrile or glutarodinitrile rather than acetonitrile, the previously required use of pressure is avoided, and much better yields are realized since, as soon as the cyanoacid is released, it irreversibly forms the corresponding cyclic imide, thus driving the reaction to completion. The utility of this reaction can be appreciated by the preparation of dodecanedinitrile in 97% yield from dodecanoic acid and α-methylglutarodinitrile, as shown in Scheme 39a.

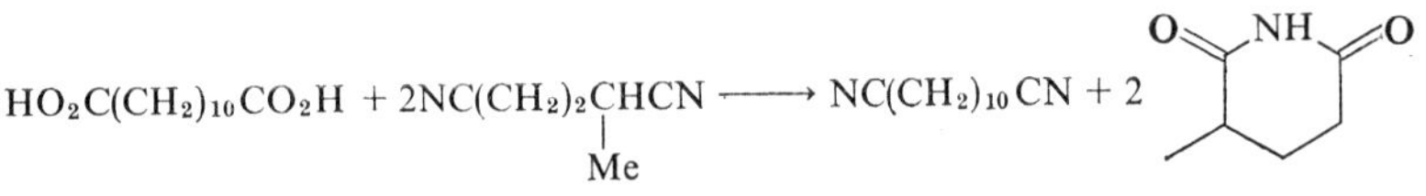

$$HO_2C(CH_2)_{10}CO_2H + 2NC(CH_2)_2\underset{\underset{Me}{|}}{C}HCN \longrightarrow NC(CH_2)_{10}CN + 2 \text{ [cyclic imide]}$$

Scheme 39a

The oxidative fragmentation of 1-amino-2,5-dialkyl-1,3,4-triazoles has been suggested[183] as a potential route to difficulty accessible nitriles; the intermediacy of the derived 1-nitrene is postulated.

[180] W. Lehnert, *Tetrahedron Letters*, 1971, 559.
[181] J. M. Lalancette and J. R. Brindle, *Canad. J. Chem.*, 1971, **49**, 2990.
[182] D. A. Klein, *J. Org. Chem.*, 1971, **36**, 3050.
[183] K. Sakai and J.-P. Anselme, *Tetrahedron Letters*, 1970, 3851.

Functional Groups other than Acetylenes, Allenes, and Olefins 113

The transformation of ketones into nitriles, α-methoxycarbonylnitriles, and α-methylnitriles from a common diazene intermediate has been reported by an American group.[184] Oxidation of the hydrazine (111), obtained from a ketonic precursor by addition of hydrogen cyanide to the corresponding methoxycarbonylhydrazone, gives the methyl dialkylcyanodiazenecarboxylate (112), which undergoes base-induced decomposition to afford the range of products in the manner shown (Scheme 40).

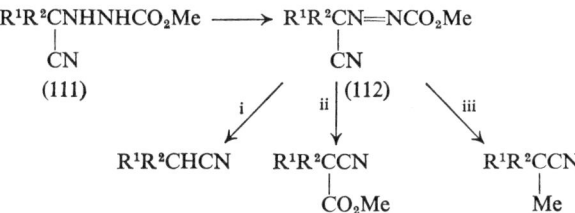

Reagents: i, NaOMe–MeOH; ii, LiOMe–(MeO)$_2$CO; iii, LiOMe–MeI

Scheme 40

A method for the *in situ* preparation of the cyanomethylene ylide (113) is described.[185] Reaction of triphenylphosphine with dibromoacetonitrile gives the bisphosphonium salt (114); this salt reacts with aldehydes in the presence of aqueous base to give the corresponding nitriles, *via* the ylide (113).

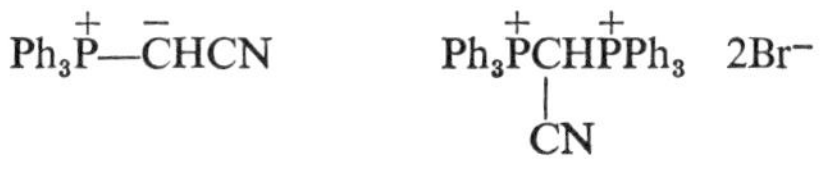

$$\text{Ph}_3\overset{+}{\text{P}}\text{—}\overset{-}{\text{CHCN}} \qquad \underset{\underset{\text{CN}}{|}}{\text{Ph}_3\overset{+}{\text{P}}\text{CH}\overset{+}{\text{P}}\text{Ph}_3} \;\; 2\text{Br}^-$$

(113)   (114)

Organoboranes are reported[186] to react with dichloroacetonitrile under the influence of potassium 2,6-di-t-butylphenoxide, giving α-chloronitriles, which can in turn react with a second organoborane to afford dialkylacetonitriles, as outlined in Scheme 41.

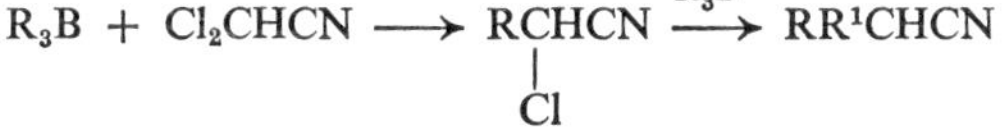

$$R_3B + Cl_2CHCN \longrightarrow \underset{\underset{Cl}{|}}{RCHCN} \xrightarrow{R^1_3B} RR^1CHCN$$

Scheme 41

Nitriles are produced from aldehydes in high yield by reaction of the latter with hydrazine hydrate under cyanide-ion catalysis.[187] A method of limited utility[188] involves the dehydrochlorination of α-nitrosophosphonium

---

[184] F. E. Ziegler and P. A. Wendler, *J. Amer. Chem. Soc.*, 1971, **93**, 4318.
[185] J. W. Wilt and A. J. Ho, *J. Org. Chem.*, 1971, **36**, 2026.
[186] H. Nambu and H. C. Brown, *J. Amer. Chem. Soc.*, 1970, **92**, 5790.
[187] W. Köhler, *Z. Chem.*, 1971, **11**, 343.
[188] M. I. Shevchuk, E. M. Volynskaya, and A. M. Dombrovskii, *J. Gen. Chem. (U.S.S.R.)*, 1971, **41**, 2019.

chlorides; nitriles and triphenylphosphine oxide are produced. A simple preparative procedure[189] utilizes the sulphonyl nitrile (115), which reacts with Grignard reagents to furnish the corresponding alkyl nitriles. The synthesis of nitriles from aldehydes using cobaltammine complexes has been reported;[190] oxidation of the intermediate imine complexes affords nitriles in high yield.

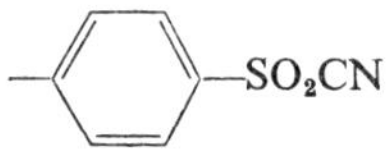

$$\text{C}_6\text{H}_5-\text{SO}_2\text{CN}$$
(115)

Reactions.—Nitriles can be reduced to aldehydes in aqueous media by photochemically generated hydrated electrons.[191] The reduction of nitriles to amines by sodium borohydride is catalysed by Raney nickel.[192] The transition metal promoted reductive decyanation of alkyl nitriles to homologous hydrocarbons with one less carbon atom is reported, and has been shown by van Tamelen[193] to proceed by a different mechanism from that of a similar reduction by alkali metals in ammonia, where a process of stepwise two-electron transfer has been proposed.[194] In the former reaction, where use is made of ferric acetylacetonate and sodium, the proton required for alkane production is derived exclusively from the acetylacetonate ligand, alkane being formed before a proton source is added; this suggests a mechanism which involves initial co-ordination of the nitrile to iron, followed by reductive cleavage of the nitrile 1,2-carbon bond and proton transfer to this area, as pictured in Scheme 42.

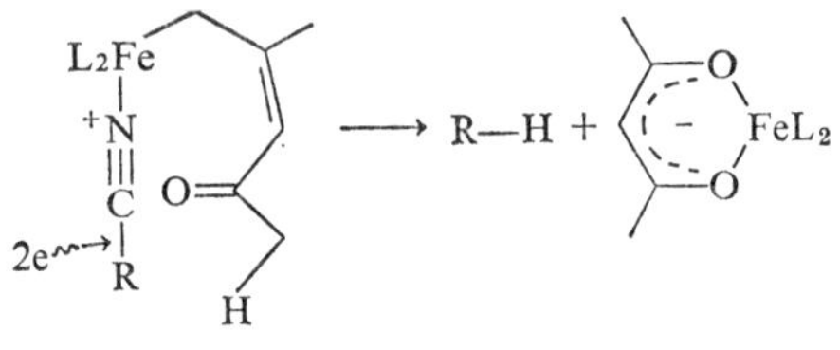

Scheme 42

A high-yield conversion[195] of nitriles into amides is effected by treatment of the nitrile with formic acid and hydrogen chloride or hydrogen bromide, elevated pressure or temperature being unnecessary.

[189] A. M. van Leusen and J. C. Jagt, *Tetrahedron Letters*, 1970, 967.
[190] I. Rhee, M. Ryang, and S. Tsutsumi, *Tetrahedron Letters*, 1970, 3419.
[191] J. D. Ferris and F. R. Antonucci, *Chem. Comm.*, 1971, 1294.
[192] R. A. Egli, *Helv. Chim. Acta*, 1970, **53**, 47.
[193] E. E. van Tamelen, H. Rudler, and C. Bjorklund, *J. Amer. Chem. Soc.*, 1971, **93**, 7113.
[194] P. G. Arapakos, M. K. Scott, and F. E. Huber, jun., *J. Amer. Chem. Soc.*, 1969, **91**, 2059.
[195] F. Becke, H. Fleig, and P. Pässler, *Annalen*, 1971, **749**, 198.

Two procedures for the preparation of α-oxocarboxylic acids via nitrile intermediates have been described. The autoxidation of 3,3-disubstituted 2-cyanocarboxylate ester anions is reported[196] to provide the related α-oxoesters in fair yield; a new general procedure[197] for the conversion of aldehydes into the homologous α-oxoacids is based on the oxidation of α-hydroxy-N-t-butyl-carboxamides (116), prepared by the acid-catalysed alcoholysis of aldehyde cyanohydrins with t-butanol, to the hitherto unknown α-oxocarboxamides, which are in turn readily converted into the parent α-oxoacids on acid-catalysed hydrolysis.

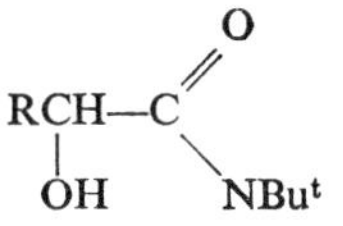

$$\begin{array}{c} \text{O} \\ \| \\ \text{RCH—C} \\ | \quad \backslash \\ \text{OH} \quad \text{NBu}^t \end{array}$$

(116)

A study[198] has been made of the asymmetric 1,4-reduction of alkylidene cyanoacetic esters and alkylidene malononitriles by chiral Grignard reagents; the stereochemical results are opposite to those predicted from a cyclic mechanism (117), suggesting the alternative acyclic transition state (118) to produce the observed S-configuration in the product (Scheme 43).

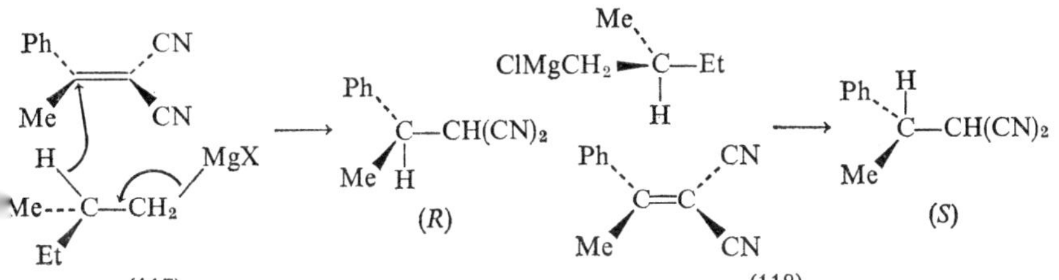

Scheme 43

In the probable prebiotic chemical evolution of adenine by pentamerization of hydrogen cyanide, the intermediacy of a dimer of the latter has been assumed; although it remains undetected, iminoacetonitrile (119) has been suggested as a plausible structure for the dimer, and the synthesis of N-t-butyliminoacetonitrile, the first known member of this family, has been reported by a Chicago school (Scheme 44).[199]

Vilsmeier formylation of acetonitrile gives the dimethylamino-enamine of cyanomalondialdehyde (120).[200] A study has been carried out on the in-

---

[196] N. A. Rabjohn and C. A. Harbert, *J. Org. Chem.*, 1970, **35**, 3240, 3726.
[197] J. Anatol and A. Medète, *Synthesis*, 1971, 538.
[198] D. Cabaret and Z. Welvart, *Chem. Comm.*, 1970, 1064.
[199] J. H. Boyer and H. Dabek, *Chem. Comm.*, 1970, 1204.
[200] C. Reichardt and W.-D. Kermer, *Synthesis*, 1970, 538.

HN=CHCN
(119)

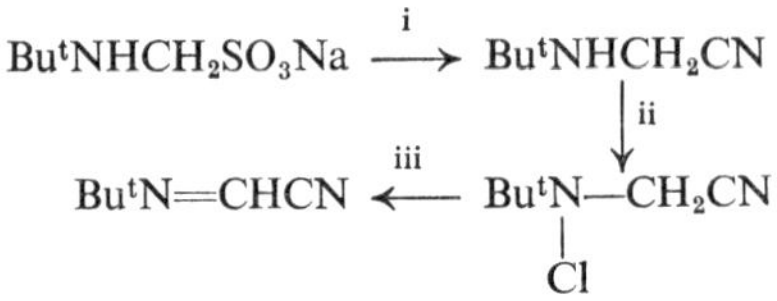

$$Bu^tNHCH_2SO_3Na \xrightarrow{i} Bu^tNHCH_2CN$$
$$\downarrow ii$$
$$Bu^tN=CHCN \xleftarrow{iii} Bu^tN-CH_2CN$$
$$\qquad\qquad\qquad\qquad |$$
$$\qquad\qquad\qquad\qquad Cl$$

Reagents: i, KCN; ii, Bu^{t}OCl; iii, Et$_3$N

Scheme 44

fluence of α-substituents in the α-halogenation of nitriles[201] with carbon tetrachloride or tetrabromide.

A correction has been made[202] to the proposed structure[203] for the product of Knoevenagel condensation of malononitrile with 2-ethoxycarbonylcyclohexanone. The product is not the alkylidene malononitrile (121), but is the pyridine (122), although the malonitrile (121) is definitely a precursor.

$$Me_2NCH=\overset{CN}{\underset{|}{C}}CHO$$
(120)

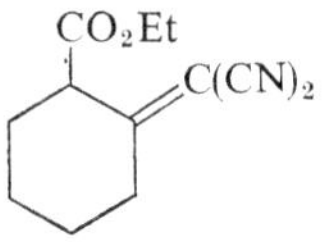

(121)

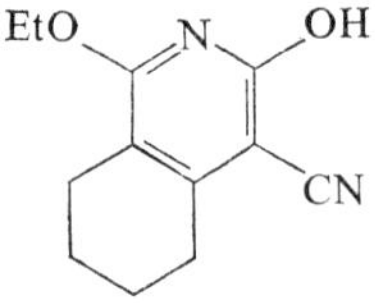

(122)

6 Aldehydes and Ketones

Preparation.—The conversion of acid chlorides into ketones using lithium dialkyl cuprates has been studied by two groups;[204] the reagents are sufficiently mild as to allow obtention of α-chloroketones from α-chloroacid chlorides.[205] The reaction of α-bromoketones with such reagents provides a new synthesis of hindered ketones.[206]

Whereas platinum-catalysed reduction of acid chlorides with triethylsilane gives aldehydes,[207] Eaborn[208] has shown that rhodium catalysis can give predominantly ketones. α-Ketophosphonate esters, readily available from the action of carboxylic acid chlorides on trialkyl phosphites, are reduced by sodium borohydride to the ester (123), which fragments to the

[201] G. Morel, R. Seux, and A. Foucaud, *Tetrahedron Letters*, 1971, 1031.
[202] J. L. van der Baan and F. L. Bickelhaupt, *Chem. Comm.*, 1970, 326.
[203] J. L. van der Baan and F. L. Bickelhaupt, *Chem. Comm.*, 1968, 1661.
[204] G. H. Posner and C. E. Whitten, *Tetrahedron Letters*, 1970, 4647; J.-E. Dubois, M. Boussu, and C. Lion, *Tetrahedron Letters*, 1971, 829.
[205] C. Jallabert, N.-T. Luong-Thi, and H. Rivière, *Bull. Soc. chim. France*, 1970, 797.
[206] J. E. Dubois, C. Lion, and C. Moulineau, *Tetrahedron Letters*, 1971, 177.
[207] J. D. Citron, *J. Org. Chem.*, 1969, **34**, 1977.
[208] S. P. Dent, C. Eaborn, and A. Pidcock, *Chem. Comm.*, 1970, 1703.

Functional Groups other than Acetylenes, Allenes, and Olefins 117

aldehyde on treatment with base,[209] as shown in Scheme 45. Aldehydes are obtained by reduction of carboxylic acids with lithium in methylamine on hydrolysis of the readily isolated imine intermediate.[210]

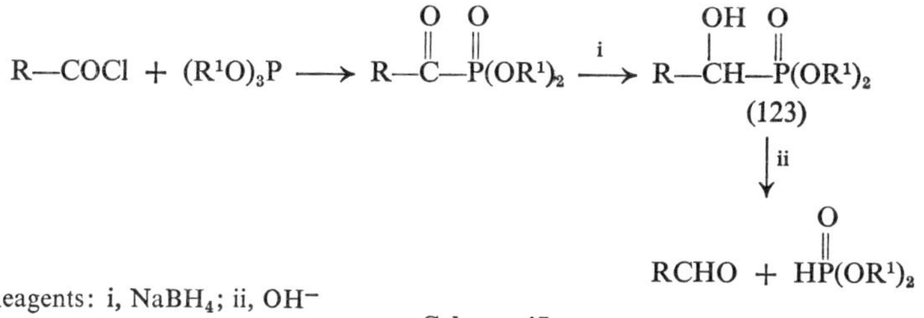

$$R\text{---}COCl + (R^1O)_3P \longrightarrow R\text{---}\overset{O}{\underset{\|}{C}}\text{---}\overset{O}{\underset{\|}{P}}(OR^1)_2 \xrightarrow{i} R\text{---}\overset{OH}{\underset{|}{C}H}\text{---}\overset{O}{\underset{\|}{P}}(OR^1)_2$$
(123)

$$\downarrow ii$$

$$RCHO + H\overset{O}{\underset{\|}{P}}(OR^1)_2$$

Reagents: i, $NaBH_4$; ii, OH^-

Scheme 45

Acyl carbonyl ferrates are involved as intermediates in two preparations of aldehydes. Alkyl bromides react with sodium tetracarbonyl ferrate($-$II) (from iron pentacarbonyl and sodium) by a process of oxidative addition to furnish the ferrate($-$I) (124); this species rearranges on treatment with triphenylphosphine to a phosphonium-substituted acyl ferrate($-$I) (125), which is subsequently cleaved by acetic acid to the homologous aldehyde,[211] as outlined in Scheme 46. A related procedure employs reaction of an acid halide with sodium tetracarbonyl ferrate($-$II) to afford the acyl ferrate($-$I) (126) directly;[212] this species is also cleaved by acetic acid, and affords yet another method for the reduction of acid halides to aldehydes.

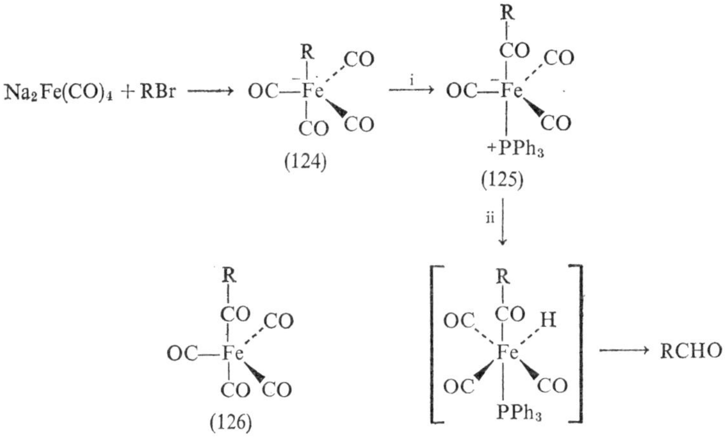

$$Na_2Fe(CO)_4 + RBr \longrightarrow \underset{(124)}{OC\text{---}Fe\overset{R}{\underset{CO}{\overset{CO}{\diagdown}}}\overset{}{\underset{CO}{\diagdown}}} \xrightarrow{i} \underset{(125)}{\underset{+PPh_3}{OC\text{---}Fe\overset{R}{\underset{CO}{\overset{CO}{\diagdown}}}\overset{CO}{\underset{CO}{\diagdown}}}}$$

$$\downarrow ii$$

$$\underset{(126)}{OC\text{---}Fe\overset{R}{\underset{CO}{\overset{CO}{\diagdown}}}\overset{CO}{\underset{CO}{\diagdown}}} \qquad \left[\underset{PPh_3}{OC\overset{R}{\underset{}{\overset{CO}{\diagdown}}Fe\overset{H}{\underset{CO}{\diagdown}}}} \right] \longrightarrow RCHO$$

Reagents: i, Ph_3P; ii, AcOH

Scheme 46

[209] L. Horner and H. Röder, *Chem. Ber.*, 1970, **103**, 2984.
[210] A. O. Bedenbaugh, J. H. Bedenbaugh, W. A. Bergin, and J. D. Adkins, *J. Amer. Chem. Soc.*, 1970, **92**, 5774.
[211] M. P. Cooke, jun., *J. Amer. Chem. Soc.*, 1970, **92**, 6080.
[212] T. Mitsudo, T. Okajima, Y. Takegami, M. Tanaka, Y. Watanabe, and K. Yamamoto, *Bull. Chem. Soc. Japan*, 1971, **44**, 2569.

Olefins are converted quantitatively into ketones by oxymercuration[213] followed by transmetallation with palladium(II) chloride; the organopalladium (127) formed is unstable, undergoing spontaneous reductive elimination to the ketone (Scheme 47).

$$R-\underset{\underset{OH}{|}}{CH}-\overline{CH_2}PdCl_2 \longrightarrow RCOMe + Pd^0 + HCl + Cl^-$$
(127)

Scheme 47

α-Diketones are obtained on oxidation of large-ring or acyclic disubstituted olefins with potassium permanganate in acetic anhydride;[214] this reaction succeeds only when the carbonyls of the product diketone are not constrained to be *cisoid:* terminal olefins give the corresponding ketoacetates. Disubstituted acetylenes are oxidized to α-diketones with ruthenium tetroxide;[215] terminal acetylenes give carboxylic acids with one less carbon atom.

Thallium(III) nitrate shows impressive qualities as an oxidizing agent.[216] Acetylenes are oxidized to give a range of products,[217] the nature of which is strikingly dependent on the structure of the acetylene (Scheme 48). Olefins

$$R-C\equiv CH \xrightarrow{i} RCO_2H$$
$$R-C\equiv C-R \xrightarrow{i} RCOCH(OH)R$$
$$Ar-C\equiv C-Ar \xrightarrow{i} ArCOCOAr$$
$$Ar-C\equiv C-R \xrightarrow{ii} Ar\underset{\underset{R}{|}}{CH}CO_2Me$$

Reagents: i, Tl(NO$_3$)$_3$–H$_2$O; ii, Tl(NO$_3$)$_3$–MeOH

Scheme 48

are efficiently oxidized to rearranged aldehydes or ketones by thallium(III) nitrate in methanol, the product being formed as the acetal;[218] an example is the oxidative ring-contraction of cyclic olefins to aldehydes (Scheme 49).

Olefins are oxidatively cleaved to aldehydes or ketones by bistriphenylsilylchromate (128).[219]

[213] D. F. Hunt and G. T. Rodeheaver, *Chem. Comm.*, 1971, 818.
[214] R. F. Lauer, O. Repič, K. B. Sharpless, A. Y. Teranishi, and D. R. Williams, *J. Amer. Chem. Soc.*, 1971, **93**, 3303.
[215] H. Gopal and A. J. Gordon, *Tetrahedron*, 1971, **31**, 2941.
[216] A. McKillop and E. C. Taylor, *Accounts Chem. Res.*, 1970, **3**, 338.
[217] A. McKillop, O. H. Oldenziel, R. L. Robey, B. P. Swann, and E. C. Taylor, *J. Amer. Chem. Soc.*, 1971, **93**, 7331.
[218] A. McKillop, J. D. Hunt, F. Kienzle, and E. C. Taylor, *Tetrahedron Letters*, 1970, 5275.
[219] L. M. Baker and W. L. Carrick, *J. Org. Chem.*, 1970, **35**, 774.

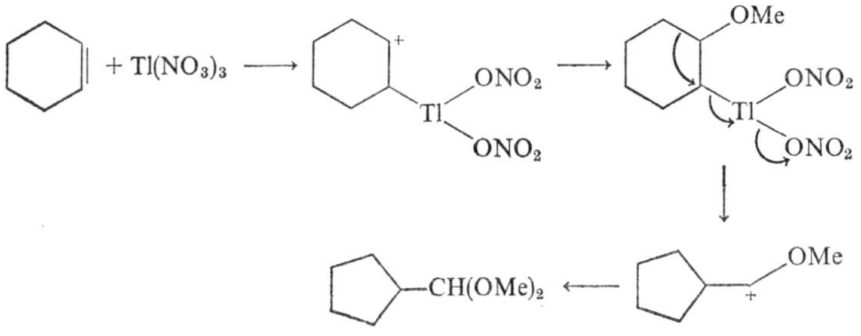

Scheme 49

Reductive electrolysis of β-ketosulphones[220] provides a convenient ketone synthesis (Scheme 50); the readily accessible sulphones give better yields of

$$(Ph_3SiO)_2CrO_2$$
(128)

ketone than do the corresponding sulphoxides.[221] Formation of γ-diketones by radical dimerization can be minimized by use of lithium salts in the supporting electrolyte.[222]

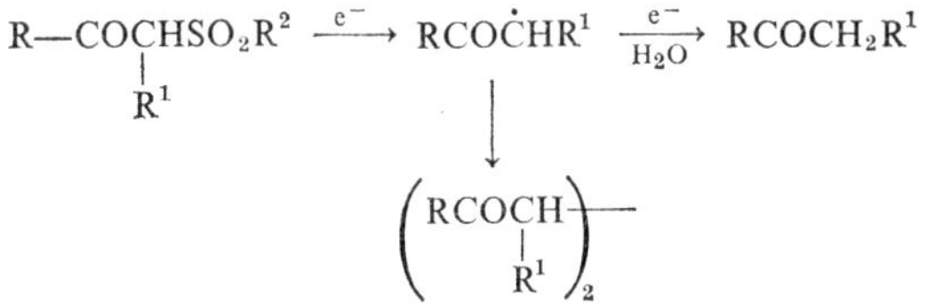

$$R-\underset{\underset{R^1}{|}}{C}OCHSO_2R^2 \xrightarrow{e^-} RCO\dot{C}HR^1 \xrightarrow[H_2O]{e^-} RCOCH_2R^1$$

$$\downarrow$$

$$\left(\underset{\underset{R^1}{|}}{RCOCH}\right)_2$$

Scheme 50

Trialkyl cyanoborates (129), formed by reaction of trialkylboranes with sodium cyanide, are attacked by electrophiles in a process involving migration of two alkyl groups from boron to carbon;[223] the cyclic species (130), formed with trifluoroacetic anhydride, is hydrolysed oxidatively to the ketone in excellent yield (Scheme 51). The effective loss of one alkyl group can be obviated by initial hydroboration with thexyl borane; as has been reported previously, the thexyl group does not undergo competitive migration: this procedure also allows the preparation of unsymmetrical ketones via sequential hydroboration.[224] In the presence of excess electrophile, a third migration[225] of an alkyl group occurs in the intermediate (130), affording on oxidation high yields of trialkylcarbinols (*loc. cit.*).

[220] B. Lamm and B. Samuelsson, *Acta Chem. Scand.*, 1970, **24**, 561, 3070.
[221] B. Lamm and B. Samuelsson, *Acta Chem. Scand.*, 1969, **23**, 691.
[222] B. Lamm and B. Samuelsson, *Chem. Comm.*, 1970, 1010.
[223] M. G. Hutchings, A. Pelter, and K. Smith, *Chem. Comm.*, 1970, 1529.
[224] M. G. Hutchings, A. Pelter, and K. Smith, *Chem. Comm.*, 1971, 1048.
[225] M. G. Hutchings, A. Pelter, and K. Smith, *Chem. Comm.*, 1971, 1048.

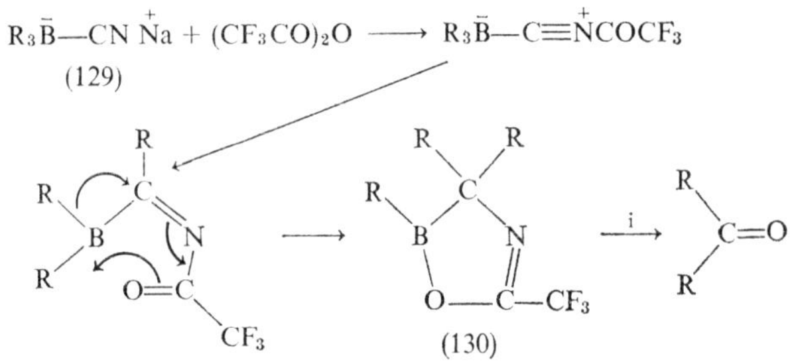

$R_3\bar{B}$—CN $\overset{+}{\text{Na}}$ + $(CF_3CO)_2O$ ⟶ $R_3\bar{B}$—C≡$\overset{+}{\text{N}}$COCF$_3$
(129)

Reagent: i, NaOH–H$_2$O$_2$

Scheme 51

α-Diazocarbonyl compounds react with trialkylboranes with loss of nitrogen to furnish homologated ketones[226] and aldehydes[227] in good yield. Vinyloxyboranes[228] have been detected as intermediates in this process. α-Diazoketones are conveniently obtained by interaction of diazoalkanes with carboxylic acid anhydrides, prepared from the acid with dicyclohexylcarbodi-imide; this offers a useful alternative for those cases where acid chloride formation is not practicable.[229] A new synthesis of α-diazoaldehydes is described;[230] reaction of an α-alkyl-β-dimethylaminoacrolein (131) with toluene-p-sulphonyl azide gives the desired aldehyde (132), as shown in Scheme 52. α-Diazoacetaldehyde prepared in this manner is found to be very unstable, decomposing in the presence of olefins to yield formylcyclopropanes. A synthesis of simple α-alkyl-β-aminoacroleins has been described.[231]

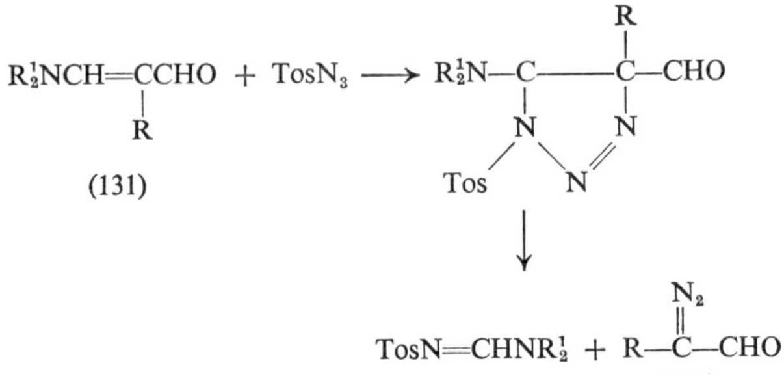

R_2^1NCH=CCHO + TosN$_3$ ⟶ R_2^1N—C——C—CHO
  | | |
  R   N   N
(131)   \ //
                                          Tos N

↓

TosN=CHNR$_2^1$ + R—C̈—CHO
                    ‖
                    N$_2$
                    (132)

Scheme 52

The synthetic utility of lithium salts of (1-methylthio)alkylphosphonate esters for the preparation of ketones is reported in an extension of the Wadsworth–Emmons–Horner modification of the Wittig reaction.[232]

[226] D. M. Gunn, J. Hooz, and H. Kono, *Canad. J. Chem.*, 1971, **49**, 2371.
[227] J. Hooz and G. F. Morrison, *Canad. J. Chem.*, 1970, **48**, 868.
[228] D. J. Pasto and P. W. Wojtkowski, *Tetrahedron Letters*, 1970, 215.
[229] D. Hodson, G. Holt, and D. K. Wall, *J. Chem. Soc. (C)*, 1970, 971.
[230] Z. Arnold, Z. Janoušek, and J. Kučera, *Coll. Czech. Chem. Comm.*, 1970, **35**, 3618.
[231] E. Breitmaier and S. Gassenmann, *Chem. Ber.*, 1971, **104**, 665.
[232] E. J. Corey and J. I. Schulman, *J. Org. Chem.*, 1970, **35**, 777.

Functional Groups other than Acetylenes, Allenes, and Olefins

Diethyl methylthiomethylphosphonate can be alkylated by successive treatment with n-butyl-lithium and an alkyl iodide; the lithio-derivatives of such alkylated esters react with carbonyl compounds to give β-alkoxyphosphonate adducts. These products decompose readily to vinyl methyl sulphides, which undergo mercury(II)-promoted hydrolysis to the corresponding ketone. These transformations are outlined in Scheme 53.

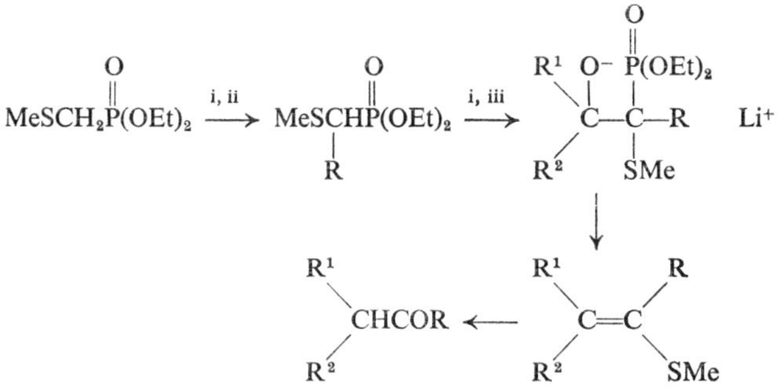

$$MeSCH_2\overset{O}{\overset{\|}{P}}(OEt)_2 \xrightarrow{i, ii} MeSCH\overset{O}{\overset{\|}{\underset{R}{P}}}(OEt)_2 \xrightarrow{i, iii} \underset{R^2}{\overset{R^1}{\underset{|}{C}}}\overset{O^-}{\underset{SMe}{\overset{|}{C}}}\overset{\overset{O}{\|}}{\underset{}{P(OEt)_2}}-R \quad Li^+$$

$$\underset{R^2}{\overset{R^1}{\diagdown}} CHCOR \longleftarrow \underset{R^2}{\overset{R^1}{\diagdown}}C=C\overset{R}{\underset{SMe}{\diagup}}$$

Reagents: i, Bu^{n}Li; ii, RX; iii, R^{1}COR^2

Scheme 53

The betaines (133), formed as intermediates in the reaction of aldehydes with phosphonium ylides, afford the useful ylide alkoxides (134) on treatment with n-butyl-lithium. Corey[233] has described the transformations of these latter compounds into ketones, halogeno-olefins, or acetylenes, as illustrated in Scheme 54.

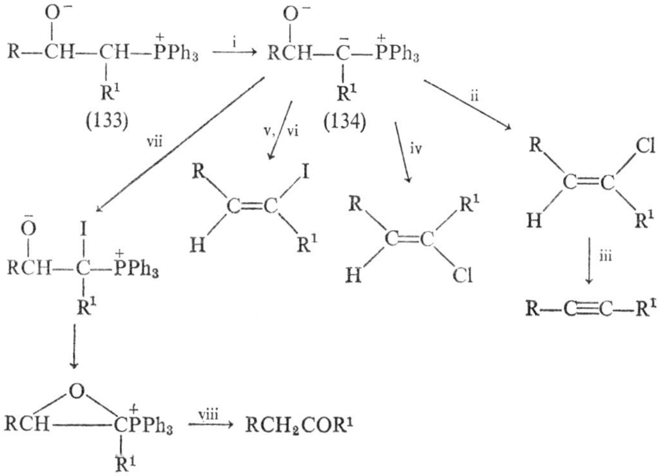

Reagents: i, Bu^{n}Li; ii, PhICl$_2$; iii, NaNH$_2$; iv, N-chlorosuccinimide; v, Hg(OAc)$_2$; vi, LiI-I$_2$; vii, I$_2$; viii, H$_2$O

Scheme 54

[233] E. J. Corey, J. I. Schulman, and H. Yamamoto, *Tetrahedron Letters*, 1970, 447.

An improved method for the conversion of alkyl halides into aldehydes via their Grignard derivatives is reported.[234] Grignard reagents react with the commercially available diethyl phenyl orthoformate to give the expected aldehyde acetals in high yield; this orthoester is more satisfactory than the triethyl analogue. NN'-Disubstituted carbamates seem ideal reagents for the preparation of ketones from Grignard reagents; the intermediate carbinolamines decompose spontaneously to the desired ketones (Scheme 55).[235]

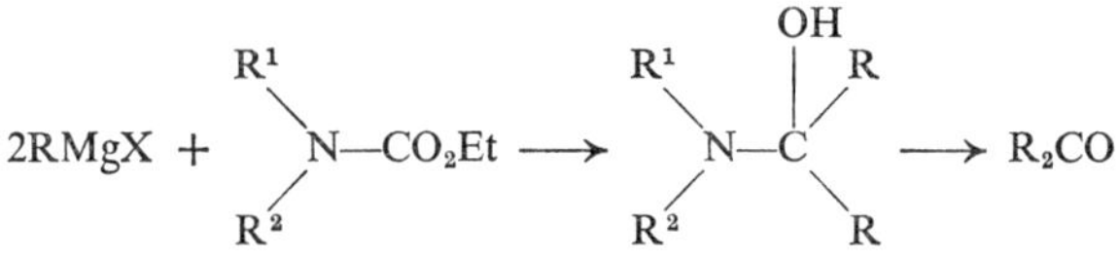

$$2RMgX + \begin{array}{c}R^1\\ \\ R^2\end{array}\!\!\!N\!-\!CO_2Et \longrightarrow \begin{array}{c}R^1\\ \\ R^2\end{array}\!\!\!N\!-\!\!\underset{R}{\overset{\overset{\displaystyle OH}{|}}{C}}\!\!-\!R \longrightarrow R_2CO$$

Scheme 55

α-Silyl organometallic reagents react with acid chlorides to give α-silylketones, which are readily hydrolysed to the parent ketones.[236] The preparation of α-epoxysilanes, such as (135), from acetylenes, and their conversion into aldehyde acetals[237] has been noted (Scheme 56).

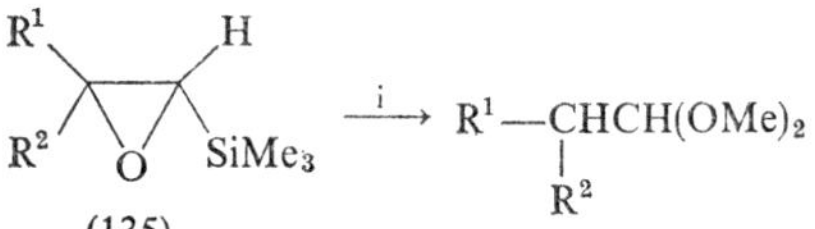

$$\underset{(135)}{\overset{R^1H}{\underset{R^2\overset{}{O}SiMe_3}{\triangle}}} \xrightarrow{i} R^1\!-\!\underset{R^2}{\overset{|}{C}}\!HCH(OMe)_2$$

Reagent: i, H_2SO_4–MeOH

Scheme 56

Vapour-phase thermolysis of the trienol (136) affords allyl vinyl ketone (137) as major product in a process of β-hydroxyolefin cleavage;[238] the alternative, oxy-Cope product (138) is formed in much lower yield.

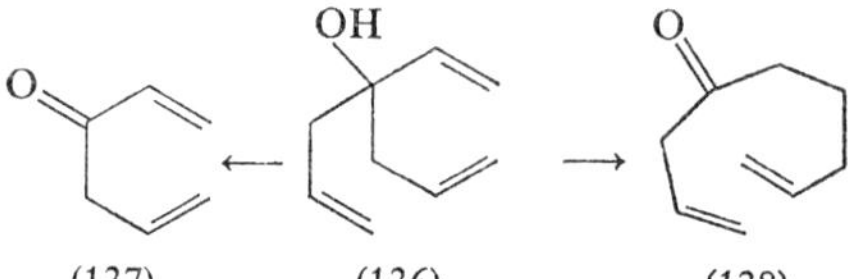

(137)   (136)   (138)

The synthesis of 1,4-diketones by hydrogenation of diacylethylenes is described; annelation of these diketones is reviewed.[239] A new synthesis of 1,4-diketones is reported[240] via reaction of diazoketones with enol acetates in the presence of copper salts; treatment of the cyclopropyl ketone so formed

[234] E. Reske and H. Stetter, *Chem. Ber.*, 1970, **103**, 643.
[235] A.-B. Hörnfeldt and U. Michael, *Tetrahedron Letters*, 1970, 5219.
[236] T. H. Chan, E. Chang, and E. Vinokur, *Tetrahedron Letters*, 1970, 1137.
[237] E. Colvin and G. Stork, *J. Amer. Chem. Soc.*, 1971, **93**, 2080.
[238] E. J. Iorio and A. Viola, *J. Org. Chem.*, 1970, **35**, 856.
[239] E. Ritchie and W. C. Taylor, *Austral. J. Chem.*, 1971, **24**, 2137.
[240] T. E. Glass and J. E. McMurry, *Tetrahedron Letters*, 1971, 2575.

Functional Groups other than Acetylenes, Allenes, and Olefins 123

with base effects hydrolysis and retroaldolization to the desired diketone, which was utilized further in a synthesis of *cis*-jasmone, as shown in Scheme 57. Of all copper catalysts investigated, copper(I) acetylacetonate proved to be the best.

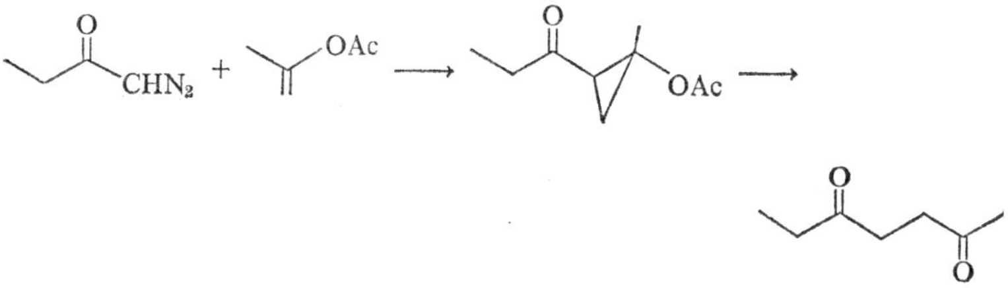

Scheme 57

A full paper[241] has been published on a general cycloalkanone synthesis involving intramolecular electrophilic alkylation of 2-chloro-1-olefins.

Oxidation of α-amino-acids with silver(II) picolinate gives almost quantitative yields of nor-aldehydes; silver oxide causes further oxidation to the nor-acid.[242] The possibility is suggested of the simultaneous operation of a radical mechanism in solution and a two-electron shift process at the solid silver salt surface. Aryl alkanes and aryl alkanols are oxidized by silver(II) picolinate to aldehydes and ketones.[243] α-Aminoketones are dehydrogenated by mercury(II) salts to ketones; iminium ions, such as (139), are postulated[244] as intermediates (Scheme 58).

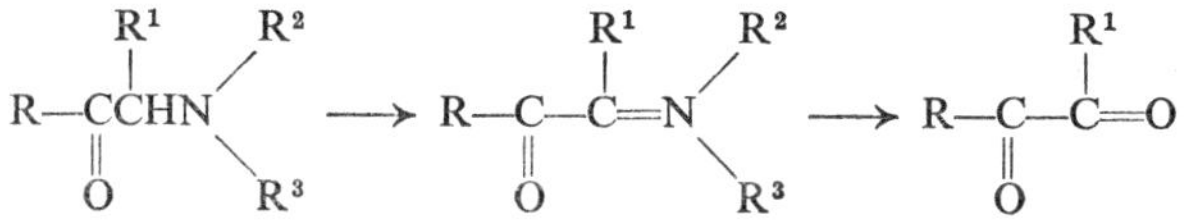

Scheme 58

The mechanism of the Serini reaction has been studied further by Ghera,[245] and the pathway shown in Scheme 59 proposed for this general reaction of 1,2-trisubstituted *vic*-glycol monoesters. A related process is seen in the formation of olefins by zinc-induced fragmentation of 1,3-diol monoesters;[246] a

---

[241] F. R. Hilfiker, P. T. Lansbury, E. J. Niehouse, and D. J. Scharf, *J. Amer. Chem. Soc.*, 1970, **92**, 5649.
[242] T. G. Clarke, N. A. Hampson, J. B. Lee, J. R. Morley, and B. Scanlon, *J. Chem. Soc.* (C), 1970, 815.
[243] J. A. Azoo, R. G. Bacon, and K. K. Gupta, *J. Chem. Soc.* (C), 1970, 1975.
[244] H. Möhrle and D. Schittenhelm, *Chem. Ber.*, 1971, **104**, 2475.
[245] E. Ghera, *J. Org. Chem.*, 1970, **35**, 660.
[246] E. Ghera, *Tetrahedron Letters*, 1970, 1539.

six-centred transition state is postulated for this reaction, which affords products opposite to those expected from a retro-Prins fragmentation (Scheme 59a).

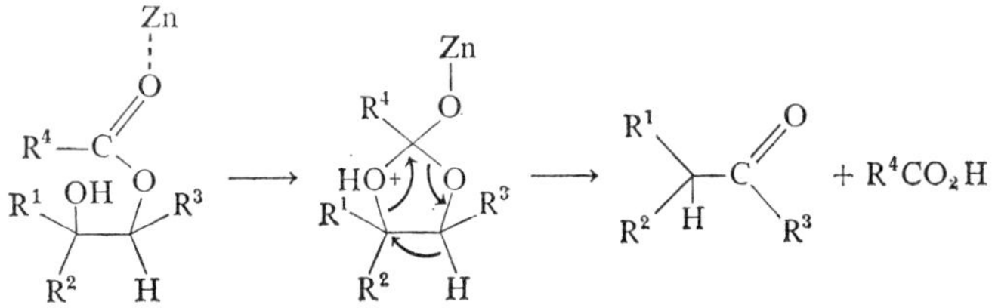

Scheme 59

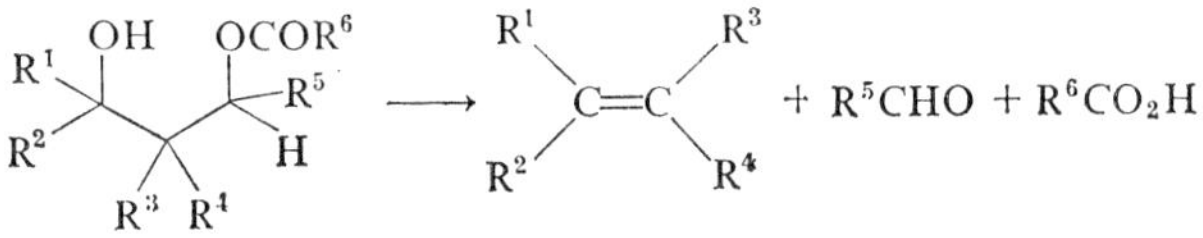

Scheme 59a

Reaction of dimethylcopperlithium with the diethyl acetal of ethyl 2-oxobutyrate gives the synthetically useful methyl ketone (139a) in nearly

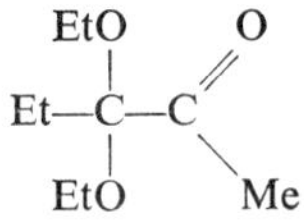

$$\text{Et}-\underset{\underset{\text{EtO}}{|}}{\overset{\overset{\text{EtO}}{|}}{\text{C}}}-\text{C}\overset{\diagup\!\!\!\!\text{O}}{\diagdown\text{Me}}$$

(139a)

quantitative yield; similarly, acetophenone is formed in good yield from the reaction of ethyl benzoate and dimethylcopperlithium.[247]

In situ preparation of dipyridine–chromium trioxide is recommended[248] as a simplification of the previous procedure[249] for the obtention of this excellent oxidizing reagent: however, the isolated reagent is preferred for the oxidation of disubstituted acetylenes to conjugated acetylenic ketones (Scheme 60).[250] The use of acetic acid as solvent for dipyridine–chromium

$$\text{RCH}_2\text{C}{\equiv}\text{CR} \longrightarrow \text{RCOC}{\equiv}\text{CR}$$

Scheme 60

trioxide oxidations has been described;[251] less selectivity is observed than when dichloromethane is the solvent. A study[252] of the oxidation of primary

---

[247] S. A. Humphrey, J. L. Herrmann, and R. H. Schlessinger, *Chem. Comm.*, 1971, 1244.
[248] R. Ratcliffe and R. Rodehorst, *J. Org. Chem.*, 1970, **35**, 4000.
[249] J. C. Collins, F. J. Frank, and W. W. Hess, *Tetrahedron Letters*, 1968, 3363.
[250] J. E. Shaw and J. J. Sherry, *Tetrahedron Letters*, 1971, 4379.
[251] K.-E. Stensiö, *Acta Chem. Scand.*, 1971, 1125.
[252] D. G. Lee and U. A. Spitzer, *J. Org. Chem.*, 1970, **35**, 3589.

Functional Groups other than Acetylenes, Allenes, and Olefins 125

alcohols with aqueous sodium dichromate has been made; neutral conditions were employed to hinder acid-catalysed formation of the aldehyde hydrate, a prerequisite for further oxidation to the carboxylic acid. Good yields of aromatic aldehydes were obtained, but the results with aliphatic alcohols were disappointing. Secondary alcohols are cleanly oxidized to ketones with no loss of epimeric purity when a two-phase ether–aqueous chromic acid system is used;[253] the alcohol passes from the ether layer as its chromic acid ester, and is returned to the organic layer as soon as oxidation to the ketone has occurred.

Potassium ferrate, K_2FeO_4, is a mild reagent for the oxidation of primary alcohols and primary alkylamines to aldehydes, and of secondary alcohols to ketones.[254] Iodonium nitrate, INO_3, is reported[255] to oxidize alcohols to carbonyl compounds; electron-withdrawing groups adjacent to the alcohol carbon cause a marked decrease in reactivity: olefinic alcohols undergo preferential electrophilic addition to the double bond.[256]

Alkyl hydroperoxides oxidize some sugar alcohols to the corresponding ketones under molybdenum-salt catalysis.[257] Ruthenium tetroxide oxidizes secondary alcohols to ketones in neutral or basic media permitting, under the latter conditions, direct conversion of γ-lactones into γ-ketoacids.[258]

An improved preparation[259] of non-enolizable α-diketones *via* acyloin condensation is described; the acyloin, trapped as its bistrimethylsilyl enol ether, undergoes mild oxidative regeneration by treatment with bromine in dichloromethane or carbon tetrachloride, as shown in Scheme 61.

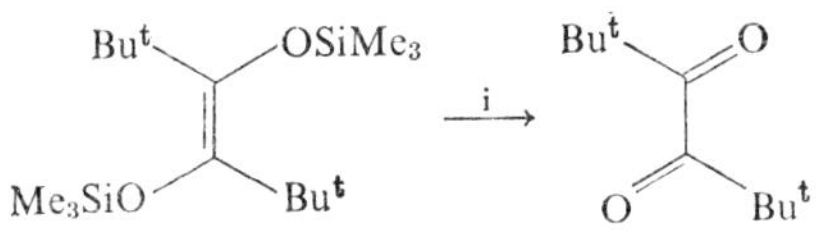

Reagent: i, Br_2–CCl_4

Scheme 61

Oxazines. The synthetic utility of dihydro-1,3-oxazines[260] has been considerably extended by Meyers. The reactivity of the double bond in dihydro-oxazines is enhanced on quaternization with methyl iodide; the resulting salt readily adds a range of organometallic reagents,[261] affording the versatile ketone synthesis summarized in Scheme 62.

[253] H. C. Brown, C. P. Garg, and K.-T. Liu, *J. Org. Chem.*, 1971, **36**, 387.
[254] R. J. Audette, J. W. Quail, and P. J. Smith, *Tetrahedron Letters*, 1971, 279.
[255] U. E. Diner, *J. Chem. Soc., (C)*, 1970, 676.
[256] U. E. Diner, J. W. Lown, and M. Worsley, *J. Chem. Soc. (C)*, 1971, 3131.
[257] U. M. Dzhemilev, G. A. Tolstikov, and V. P. Yur'ev, *J. Gen. Chem. (U.S.S.R.)*, 1971, **41**, 943.
[258] T. Adams, H. Gopal, and R. M. Moriarty, *Tetrahedron Letters*, 1970, 4003.
[259] J. Strating, S. Reiffers, and H. Wynberg, *Synthesis*, 1971, 209.
[260] *Org. Synth.*, 1971, **51**, 24.
[261] A. I. Meyers and E. M. Smith, *J. Amer. Chem. Soc.*, 1970, **92**, 1084.

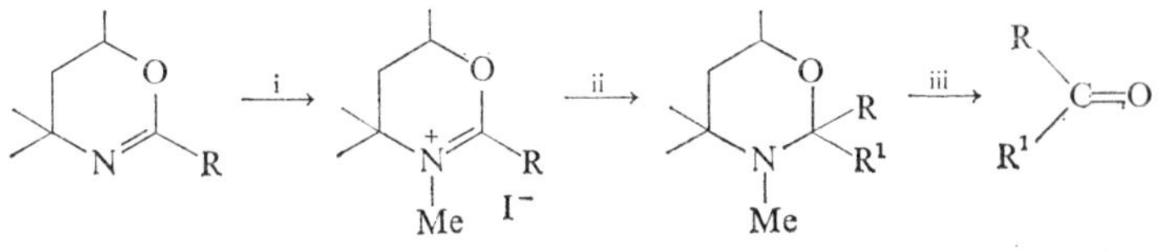

Reagents: i, MeI; ii, R¹M; iii, H₃O⁺

Scheme 62

The potential of dihydro-oxazines is displayed to advantage in a high-yield synthesis[262] of the male bollworm moth pheromone, *trans*-1-acetoxy-10-n-propyltrideca-5,9-diene. Alkylation of anions from dihydro-oxazines with α,ω-dihalogenoalkanes[263] and subsequent elaboration of the terminal halide provides a route to variously functionalized aldehydes (Scheme 63).

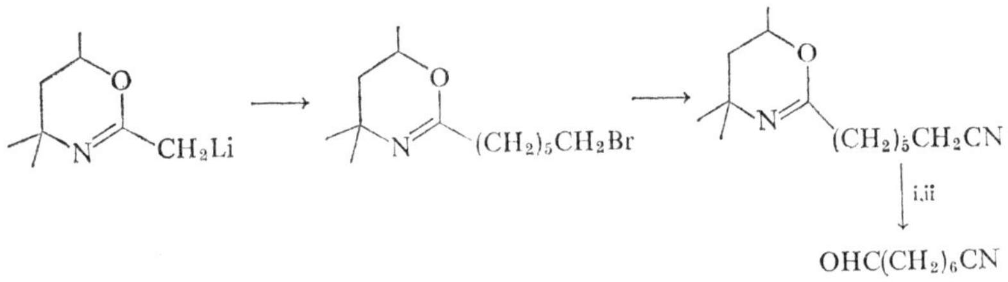

Reagents: i, NaBH₄; ii, H₃O⁺

Scheme 63

The failure of all but 2-methyl- or 2-benzyl-dihydro-1,3-oxazine anions to alkylate has been turned to ingenious advantage. Meyers[264] has reported that the above two anions are stable at the generation temperature, −78 °C, but, on warming to room temperature, they rapidly rearrange to the ketenimines (140); higher alkyl dihydro-oxazine anions form the ketenimine spontaneously at −78 °C. Thus, dihydro-oxazine anions can function as nucleophiles or electrophiles, depending on the temperature employed. This has been used[265] for the assembly of ketones with α-quaternary carbon atoms of varied structure, but specifically alkylated in one of two available sites (Scheme 64).

Aromatic Grignard reagents can be formylated[266] as their complexes with hexamethylphosphoramide by reaction with the quaternized 2-oxazoline

[262] A. I. Meyers and E. W. Collington, *Tetrahedron*, 1971, **27**, 5979.
[263] H. W. Adickes, G. R. Malone, and A. I. Meyers, *Tetrahedron Letters*, 1970, 3715.
[264] A. I. Meyers and E. M. Smith, *Tetrahedron Letters*, 1970, 4355.
[265] A. F. Jurjevich, A. I. Meyers, and E. M. Smith, *J. Amer. Chem. Soc.*, 1971, **93**, 2314
[266] E. W. Collington and A. I. Meyers, *J. Amer. Chem. Soc.*, 1970, **92**, 6676.

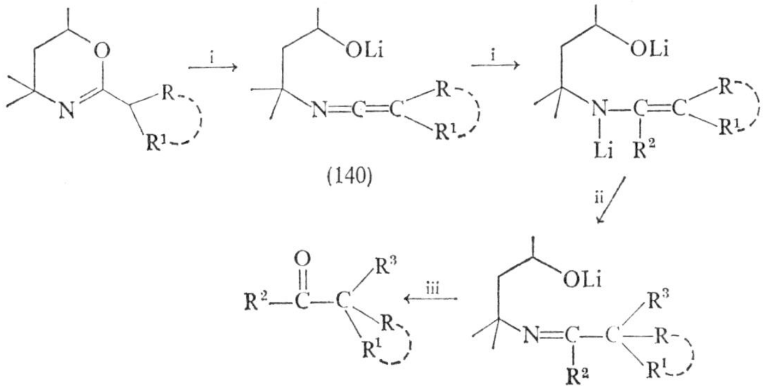

Scheme 64

Reagents: i, R²Li; ii, R³I; iii, H₃O⁺

Scheme 64

(141); aliphatic Grignard reagents are too basic, and merely abstract a proton.

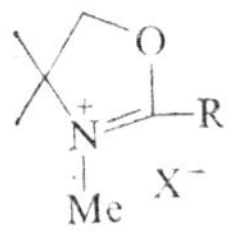

(141) R = H or D

A conceptually similar method of aldehyde synthesis utilizes the 1,3-thiazole (142); this route is distinguished by a final mild hydrolysis step.[267]

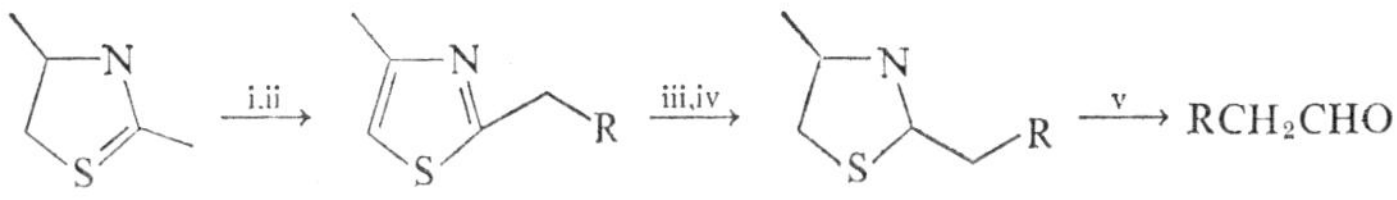

Reagents: i, Bu^{n}Li; ii, RX; iii, Me₃OBF₄–SO₂; iv, NaBH₄; v, HgCl₂–H₂O

Scheme 65

Enone Formation. The transformation of saturated carbonyl compounds into their αβ-unsaturated analogues by a process of homogeneous liquid-phase oxidative dehydrogenation has been described;[268] use of oxygen or air in the presence of a palladium(II) catalyst and a co-catalyst, usually copper(II) salts or quinone, gives good yields based on ketone consumed, though overall conversion is low.

Tetramethylammonium salts of phosphoric acid derivatives[269] convert

---

[267] L. Altman and S. L. Richheimer, *Tetrahedron Letters*, 1971, 4709.
[268] R. J. Theissen, *J. Org. Chem.*, 1971, **36**, 752.
[269] G. Sturtz and A. Raphalen, *Tetrahedron Letters*, 1970, 1529; A. Raphalen and G. Sturtz, *Bull. Soc. chim. France*, 1971, 2962; J. L. Kraus and G. Sturtz, *Bull. Soc. chim. France*, 1971, 4012.

α-halogenocarbonyl compounds into αβ-unsaturated ketones; varying amounts of α-ketophosphate, formed by direct nucleophilic displacement, are also observed.

Aqueous palladium(II) chloride oxidizes allylic alcohols to the corresponding enones.[270] Silver carbonate on Celite is recommended[271] for the oxidation of sugar allylic alcohols; D-glucal (143) affords the enone (144) in 60% yield, a considerable improvement on previous methods.

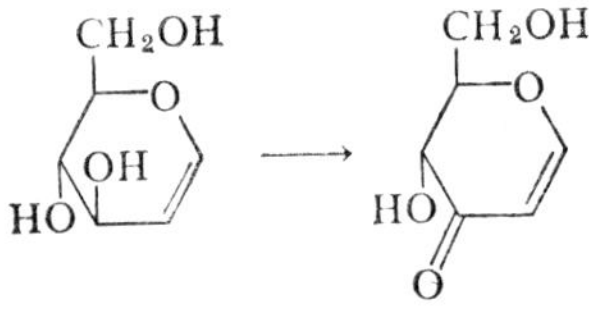

(143)   (144)

Conjugate Addition to Enones. Following the recognition that the conjugate addition of the alkyl groups of trialkylboranes to αβ-unsaturated carbonyl compounds was free-radical in nature,[272] peroxides, light,[273] and oxygen[274] have been studied as potential initiators; this reaction is facile in most cases, highly hindered alkyl groups being introduced by *B*-alkylboracyclanes;[275] organoboranes can be induced to react with ethynyl methyl ketone, affording a new synthesis[276] of αβ-unsaturated ketones (Scheme 66).

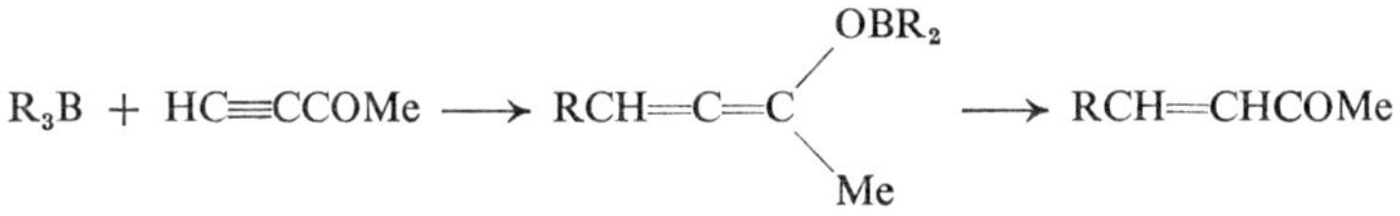

$$R_3B + HC{\equiv}CCOMe \longrightarrow RCH{=}C{=}C\genfrac{}{}{0pt}{}{OBR_2}{Me} \longrightarrow RCH{=}CHCOMe$$

Scheme 66

Diethylalkynylalanes[277] effect conjugate addition of acetylenic units to simple αβ-unsaturated ketones; this reaction is limited to *cisoid* enones, suggesting intramolecular delivery of the alkynyl group in a six-membered transition state such as (145): *transoid* enones give products of 1,2-addition. The conjugate addition of lithium divinyl copper to αβ-unsaturated ketones affords a route to the γδ-unsaturated homologues.[278]

Nagata[279] has described a method for effecting overall conjugate hydrocyanation of αβ-unsaturated aldehydes; normally, organoaluminium cyanides

---

[270] R. Jira, *Tetrahedron Letters*, 1971, 1225.
[271] J. M. J. Tronchet, J. Tronchet, and A. Birkhäuser, *Helv. Chim. Acta*, 1970, **53**, 1489.
[272] H. C. Brown, G. W. Kabalka, A. Suzuki, S. Honma, A. Arase, and M. Itoh, *J. Amer. Chem. Soc.*, 1970, **92**, 710.
[273] H. C. Brown and G. W. Kabalka, *J. Amer. Chem. Soc.*, 1970, **92**, 712.
[274] H. C. Brown and G. W. Kabalka, *J. Amer. Chem. Soc.*, 1970, **92**, 714.
[275] H. C. Brown and E. Negishi, *J. Amer. Chem. Soc.*, 1971, **93**, 3777.
[276] A. Suzuki, S. Nozawa, M. Itoh, H. C. Brown, G. W. Kabalka, and G. W. Holland, *J. Amer. Chem. Soc.*, 1970, **92**, 3503.
[277] J. Hooz and R. B. Layton, *J. Amer. Chem. Soc.*, 1971, **93**, 7320.
[278] J. Hooz and R. B. Layton, *Canad. J. Chem.* 1970, **48**, 1626.
[279] W. Nagata, M. Yoshioka, T. Okumura, and M. Murakami, *J. Chem. Soc.* (C), 1970, 2355.

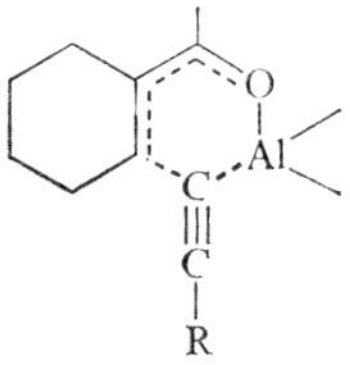

(145)

react with such substrates in a process of 1,2-addition, but reaction with the derived alkylideneamines (146), with a bulky substituent on nitrogen, gives a product of double addition, (147), which is readily hydrolysed to the desired β-cyanoaldehyde (Scheme 67).

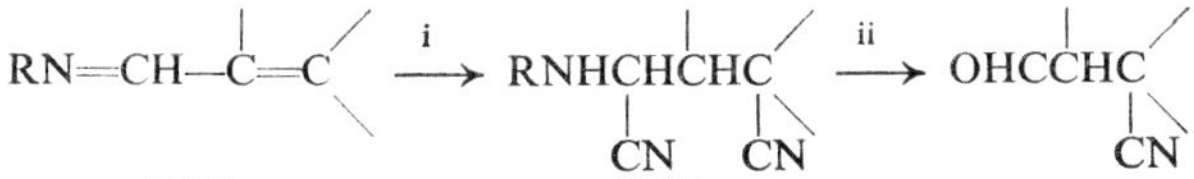

RN=CH—C=C →ⁱ RNHCHCHC →ⁱⁱ OHCCHC
                    | | |
                    CN CN   CN
(146)   (147)

Reagents: i, R_3^iAl–HCN; ii, $(CO_2H)_2$

Scheme 67

Regeneration from Derivatives. The need for mild methods of effecting aldehyde and ketone regeneration from derivatives has stimulated the efforts of a number of workers. Thallium(III) nitrate[280] efficiently regenerates aldehydes and ketones from oximes. Chromium(II) acetate[281] reductively deoximates ketoxime O-acetates under mild conditions. Selectivity is attainable with this reagent, as conjugated and aromatic ketoxime O-acetates react considerably faster than the saturated analogues; this is exemplified in the conversion of progesterone 3,20-dioxime diacetate into progesterone 20-oxime acetate. This method has been extended to the conversion of olefins into ketones and, much more significantly, for the transposition of the ketone carbonyl group, as shown in Scheme 68.

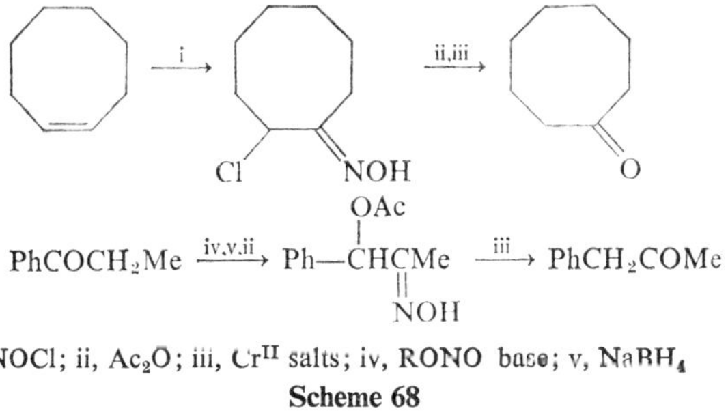

PhCOCH$_2$Me $\xrightarrow{iv,v,ii}$ Ph—CHCMe $\xrightarrow{iii}$ PhCH$_2$COMe
                                   | ||
                                   OAc   NOH

Reagents: i, NOCl; ii, Ac$_2$O; iii, Cr^{II} salts; iv, RONO base; v, NaBH$_4$

Scheme 68

[280] J. D. Hunt, A. McKillop, R. D. Naylor, and E. C. Taylor, *J. Amer. Chem. Soc.*, 1971, **93**, 4918.
[281] E. J. Corey and J. E. Richman, *J. Amer. Chem. Soc.*, 1970, **92**, 5276.

The reduction of ketoximes with titanium(III) chloride is noted in a partial synthesis of erythromycylamine;[282] in this case, the intermediate imine is stable and can be further reduced, though normally imine hydrolysis is rapid and ketones are obtained in high yield. This observation has been utilized to effect conversion of a nitro-compound into a ketone. In search of a route to 1,4-diketones *via* conjugate addition of a masked carbonyl anion to an $\alpha\beta$-unsaturated ketone, McMurry considered Michael addition with a nitroalkane.[283] Treatment of the resulting γ-nitroketone with titanium(III) chloride afforded the required 1,4-diketone (148) in excellent yield, presumably by sequential intermediacy of the oxime and the imine; this mild conversion avoids the harsh conditions of the alternative Nef procedure (Scheme 69).

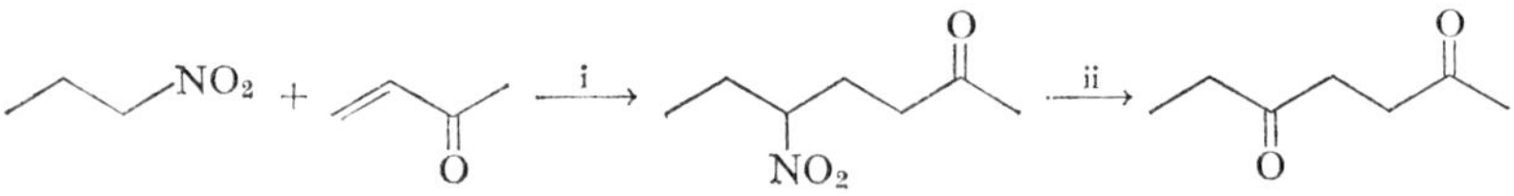

Reagents: i, $Pr_2^i NH$; ii, Ti^{III} salts

Scheme 69

Toluene-*p*-sulphonic acid monohydrate promotes hydration of imine double bonds, providing a method for the ready regeneration of ketones from their 2,4-dinitrophenylhydrazones;[284] this method is unsuitable for regeneration of $\alpha\beta$-unsaturated ketones, where extensive retro-aldol fragmentation occurs. Imines can be oxidized photochemically to ketones.[285] In a re-investigation of the acid-catalysed conversion of *gem*-bromonitro-compounds into ketones, loss of a bromonium ion followed by a Nef reaction is postulated (Scheme 70).[286]

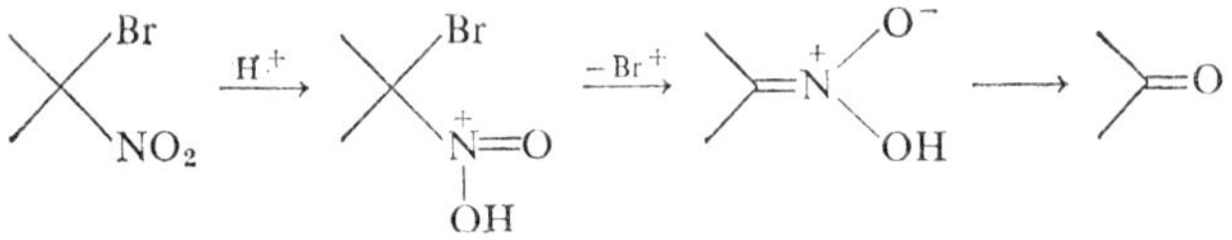

Scheme 70

Synthons. More ingenious functional equivalents for the carbonyl moiety are reported, where the normal electrophilicity of this group is inverted to exhibit nucleophilic properties.

Metallation of a protected aldehyde cyanohydrin gives the formal acyl carbanion equivalent (149), which undergoes ready alkylation; this forms the

---

[282] G. H. Timms and E. Wildsmith, *Tetrahedron Letters*, 1971, 195.
[283] J. Melton and J. E. McMurry, *J. Amer. Chem. Soc.*, 1971, **93**, 5309.
[284] M. E. N. Nambudiry and G. S. K. Rao, *Austral. J. Chem.*, 1971, **25**, 2183.
[285] H. Hirai and N. Toshima, *Tetrahedron Letters*, 1970, 433.
[286] H. Raman and S. Ranganathan, *Tetrahedron Letters*, 1970, 3331.

Functional Groups other than Acetylenes, Allenes, and Olefins 131

basis of a new synthesis of ketones reported by Stork,[287] and outlined in Scheme 71.

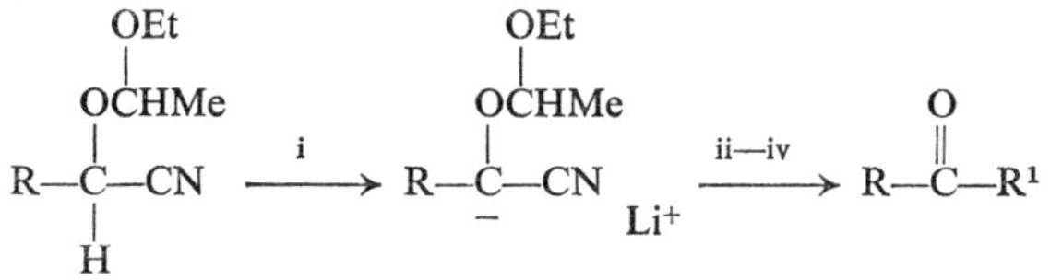

$$R-\underset{\underset{H}{|}}{\overset{\overset{OEt}{|}}{\underset{|}{C}}}\text{—CN} \xrightarrow{i} R-\underset{\underset{-}{|}}{\overset{\overset{OEt}{|}}{\underset{|}{C}}}\text{—CN} \xrightarrow{ii-iv} R-\overset{\overset{O}{\|}}{C}-R^1$$
$$\text{Li}^+$$
$$(149)$$

Reagents: i, $LiNR_2$; ii, R^1X; iii, H_3O^+; iv, OH^-

Scheme 71

Aliphatic organomagnesium or organolithium reagents add to 1,1,3,3-tetramethylbutyl isocyanide to give the corresponding metallo-aldimine (150), another acyl carbanion equivalent,[288] which reacts with a range of electrophiles in the normal manner.

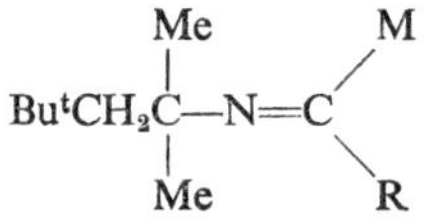

$$Bu^tCH_2\underset{\underset{Me}{|}}{\overset{\overset{Me}{|}}{C}}-N=C\underset{R}{\overset{M}{\diagup}}$$

(150)   M = MgX or Li

The electrophilic reactivity of the β-carbon in an $\alpha\beta$-unsaturated aldehyde is reversed in 1,3-bis(methylthio)allyl-lithium (151), derived by metallation of 1,3-bis(methylthio)-2-methoxypropane with lithium di-isopropylamide; this reagent functions as the synthetic equivalent[289] of the unknown β-formylvinyl anion (152) in the preparation of a number of $\alpha\beta$-unsaturated aldehydes, as exemplified in Scheme 72. Corey[290] has utilized this reagent further in a total synthesis of prostaglandin $F_{2\alpha}$.

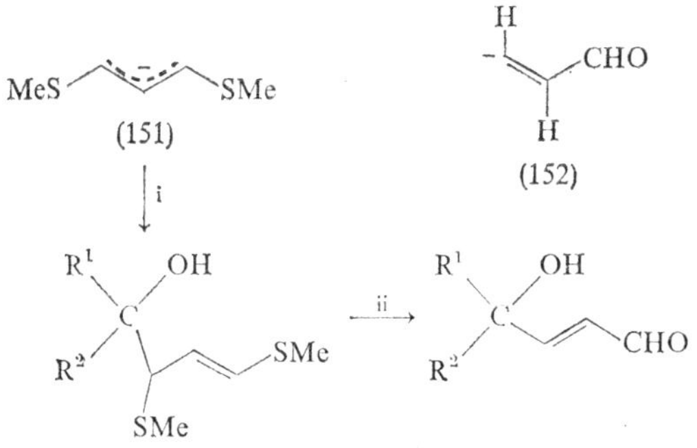

Reagents: i, R^1COR^2; ii, $HgCl_2$–H_2O–MeCN

Scheme 72

[287] L. Maldonado and G. Stork, *J. Amer. Chem. Soc.*, 1971, **93**, 5286.
[288] W. H. Morrison, G. E. Niznik, and H. M. Walborsky, *J. Amer. Chem. Soc.*, 1970, **92**, 6675; 1969, **91**, 7778; *Org. Synth.*, 1971, **51**, 31.
[289] E. J. Corey, B. W. Erickson, and R. Noyori, *J. Amer. Chem. Soc.*, 1971, **93**, 1724.
[290] E. J. Corey and R. Noyori, *Tetrahedron Letters*, 1971, 311.

Full details for the preparation of 1,3-dithian have been published.[291] The anion of methylthiomethyl sulphoxide functions[292] as a new formyl anion equivalent (Scheme 73); acid-catalysed hydrolysis is very rapid, perhaps explaining the low yields observed in the attempted oxidation of some thioacetals to disulphoxides.[293]

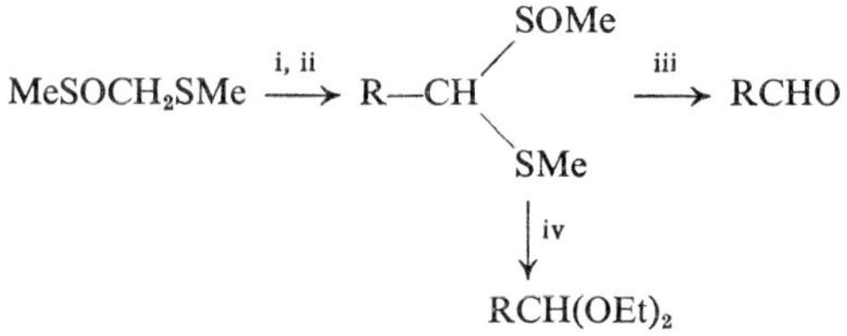

$$\text{MeSOCH}_2\text{SMe} \xrightarrow{\text{i, ii}} \text{R--CH}\begin{array}{c}\text{SOMe}\\ \text{SMe}\end{array} \xrightarrow{\text{iii}} \text{RCHO}$$

$$\downarrow \text{iv}$$

$$\text{RCH(OEt)}_2$$

Reagents: i, NaH–THF; ii, RX; iii, H_3O^+; iv, $HC(OEt)_3$

Scheme 73

A method for the homologation of aldehydes *via* the derived keten thioacetals is described.[294] Aldehydes react with the phosphorus ylide (153) in the normal manner to give the keten thioacetals (154); these latter compounds are converted into aldehyde thioacetals by a protonation–hydride-transfer process using triethylsilane in an acidic medium, the cationic intermediates being stabilized by electron donation from sulphur (Scheme 74). One might predict a considerable increase in the use of this reducing system.

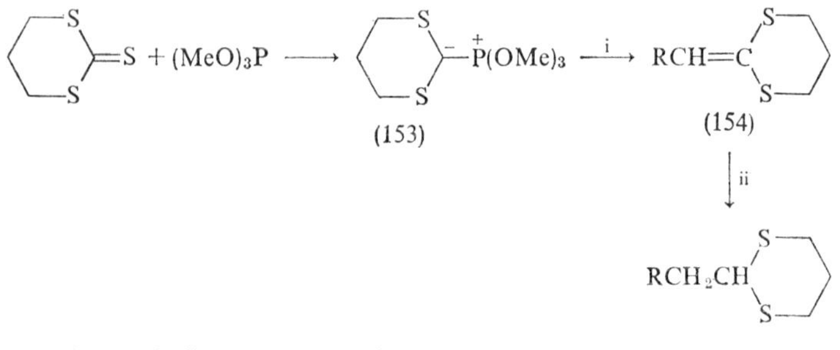

Reagents: i, RCHO; ii, CF_3CO_2H–Et_3SiH

Scheme 74

Isocyanatomethyl-lithium,[295] from methyl isocyanide and n-butyl-lithium, introduces aminomethyl groups into carbonyl compounds, allowing the

---

[291] *Org. Synth.*, 1970, **50**, 72.
[292] K. Ogura and G. Tsuchihashi, *Tetrahedron Letters*, 1971, 3151.
[293] R. Louw and H. Niewenhuyse, *Tetrahedron Letters*, 1971, 4141.
[294] F. A. Carey and J. R. Neergaard, *J. Org. Chem.*, 1971, **36**, 2731.
[295] P. Böhme and U. Schöllkopf, *Angew. Chem. Internat. Edn.*, 1971, **10**, 491.

homologation of cyclic ketones (Scheme 75).

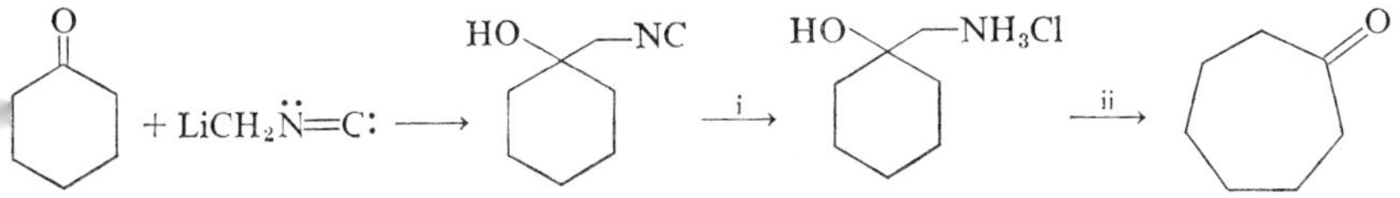

Reagents: i, MeOH–HCl; ii, HNO$_2$

Scheme 75

Ketone trimethylsilyl enol ethers have been employed in an improvement[296] to the cyanohydrin route for the generation of aminomethyl groups (Scheme 76); this modification increases the generality of cyanohydrin formation.

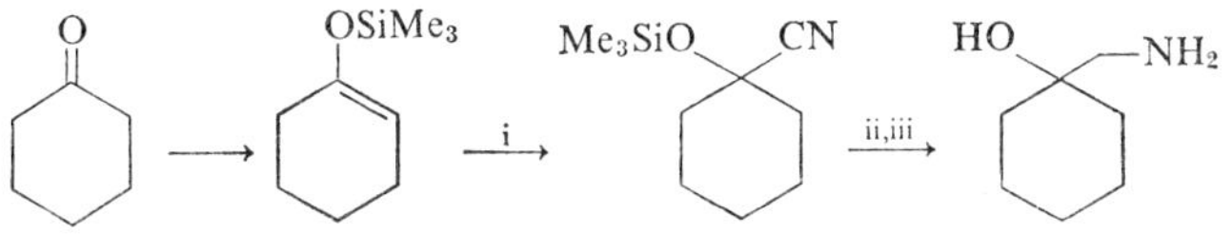

Reagents: i, HCN; ii, LiAlH$_4$; iii, H$_2$O

Scheme 76

Corey[297] has reported a novel method for the introduction of two alkyl appendages at the carbonyl carbon of ketones. Ketones react with the anion of diethyl allylthiomethylphosphonate (155) to give vinyl allyl sulphides such as (156); heating in the presence of mercury(II) oxide induces a thio-Claisen rearrangement to yield the aldehyde (157), which can be further elaborated by various oxidation–reduction sequences to give, for example, the spiro-enone (158).

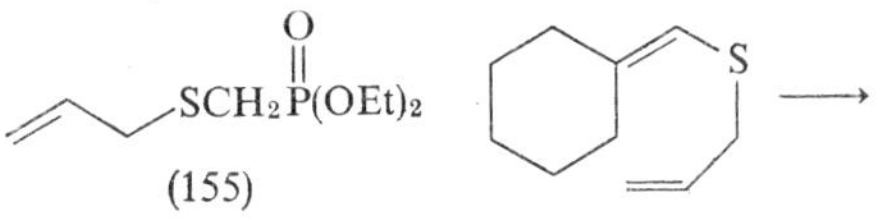

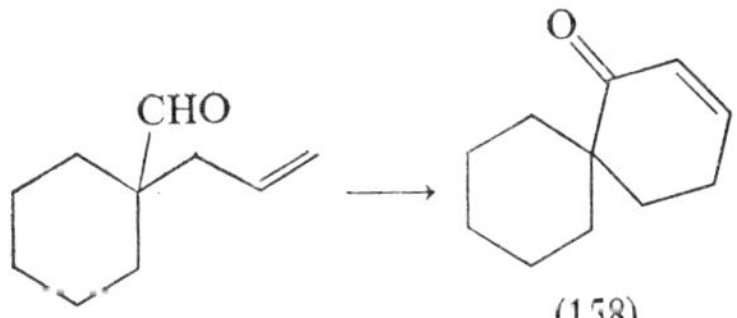

[296] W. E. Parham and C. S. Roosevelt, *Tetrahedron Letters*, 1971, 923.
[297] E. J. Corey and J. I. Shulman, *J. Amer. Chem. Soc.*, 1970, **92**, 5522.

Enamine phosphine oxides (159) seem suitable reagents for the preparation of αβ-unsaturated ketones,[298a] and of cyclopropyl ketones[298b] (Scheme 77).

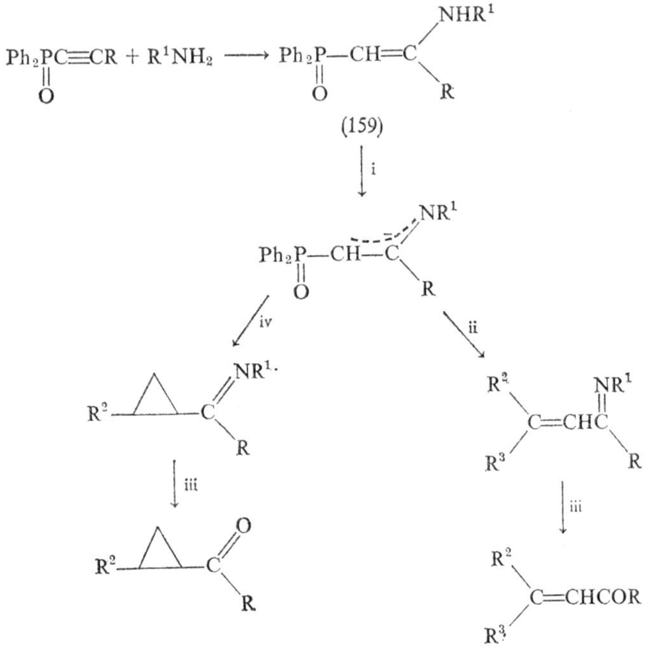

$$Ph_2PC\equiv CR + R^1NH_2 \longrightarrow Ph_2P(O)-CH=C(NHR^1)(R)$$
(159)
$$\downarrow i$$
$$Ph_2P(O)-CH\cdots C(NR^1)(R)$$

iv ↙   ↘ ii

$$R^2-\triangle-C(=NR^1)(R)$$ $$R^2R^3C=CHC(=NR^1)(R)$$

↓ iii   ↓ iii

$$R^2-\triangle-C(=O)(R)$$ $$R^2R^3C=CHCOR$$

Reagents: i, Bu^{n}Li; ii, R^{2}COR^3; iii, H$_3$O$^+$; iv, R^2–CH(O)–

Scheme 77

The synthetic flexibility of 1,3-dithians as masked carbonyl synthons has been frequently marred by difficulty in regeneration of the desired carbonyl group; 2,2-dialkyl-1,3-dithians are readily hydrolysed to ketones by mercury(II) salts, but the reactions of the 2-alkyl (to aldehydes) and 2-acyl (to α-dicarbonyl compounds) analogues are very sluggish, in accord with the ease of carbon–sulphur bond heterolysis, which is in turn determined by the electron-supplying ability of substituents at C-2. Vedejs[299] has now found that a combination of mercury(II) acetate and boron trifluoride etherate in acetic acid effects rapid transacetalization of 2-alkyl-1,3-dithians to the corresponding acetal diacetates; alternatively, mercury(II) oxide and boron trifluoride etherate in aqueous THF give the aldehyde directly (Scheme 78):

[298] (a) A. M. Aguiar, M. S. Chatta, C. J. Morrow, N. A. Portnoy, and J. C. Williams, *Tetrahedron Letters*, 1971, 1397, 1401; A. M. Aguiar and M. S. Chatta, *ibid*., p. 1419; (b) A. M. Aguiar, N. A. Portnoy, and K. S. Yong, *ibid*., p. 2559.
[299] P. L. Fuche and E. Vedejs, *J. Org. Chem*., 1971, **36**, 366.

Functional Groups other than Acetylenes, Allenes, and Olefins 135

both methods proceed in very high yield. Unfortunately, they are inapplicable to 2-acyl-1,3-dithians and to the regeneration of enolizable ketones such as cyclohexanone and cyclopentanone.

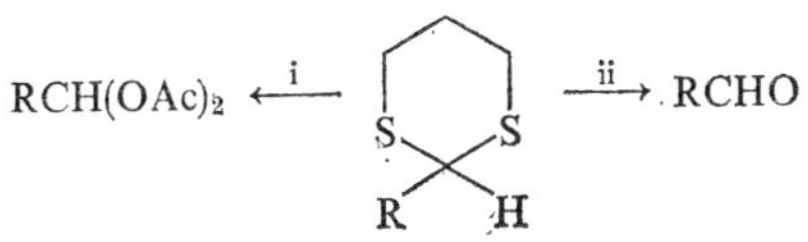

$$RCH(OAc)_2 \xleftarrow{i} \underset{R\ H}{\underset{S\diagdown\diagup S}{\bigcirc}} \xrightarrow{ii} RCHO$$

Reagents: i, $Hg(OAc)_2$–BF_3, Et_2O–AcOH; ii, HgO–BF_3, Et_2O–THF–H_2O

Scheme 78

Oxidative hydrolysis of 2-alkyl- and 2-acyl-1,3-dithians by *N*-halogenosuccinimides is recommended[300] as a high-yield method of carbonyl regeneration; the combination of *N*-chlorosuccinimide and silver(I) ion is suitable for reaction with unsaturated 1,3-dithians. In a related oxidative procedure, ketones and aldehydes can be readily regenerated from their ethylene thio-[301] or hemithio-acetals[302] by treatment with aqueous chloramine-T.

The difficulties encountered in the regeneration of steroidal ketones from their ethylene or trimethylene thioacetals by the mercury(II) salt procedure have been circumvented;[303] 1-chlorobenzotriazole smoothly oxidizes thioacetals to the corresponding disulphoxides, which, without isolation, furnish the desired ketones in satisfactory yield on treatment with aqueous base.

The trityl carbonium ion is well documented as a hydride-ion acceptor; indeed it can be reduced by aldehyde acetals in an intramolecular hydride-transfer process.[304,305] This is pivotal in a new method for the deacetalization of ketone acetals by oxidative hydride transfer;[306] trityl fluoroborate is used as reagent, and the glycol moiety is oxidized to an α-ketol, as is shown in Scheme 79. This novel procedure is successful also with hemithioacetals, but fails with thioacetals.

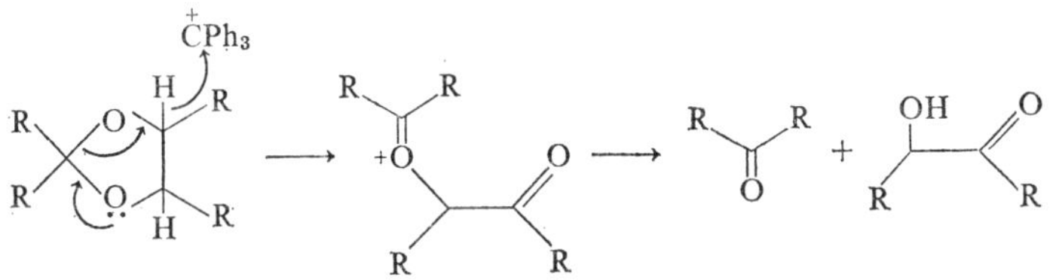

Scheme 79

[300] E. J. Corey and B. W. Erickson, *J. Org. Chem.*, 1971, **36**, 3553.
[301] D. W. Emerson, W. F. J. Huurdeman, and H. Wynberg, *Tetrahedron Letters*, 1971, 3449.
[302] D. W. Emerson and H. Wynberg, *Tetrahedron Letters*, 1971, 3445.
[303] P. R. Heaton, J. M. Midgley, and W. B. Whalley, *Chem. Comm.*, 1971, 750.
[304] V. Hederich, H. Meerwein, H. Morschel, and K. Wunderlich, *Annalen*, 1960, **635**, 1.
[305] St. Penczek and St. Słomkowski, *Chem. Comm.*, 1970, 1347.
[306] (*a*) D. H. R. Barton, P. D. Magnus, G. Smith, and D. Zurr, *Chem. Comm.*, 1971, 861; (*b*) D. H. R. Barton, P. D. Magnus, G. Smith, G. Streckert, and D. Zurr, *J. C. S. Perkin I*, 1972, 542.

Alkylation.—Position-specific alkylation of ketones can be achieved by a number of new methods. Vinyloxyboranes, formed by reaction of trialkylboranes with diazoketones or by radical addition to methyl vinyl ketone, react with alkyl-lithium reagents to give the corresponding lithium enolates, which undergo facile, site-specific alkylation;[307] this allows the formation of αα-disubstituted ketones from diazoketones and αβ-disubstituted ketones from methyl vinyl ketone (Scheme 80).

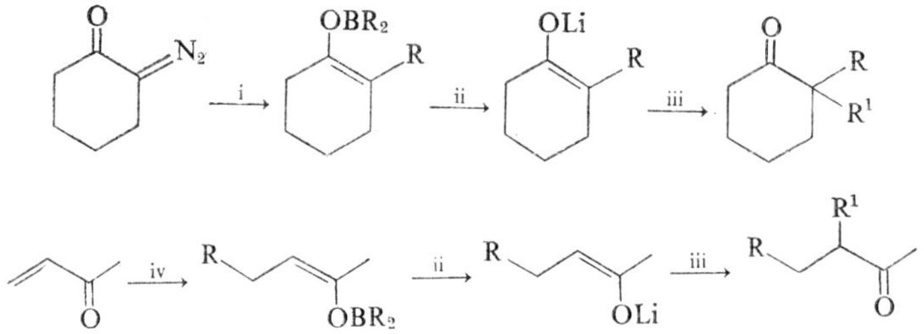

Reagents: i, R_3B; ii, Bu^nLi; iii, R^1I; iv, $R_3B-I\cdot$

Scheme 80

The reductive generation of specific lithium enolates from cyclic αβ-unsaturated ketones can be extended to acyclic analogues;[308] however, much polymethylation was observed on alkylation with methyl iodide in the latter cases. This undesirable side-reaction was ascribed to the effect of the conjugate base of the proton donor used in the reduction step, it being minimized by use of triphenylcarbinol as proton donor.

The relative utility of various enol derivatives for the generation of specific lithium enolates has been critically assessed by House;[309] the method of choice for generation of the less highly substituted enolate of an unsymmetrical ketone seems to be kinetically controlled deprotonation with lithium di-isopropylamide.

An Illinois group[310] has described a method of selective geminal alkylation of ααα'-trisubstituted ketones; the corresponding n-butylthiomethylene derivatives of such ketones undergo a process of 'double reduction' with lithium in ammonia, affording a methyl-substituted enolate anion at the original methylene position, which can be alkylated *in situ*. This permits the introduction of a methyl group and a second, variable substituent at that ketone flank which condenses with ethyl formate (Scheme 81).

The reduction of α-epoxyketones with lithium in ammonia gives rise

---

[307] D. J. Pasto and P. W. Wojtkowski, *J. Org. Chem.*, 1971, **36**, 1790.
[308] L. E. Hightower, L. R. Glasgow, D. A. Albertson, and H. A. Smith, *J. Org. Chem.*, 1970, **35**, 1881.
[309] H. O. House, M. Gall, and H. D. Olmstead, *J. Org. Chem.*, 1971, **36**, 2361.
[310] R. M. Coates and R. L. Sowerby, *J. Amer. Chem. Soc.*, 1971, **93**, 1027.

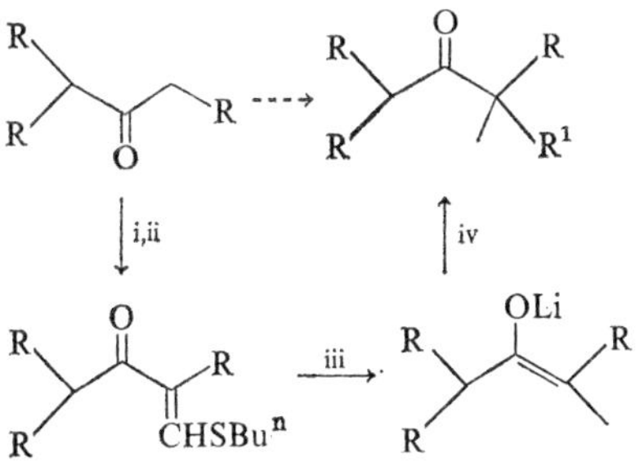

Reagents: i, HCO_2Et–NaH; ii, Bu^nSH–TosOH; iii, Li–NH_3–2ROH; iv, R^1X

Scheme 81

to a specific enolate;[311] this allows the specific α-alkylation of an αβ-unsaturated ketone as shown in Scheme 82.

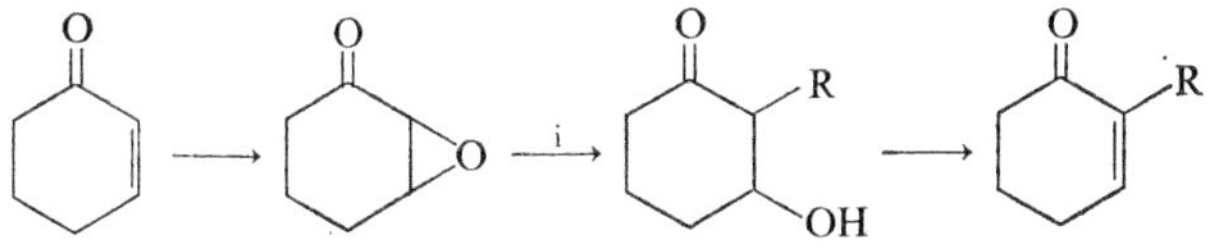

Reagents: i, Li–NH_3–RI

Scheme 82

The specific formation and alkylation of enolates from enol-phosphorylated species has been described (Scheme 83).[312]

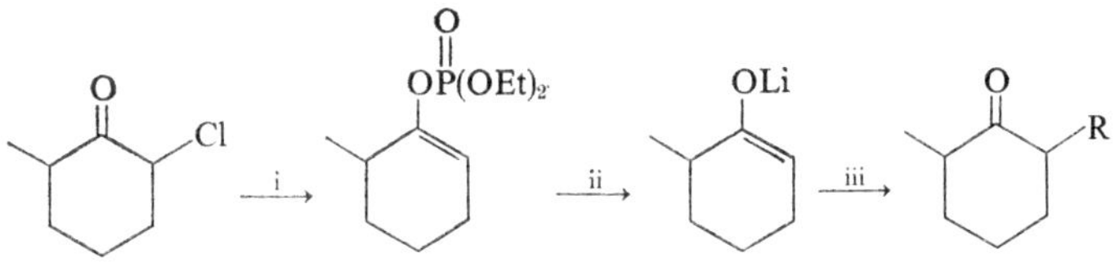

Reagents: i, $(EtO)_3P$; ii, PhLi; iii, RX

Scheme 83

Aldehydes can be alkylated indirectly via the corresponding imines; lithium dialkylamides readily deprotonate such imines at the position α to the azomethine group, and the carbanions formed can be alkylated with primary halides.[313] This route to α-substituted aldehydes avoids the use of sodium borohydride, necessary in the dihydro-1,3-oxazine alternative.

[311] J. D. McChesney and A. F. Wycpalek, *Chem. Comm.*, 1971, 542.
[312] I. J. Borowitz, E. W. R. Casper, and R. K. Crouch, *Tetrahedron Letters*, 1971, 105.
[313] Th. Cuvigny and H. Normant, *Bull. Soc. chim. France*, 1970, 3976; G. J. Heiszwolf and H. Koosterziel, *Rec. Trav. chim.*, 1970, **89**, 1217.

138 *Aliphatic, Alicyclic, and Saturated Heterocyclic Chemistry*

N-Acetylpyrrolidone can effect C-acylation of ketones[314] (Scheme 84).

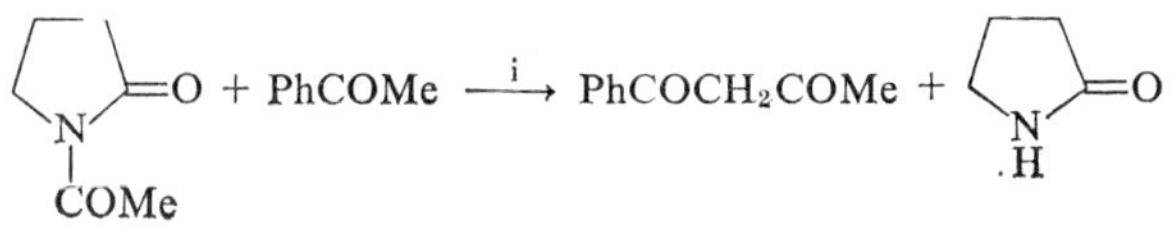

Reagent: i, NaH

Scheme 84

Triethylbenzylammonium chloride functions as a catalyst in the sodium hydroxide-induced alkylation of ketones in aqueous media, thus avoiding the use of inflammable solvents and highly reactive reagents.[315] α-Diazoketones seem to be stable to dilute aqueous base, allowing the normal alkylation shown in Scheme 85.[316]

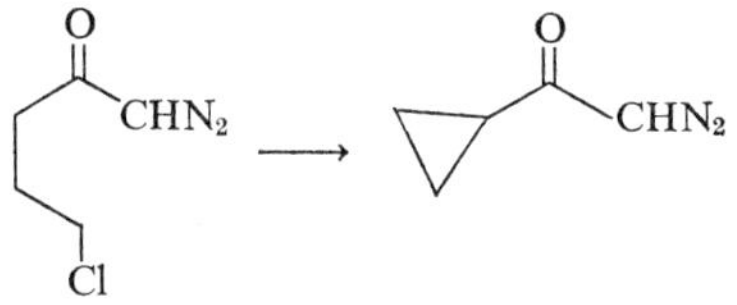

Scheme 85

An intriguing report on the alkylation of alkali-metal salts of acetylacetone and methyl acetoacetate with chiral 2-bromobutane has been published;[317] the C-alkylated product is found to retain the configuration of the alkyl group, whereas the expected inversion is shown by the O-alkylated product. The reaction of carbanions of β-ketoesters and β-diketones with 1,4-dichlorobutan-2-one leads to epoxyannelation and annelated chlorohydrins; as a route[318] to epoxyketones, it avoids the possible blocking–deblocking sequence in the normal basic peroxidation of enones (Scheme 86).

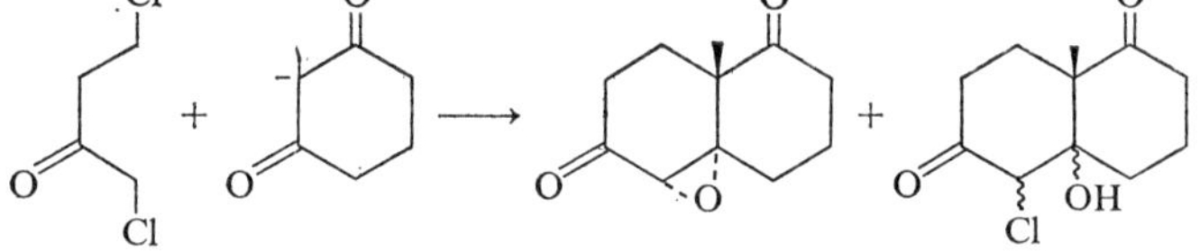

Scheme 86

[314] H. Stetter and W. Steinbeck, *Chem. Ber.*, 1970, **103**, 1088.
[315] A. Jończyk, B. Serafin, and M. Mąkosza, *Tetrahedron Letters*, 1971, 1351.
[316] N. F. Woolsey and D. D. Hammargren, *Tetrahedron Letters*, 1970, 2087; *cf.* T. L. Burkoth, *Tetrahedron Letters*, 1969, 5049.
[317] M. Suama, T. Sugita, and K. Ichikawa, *Bull. Chem. Soc. Japan*, 1971, **44**, 1999.
[318] S. Danishefsky and G. A. Koppel, *Chem. Comm.*, 1971, 367.

Functional Groups other than Acetylenes, Allenes, and Olefins 139

Reduction.—Contrary to accepted belief, epimerization can precede reduction in the reaction of α-chiral ketones with sodium borohydride in anhydrous media, such as alcohols or ethers; this undesirable event can be nullified by the addition of as little as 2% of water.[319] An overdue report has been published on the occurrence of 1,4-addition in the reduction of αβ-unsaturated ketones with sodium borohydride; no methods of avoidance are suggested.[320]

The utility of the n-butyl-lithium–pyridine adduct as a reducing agent has been rediscovered;[321] it reduces ketones under non-equilibrating conditions to the equatorial alcohol, a useful reagent when axial approach by hydride ion is sterically hindered.[322]

Sodium cyanoborohydride[323] appears to be an exceptionally selective reducing agent; it has only a weak reducing ability for most functional groups except halides[324] (*loc. cit.*) and iminium ions.[325] This latter capacity is utilized in the acid-catalysed reduction of aliphatic ketones and aldehydes to hydrocarbons *via* the corresponding toluene-*p*-sulphonylhydrazones[326] under conditions much milder than those of the analogous Caglioti procedure; prior preparation of the hydrazones is unnecessary, owing to the very slow rate of carbonyl reduction.

The reduction of carbonyl groups by chiral reagents continues to arouse much interest. Full papers describing the asymmetric reduction of ketones[327a] and imines[327b] have been published; use is made of chiral borohydride ions, prepared by reaction of di-isopinocamphenylborane with a number of alkyl-lithium reagents. A similar range of reagents was critically assessed by Corey[328] to achieve asymmetric reduction of the carbonyl group in the enone (160), a prostaglandin precursor. The most effective borohydride ion was that formed by reaction of the borane (161), which could be racemic, and t-butyl-lithium, presumably by β-hydrogen transfer from the organolithium with elimination of isobutene. Using this reagent, the desired α-epimer (162) was found to predominate over the β-epimer (163) by a ratio of 4.5:1.

Optically active beryllium and aluminium alkyls have been studied as asymmetric reducing agents;[329] they appear to be considerably more effective

---

[319] V. Hach, E. C. Fryberg, and E. McDonald, *Tetrahedron Letters*, 1971, 2629.
[320] M. R. Johnson and B. Rickborn, *J. Org. Chem.*, 1970, **35**, 1041.
[321] R. Levine and W. M. Kadunce, *Chem. Comm.*, 1970, 921.
[322] R. A. Abramovitch, W. C. Marsh, and J. G. Saha, *Canad. J. Chem.*, 1965, **43**, 2631; Fieser and Fieser, 'Reagents for Organic Synthesis', Volume 2, Wiley-Interscience, 1969, p. 351.
[323] R. F. Borch, M. D. Bernstein, and H. D. Durst, *J. Amer. Chem. Soc.*, 1971, **93**, 2897.
[324] R. O. Hutchins, B. E. Maryanoff, and C. A. Milewski, *Chem. Comm.*, 1971, 1097.
[325] R. F. Borch and H. D. Durst, *J. Amer. Chem. Soc.*, 1969, **91**, 3996.
[326] R. O. Hutchins, B. E. Maryanoff, and C. E. Milewski, *J. Amer. Chem. Soc.*, 1971, 1793.
[327] (a) M. F. Grundon, W. A. Khan, D. R. Boyd, and W. R. Jackson, *J. Chem. Soc. (C)*, 1971, 2557; (b) J. F. Archer, D. R. Boyd, W. T. Jackson, M. F. Grundon, and W. A. Khan, *J. Chem. Soc. (C)*, 1971, 2560.
[328] E. J. Corey, S. M. Albonico, U. Koelliker, T. K. Schaaf, and R. K. Varma, *J. Amer. Chem. Soc.*, 1971, **93**, 1491.
[329] G. P. Giacomelli, R. Menicagli, and L. Lardicci, *Tetrahedron Letters*, 1971, 4135.

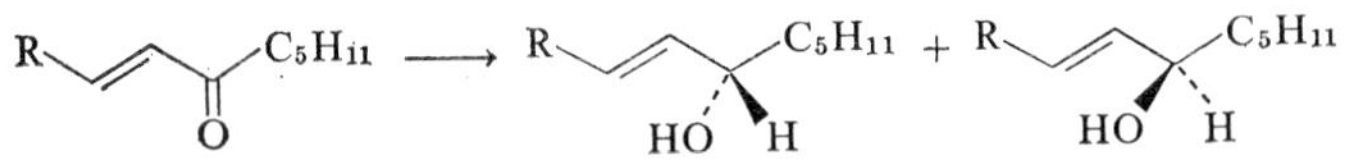

R–CH=CH–CO–C$_5$H$_{11}$ ⟶ R–CH=CH–C(OH)(H)–C$_5$H$_{11}$ + R–CH=CH–C(OH)(H)–C$_5$H$_{11}$
(160)　　　　　　　　　(162)　　　　　　　　(163)

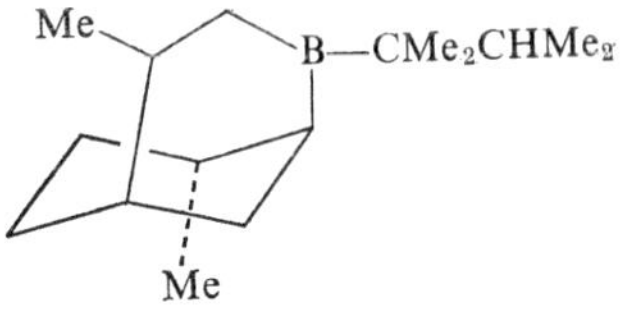

Me–[bicyclic]–B–CMe$_2$CHMe$_2$
(161)

than the analogous magnesium Grignard reagents. Ketone reduction by aluminium alkyls proceeds by an initial rapid, reversible complexation, followed by a slow, rate-determining hydride-transfer,[330] as shown in Scheme 87.

$$Bu^i_3Al + Ph_2CO \rightleftharpoons Bu^i_3Al \cdot O=CPh_2$$

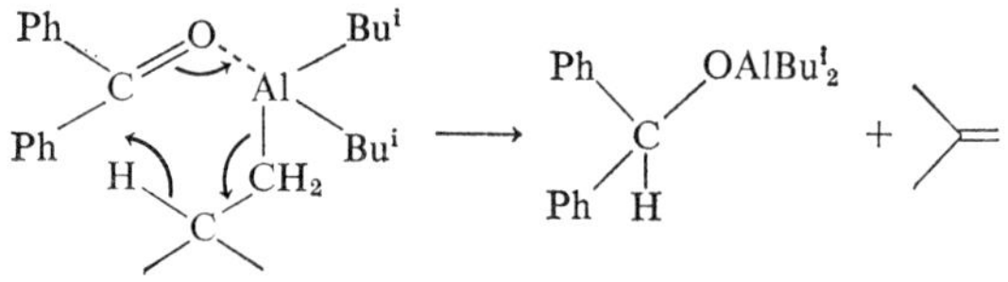

Scheme 87

Chiral [α-^{2}H]benzyl alcohol has been prepared by asymmetric reduction of [1-^{2}H]benzaldehyde; the most efficient reagent was bornyloxymagnesium bromide.[331]

Certain soluble rhodium complexes, such as (164), function as homogeneous catalysts for the hydrogenation of ketones;[332] in dry solvents, the reaction is very slow, but the addition of ca. 1% of water produces a marked rate enhancement and, as a bonus, also inhibits the rate of reduction of olefins.

$$[RhH_2(PMe_3)_2L_2]^+ClO_4^- \quad L = \text{solvent}$$
(164)

The homogeneous transfer-hydrogenation[333] of αβ-unsaturated carbonyl compounds using ruthenium catalysts has been described, and is shown in Scheme 88.

PhCH=CHCOMe + 2PhCH$_2$OH $\xrightarrow{i}$ PhCH$_2$CH$_2$COMe + 2PhCHO

Reagent: i, (Ph$_3$P)$_3$RuCl$_2$

Scheme 88

[330] E. C. Ashby and S. H. Yu, *J. Org. Chem.*, 1970, **35**, 1034.
[331] D. Nasipuri, C. K. Ghosh, and R. J. L. Martin, *J. Org. Chem.*, 1970, **35**, 657.
[332] R. R. Schrock and J. A. Osborn, *Chem. Comm.*, 1970, 567.
[333] Y. Sasson and J. Blum, *Tetrahedron Letters*, 1971, 2167.

Functional Groups other than Acetylenes, Allenes, and Olefins

The reduction of enones by electron transfer has been studied further, a comparison being drawn between ammonia and hexamethylphosphoramide as solvents.[334] In general, reduction with sodium in hexamethylphosphoramide yields more of the less stable saturated ketone epimer than is observed in ammonia; methods of optimizing this selectivity are discussed.[335]

1,3-Dicarbonyl compounds can be specifically reduced to $\alpha\beta$-unsaturated ketones via the corresponding boron difluoride complexes,[336] as exemplified in Scheme 89.

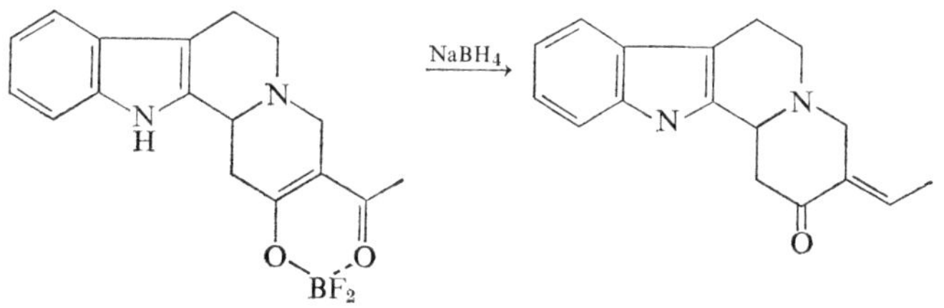

Scheme 89

Such complexes find further use in the conversion of 2-formyl ketones into their 2-alkylidene derivatives (Scheme 90), conserving half of the normal requirement of organometallic reagent.[337]

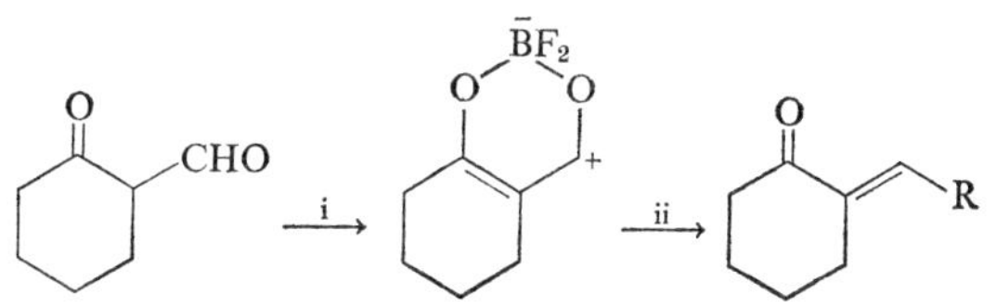

Reagents: i, BF_3, Et_2O; ii, RM

Scheme 90

Reduction of 1,3-diketones with lithium aluminium hydride under forcing conditions can afford allylic alcohols as products of elimination in addition to the expected 1,3-diols; the allylic alcohols[338] are obtained in yields corresponding to the enol content of the starting diketone, trans-olefins being produced exclusively. The two tautomers of the diketones appear to react independently, the enolic portion giving rise to allylic alcohol and a small

---

[334] (a) K. W. Bowers, R. W. Giese, J. Grimshaw, H. O. House, N. H. Kolodny, K. Kronenberger, and D. K. Roe, J. Amer. Chem. Soc., 1970, 92, 2783; (b) M. Larchevêque, Ann. Chim. (France), 1970, 5, 129.
[335] H. O. House, R. W. Giese, K. Kronenberger, J. P. Kaplan, and J. F. Simeone, J. Amer. Chem. Soc., 1970, 92, 2800.
[336] E. Winterfeldt, J. M. Nelke, and T. Korth, Chem. Ber., 1971, 104, 802.
[337] R. A. J. Smith and T. A. Spencer, J. Org. Chem., 1970, 35, 3220.
[338] J. W. Frankenfeld and W. E. Tyler, J. Org. Chem., 1971, 36, 2110.

amount of saturated alcohol and the diketo-form being reduced to the diol (Scheme 91).

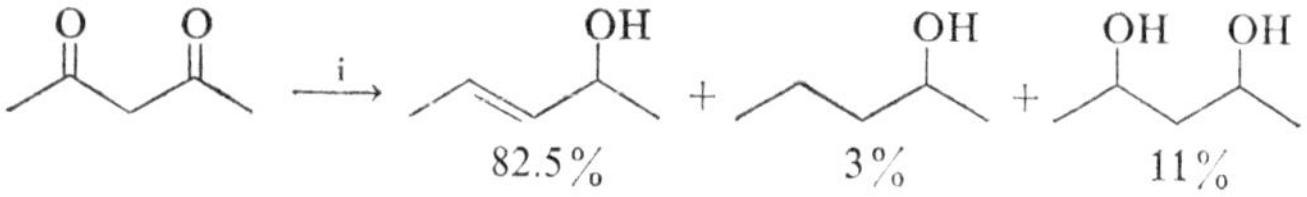

Reagent: i, LiAlH$_4$

Scheme 91

The Clemmensen reduction of 1,4-diketones has been investigated by two groups. Alcohols with unchanged carbon skeletons are obtained[339] when conformational mobility allows the two carbonyl groups to approach each other; for example, hexane-2,5-dione gives a mixture of hexan-2-ol and *cis*- and *trans*-hex-4-en-2-ol; the participation of the carbonyl oxygen in the solvolysis of intermediates such as (165) is proposed (Scheme 92). Pinacol-type prod-

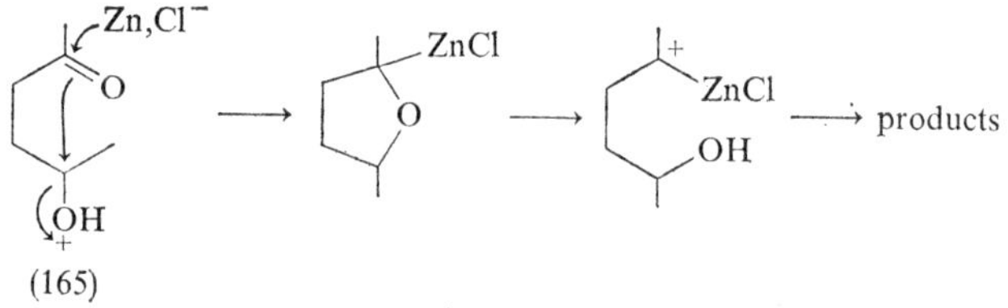

Scheme 92

ucts are formed from 1,4-diketones in seven- and eight-membered rings,[340] whereas cyclohexane-1,4-dione gives hexane-2,5-dione, perhaps by fragmentation of the diradical (166).

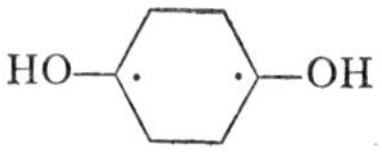

(166)

Amalgamated aluminium in dichloromethane or THF is recommended[341] for the pinacolic reduction of aliphatic or cyclo-aliphatic ketones; soluble aluminium pinacolates are formed.

The formation of the diol (167) from the reaction of di-t-butyl ketone with sodium in diethyl ether has been re-investigated, and a new mechanism (Scheme 93) proposed;[342] the involvement of ethylene, derived from diethyl

[339] D. R. Crump and B. R. Davis, *Chem. Comm.*, 1970, 768.
[340] E. Wenkert and J. E. Yoder, *J. Org. Chem.*, 1970, **35**, 2986.
[341] A. A. P. Schreibmann, *Tetrahedron Letters*, 1970, 4271.
[342] P. D. Bartlett, T. T. Tidwell, and W. P. Weber, *Tetrahedron Letters*, 1970, 2919.

ether, was proven by conducting the reaction in benzene in the presence of ethylene, when the same product, (167), was formed. The pinacolic reduction of acetone has been subjected to further study.[343]

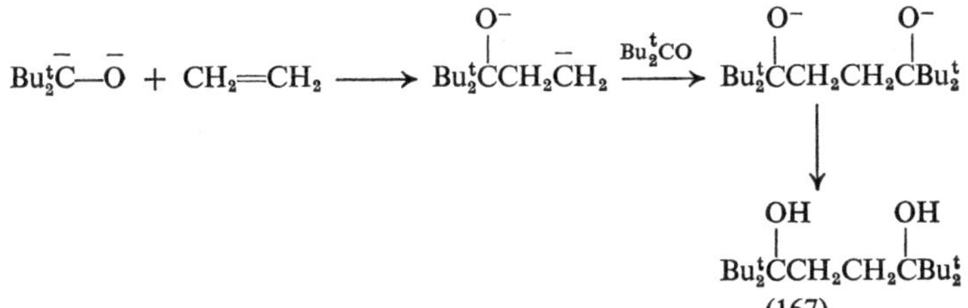

$$Bu_2^t\bar{C}-\bar{O} + CH_2=CH_2 \longrightarrow Bu_2^t\overset{O^-}{\underset{|}{C}}CH_2CH_2 \xrightarrow{Bu_2^tCO} Bu_2^t\overset{O^-}{\underset{|}{C}}CH_2CH_2\overset{O^-}{\underset{|}{C}}Bu_2^t$$

$$\downarrow$$

$$Bu_2^t\overset{OH}{\underset{|}{C}}CH_2CH_2\overset{OH}{\underset{|}{C}}Bu_2^t$$
(167)

Scheme 93

Addition of Grignard Reagents.—A sensitive measure[344] of the degree of steric hindrance in the neighbourhood of a carbonyl group is provided by reaction with t-butylallylmagnesium bromide. Two transition states are possible, on the basis of an S_E2' non-cyclic bimolecular electrophilic substitution mechanism (Scheme 94);[345] the ratio of products is extremely sensitive to the steric environment of the carbonyl group, and thus constitutes a useful measure of the effective bulk of the substituents. Felkin has found that the carbonyl group in hexadeuterioacetone is less hindered than that of acetone itself; this was taken as substantiation of the view that secondary deuterium effects are steric in origin, but an alternative explanation based on hyperconjugative inductive effects has been advanced.[346]

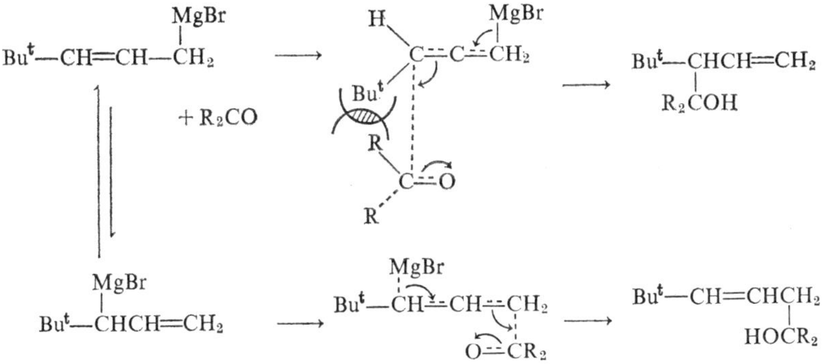

Scheme 94

In a related study,[347] the factors affecting the steric outcome of reaction of

[343] J. R. Binks and D. Lloyd, *J. Chem. Soc.* (*C*), 1971, 2641.
[344] M. Chérest, H. Felkin, and C. Frajerman, *Tetrahedron Letters*, 1971, 379.
[345] H. Felkin, Y. Gault, and G. Roussi, *Tetrahedron*, 1970, **26**, 3761.
[346] A. J. Kresge and V. Nowlan, *Tetrahedron Letters*, 1971, 4297.
[347] M. Chérest and H. Felkin, *Tetrahedron Letters*, 1971, 383.

simple cyclohexanones with hydride and Grignard reagents are separated into the net differences between the steric strain in the transition state leading to the equatorial alcohol and the torsional strain in that leading to the axial epimer.

A one-step alternative to the Grignard reaction has been described, involving reaction of an alkyl halide, lithium metal, and the ketone in THF.[348] The yield of addition of a Grignard reagent to a hindered ketone can be enhanced[349] at the expense of reduction by complexing the reagent with lithium perchlorate or tetrabutylammonium bromide; in the salt complex, the ionic character of the organometallic will be enhanced, favouring addition, since the corresponding transition state must be more polar than that for reduction.

Halogeno-derivatives.—A new general method[350] has been described for the selective halogenation of the methyl group of methyl ketones, based on treatment of the ethanolamine-derived ketimine (168) with either N-chloro- or N-bromo-succinimide in ether, followed by mild acid hydrolysis; presumably the tautomeric enamine (169) is the reactive species.

$$\underset{(168)}{\underset{R}{\diagup}\overset{N}{\underset{\|}{\diagup}}\overset{\diagdown}{}\overset{OH}{}\underset{R}{\diagup}\overset{\diagdown}{Me}} \qquad \underset{(169)}{\underset{R}{\diagup}\overset{NH}{\diagup}\overset{\diagdown}{}\overset{OH}{}\overset{\diagdown}{CH_2}}$$

Acid-catalysed bromination of methyl ketones in methanol constitutes an excellent method for the obtention of bromomethyl ketones; the addition of small amounts of methanol to other solvents, such as ether or carbon tetrachloride, leads to results comparable to those observed in methanol alone. The directive effect of methanol is ascribed by the French authors[351] to its intervention as a base in the reaction.

$\alpha\beta$-Unsaturated aldehydes can be halogenated in the γ-position *via* 1,4-addition to the corresponding dienol acetates (Scheme 95).[352a] The γ-halogeno-compounds produced can be utilized in Wittig reactions, when an interesting difference in phosphonium salt formation is observed[352b]; The halogeno-compound derived from crotonaldehyde gives the expected ylide (170), but the ylide (171) is produced by allylic displacement on the halogeno-compound from 3-methylcrotonaldehyde.

---

[348] P. J. Pearce, D. H. Richards, and N. F. Scilly, *Chem. Comm.*, 1970, 1160.
[349] M. Chastrette and R. Amouroux, *Chem. Comm.*, 1970, 470.
[350] J. F. W. Keana and R. R. Schumaker, *Tetrahedron*, 1970, **26**, 5191.
[351] M. Gaudry and A. Marquet, *Tetrahedron*, 1970, **26**, 5611, 5617.
[352] (*a*) M. J. Berenguer, J. Castells, J. Fernández, and R. M. Galard, *Tetrahedron Letters*, 1971, 493; (*b*) M. J. Berenguer, J. Castells, R. M. Galard, and M. Moreno-Mañas, *Tetrahedron Letters*, 1971, 495.

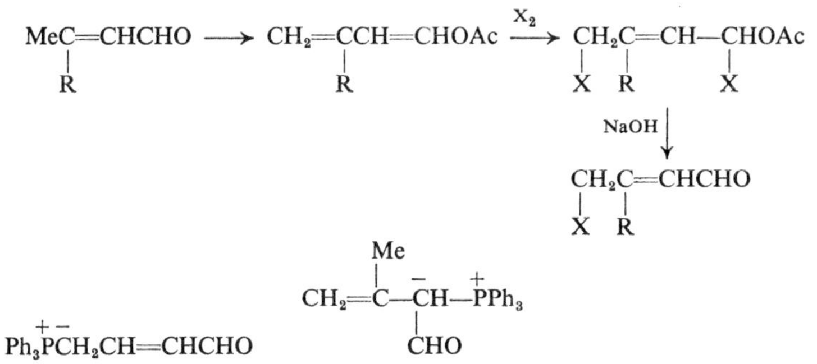

Functional Groups other than Acetylenes, Allenes, and Olefins

$$\underset{\underset{R}{|}}{MeC}=CHCHO \longrightarrow \underset{\underset{R}{|}}{CH_2}=CCH=CHOAc \xrightarrow{X_2} \underset{\underset{X \ R \ \ \ X}{| \ \ \ \ \ |}}{CH_2C}=CH-CHOAc$$

$$\downarrow NaOH$$

$$\underset{\underset{X \ \ R}{| \ \ |}}{CH_2C}=CHCHO$$

$$\overset{+}{Ph_3}\overset{-}{P}CH_2CH=CHCHO \qquad \underset{\underset{CHO}{|}}{CH_2=\overset{\overset{Me}{|}}{C}-\overset{-}{C}H-\overset{+}{P}Ph_3}$$

(170) (171)

Scheme 95

The preparation of 1-bromo-1-chloro-1-fluoroacetone has been described.[353] The combination of lithium iodide and boron trifluoride etherate provides a mild, high-yield process for the reductive deiodination[354] of iodoketones. αα'-Dibromoketones[355] react with 1,3-dienes in the presence of iron carbonyls to give troponoid compounds (Scheme 96).

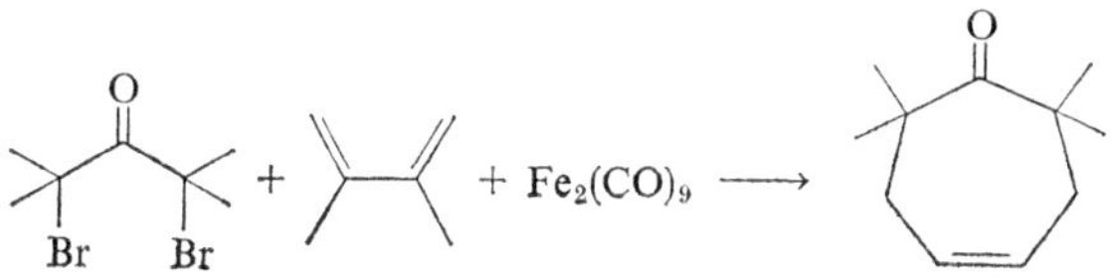

Scheme 96

Dimerization.—The dimerization of organic substrates has obvious synthetic potential, and some further attention has been directed at carbonyl compounds in this context. αα-Disubstituted aldehydes, when treated with metal oxides, undergo oxidative coupling by hydrogen abstraction and subsequent carbon–carbon and carbon–oxygen bond formation (Scheme 97); the ratio of the two products varies with the metal oxide used, but further oxidation of the products is never observed.[356] The isolation of carbon–oxygen coupled products offers an alternative explanation for the formation of α-hydroxyaldehydes by one-electron oxidants in aqueous acidic media.

Whereas electrochemical dimerization of crotonaldehyde gives the glycol (172), 3-methylcrotonaldehyde gives the lactol (173) as major product, severely limiting the generality of this method of dimerization.[357]

---

[353] G. C. Barrett, D. M. Hall, M. K. Hargreaves, and B. Modari, *J. Chem. Soc.*, (*C*), 1971, 279.
[354] J. M. Townsend and T. A. Spencer, *Tetrahedron Letters*, 1971, 137.
[355] R. Noyori, S. Makino, and H. Takaya, *J. Amer. Chem. Soc.*, 1971, **93**, 1272.
[356] J. C. Leffingwell, *Chem. Comm.*, 1970, 357.
[357] D. Miller, L. Mandell, and R. A. Day, jun., *J. Org. Chem.*, 1971, **36**, 1683.

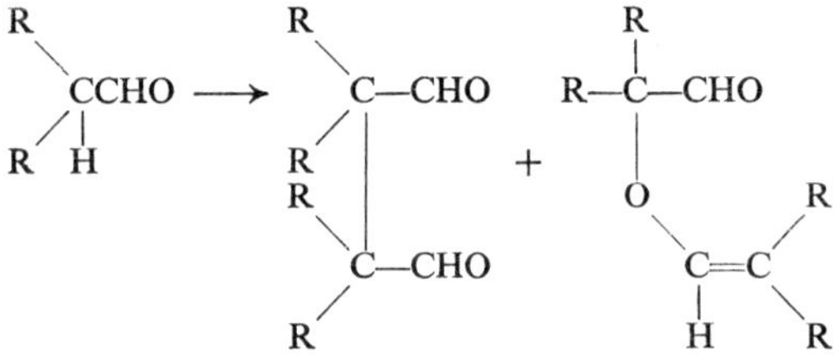

R\
 \CCHO → R\C—CHO   R—C(R)—CHO
R/H   R/| |
          R\   O   R
           C—CHO   \C=C/
          R/   H/ \R

Scheme 97

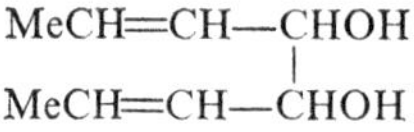

MeCH=CH—CHOH
         |
MeCH=CH—CHOH
      (172)

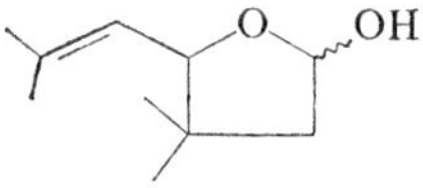

(173)

A copper(I)–alkyl isocyanide catalyst for the dimerization of $\alpha\beta$-unsaturated ketones, esters, and nitriles has been reported;[358] β-unsubstituted compounds do not dimerize, but can co-dimerize with β-substituted monomers, as shown in Scheme 98.

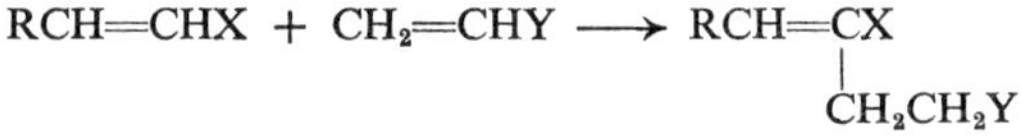

$$RCH=CHX + CH_2=CHY \longrightarrow RCH=CX$$
$$|$$
$$CH_2CH_2Y$$

Scheme 98

The oxidative coupling of pinacolone with lead dioxide in toluene affords not only the expected dimer, 2,2,7,7-tetramethyloctane-3,6-dione, but also a cross-coupled product, 4,4-dimethyl-1-phenylpentan-3-one, and bibenzyl (Scheme 99).[359]

$$Bu^tCOMe \longrightarrow Bu^tCOCH_2CH_2COMe + Bu^tCOCH_2CH_2Ph + PhCH_2CH_2Ph$$

Scheme 99

Metal-induced Bond Formation.—A satisfactory reagent for the conversion of a halogenocarbonyl compound into a reactive organometallic derivative has been discovered. Dialkylcopperlithium reagents react with vinylic iodides at a much faster rate than they react with non-conjugated ketones; this allows a new cyclization method[360] by intramolecular addition of the newly formed organometallic reagent to the carbonyl group. The reagent of choice for alkyl halides is the anion derived by reaction of nickel tetraphenylporphine and

---

[358] T. Saegusa, Y. Ito, S. Tomita, and H. Kinoshita, *J. Org. Chem.*, 1970, **35**, 670.
[359] R. Brettle, *Chem. Comm.*, 1970, 342.
[360] E. J. Corey and I. Kuwajima, *J. Amer. Chem. Soc.*, 1970, **92**, 395.

Functional Groups other than Acetylenes, Allenes, and Olefins   147

lithium naphthalene, NiTPP^{2-}. These transformations are exemplified in Scheme 100.

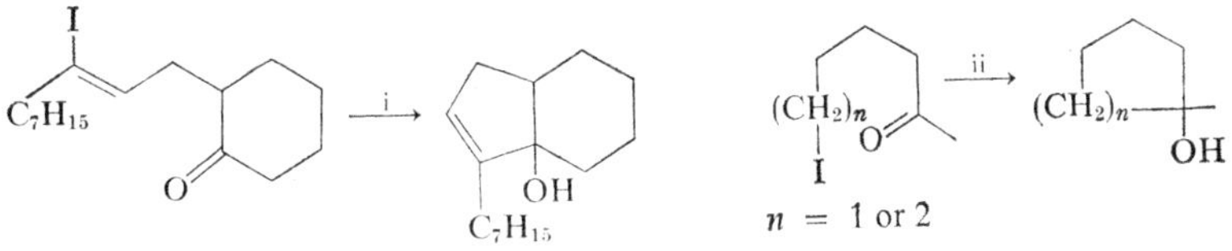

Reagents: i, Bu$_2^n$CuLi; ii, NiTPP^{2-}

Scheme 100

An electrochemical counterpart[361] to this cyclization is provided by the anodic reduction of non-conjugated olefinic ketones; cyclic tertiary alcohols are formed (Scheme 101) in a process initiated by electron addition to the carbonyl group. This method is limited to cases leading to formation of five- and six-membered rings; it provides some interesting stereochemical results, e.g. the isolation of cis-1,2-dimethylcyclopentan-1-ol, whereas methylmagnesium iodide on 2-methylcyclopentanone gives mainly the trans-dimethyl isomer.

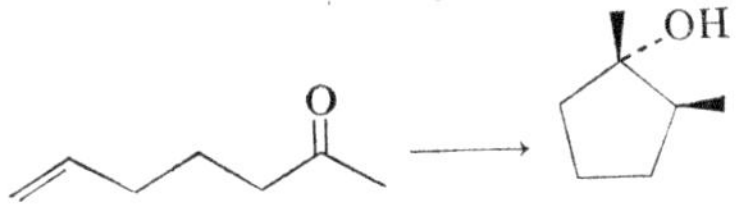

Scheme 101

Enolizable ketones react with olefins in the presence of manganese(III) or cerium(IV) oxidants to give a mixture of products,[362] including $\beta\gamma$-unsaturated ketones (Scheme 102).

MeCOMe + C$_6$H$_{13}$CH=CH$_2$ $\xrightarrow{i}$
  C$_9$H$_{19}$COMe + C$_6$H$_{13}$CH=CHCH$_2$COMe + C$_6$H$_{13}$CHCH$_2$CH$_2$COMe
                                                                    |
                                                                   OAc

Reagents: i, Mn(OAc)$_3$–AcOH

Scheme 102

Aldol Condensation.—Russian authors[363] report that in the presence of a stoicheiometric amount of methylcalcium iodide, acetone condenses rapidly to give mesityl oxide in high yield; similar reactivity was shown by a range of carbonyl compounds.

A number of methods of directing the orientation of aldol condensation have been described. In an extension to a patent,[364] a procedure is given where

[361] T. Shono and M. Mitani, *J. Amer. Chem. Soc.*, 1971, **93**, 5284.
[362] E. I. Heiba and R. M. Dessau, *J. Amer. Chem. Soc.*, 1971, **93**, 524.
[363] A. V. Bogatskii, A. E. Kozhukhova, and T. K. Chumachenko, *J. Gen. Chem. (U.S.S.R.)*, 1970, **40**, 1167.
[364] S. Satsumabayashi, K. Nakajo, R. Soneda, and S. Motoki, *Bull. Chem. Soc. Japan*, 1970, **43**, 1586.

one of the aldehyde components is used as its enol ether; the intermediate 1,3-dioxan (174) can be isolated (Scheme 103).

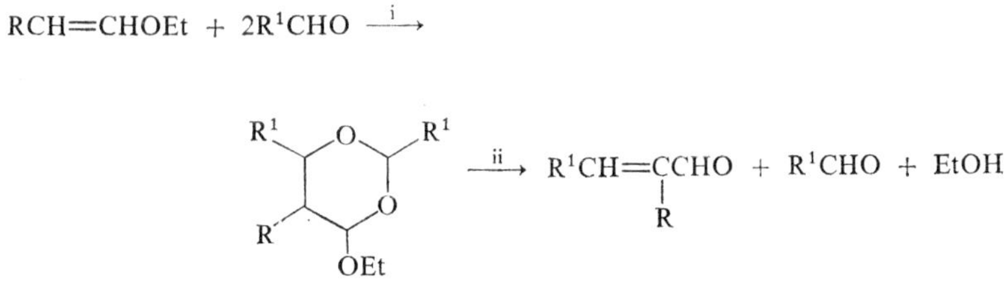

RCH=CHOEt + 2R¹CHO —ⁱ→

$$\underset{(174)}{\text{[dioxan with } R^1, R^1, R, \text{OEt substituents]}} \xrightarrow{ii} R^1CH=\underset{R}{\overset{|}{C}}CHO + R^1CHO + EtOH$$

Reagents: i, BF_3, Et_2O; ii, H_3O^+

Scheme 103

Kinetic trapping is reported[365] in the reaction of formaldehyde with unsymmetrical ketones in the presence of trifluoroacetic acid; condensation at the more substituted carbon atom predominates, the β-ketols formed being trifluoroacetylated more rapidly than they are produced. Controlled hydroxymethylation of ketones can be achieved[366] by reduction of the sodium salt of the corresponding hydroxymethylene derivative with aluminium hydride; direct base-catalysed condensation with formaldehyde normally gives products of polycondensation.

General.—A convenient method[367] for the regeneration of chlorotris(triphenylphosphine)rhodium(I) has been reported, enhancing the utility of aldehyde decarbonylation as a synthetic reaction. The mechanism of this process, which affords a high degree of retention of configuration, is now considered to involve a radical cage disproportionation followed by a product-forming cage recombination,[368] rather than the earlier proposed intermediacy of carbanions.[369]

A reagent suitable for the specific detection of aldehydes in concentrations as low as 10^{-4} mol l^{-1} is provided by the triazole (175). The immediate condensation product (176) is air-oxidized to the triazolotetrazine (177), whose

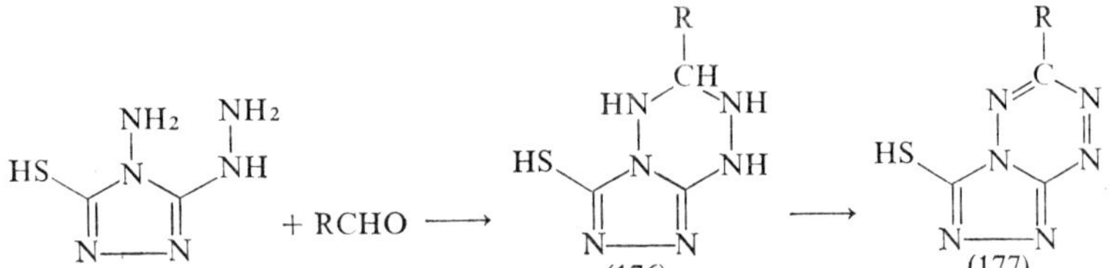

(175) + RCHO ⟶ (176) ⟶ (177)

[365] W. C. Lumma, jun. and O. H. Ma, *J. Org. Chem.*, 1970, **35**, 2391.
[366] E. J. Corey and D. E. Cane, *J. Org. Chem.*, 1971, **36**, 3070.
[367] R. W. Fries and J. K. Stille, *Synthesis Inorg. Metal-org. Chem.*, 1971, **1**, 295.
[368] H. M. Walborsky and L. E. Allen, *J. Amer. Chem. Soc.*, 1971, **93**, 5465.
[369] H. M. Walborsky and L. E. Allen, *Tetrahedron Letters*, 1970, 823; J. Tsuji and K. Ohno, *J. Amer. Chem. Soc.*, 1968, **90**, 99.

Functional Groups other than Acetylenes, Allenes, and Olefins 149

colour ranges from magenta with aliphatic to purple with aromatic aldehydes.[370]

An improved preparation of 1-deuteriated aldehydes of high isotopic purity is reported;[371] reaction of an aldehyde with triphenylphosphine dibromide in pyridine leads utimately to the salt (178), which is smoothly cleaved by deuterium oxide (Scheme 104).

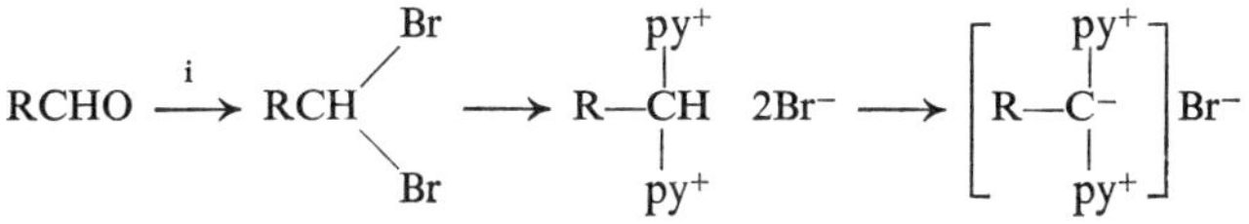

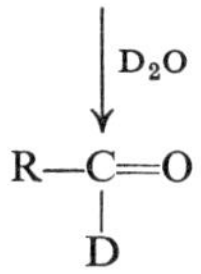

Reagents: i, Ph$_3$P, Br$_2$–py

Scheme 104

αβ-Acetylenic ketoximes undergo Beckmann rearrangement and/or fragmentation with phosphorous pentachloride in ether (Scheme 105); contrary to a previous report,[372] the products are isolable and readily characterized.[373] The immediate product of rearrangement, the acetylenic amide, is never isolated, rapidly adding hydrogen chloride to give (179).

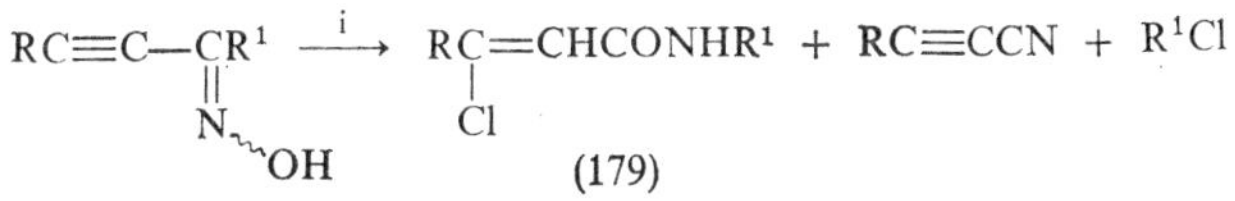

$$RC\equiv C-CR^1 \xrightarrow{i} RC=CHCONHR^1 + RC\equiv CCN + R^1Cl$$
$$\underset{OH}{\overset{N}{\|}} \quad \underset{Cl}{} \quad (179)$$

Reagents: i, PCl$_5$–Et$_2$O

Scheme 105

Non-enolizable α-diketones are cleaved to diesters with lead tetra-acetate.[374] The α-methylene group of highly enolized β-diketones is iodinated by periodic acid.[375]

[370] R. G. Dickinson and N. W. Jacobson, *Chem. Comm.*, 1970, 1719.
[371] T. Hase, *Acta Chem. Scand.*, 1970, **24**, 2263.
[372] S. Morrochi, A. Ricca, A. Zanarotti, G. Bianchi, R. Gandolfi, and P. Grunanger, *Tetrahedron Letters*, 1969, 3329.
[373] Z. Hamlet and M. Rampersad, *Chem. Comm.*, 1970, 1230.
[374] L. Canonica, B. Danieli, P. Manitto, and G. Russo, *Gazzetta*, 1970, **100**, 1026.
[375] A. J. Fatiadi, *Chem. Comm.*, 1970, 11.

Whereas the acid-catalysed rearrangement of acyloins normally gives isomeric α-hydroxycarbonyl products, the hydroxyaldehyde (180) rearranges to the αβ-unsaturated ketone (182) by some unspecified mechanism, though the expected product (181) is presumed to be an intermediate.[376]

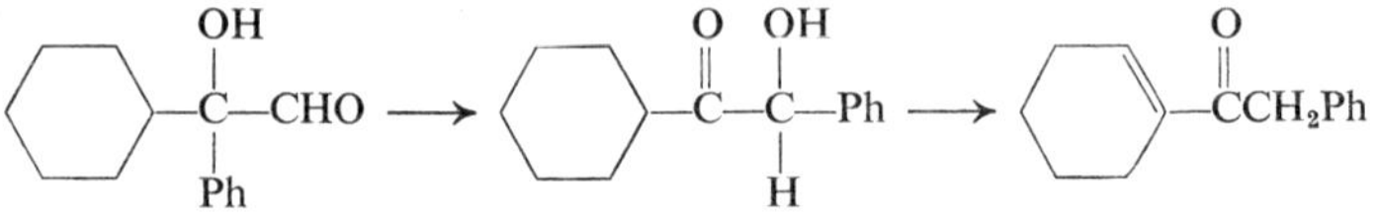

(180) → (181) → (182)

Glyoxal monohydrate seems[377] to have the dimeric structure (183), but in 40% aqueous solution the presence of the hydrated monomer (184) and two isomeric dioxolans, (185) and (186), is claimed.[378]

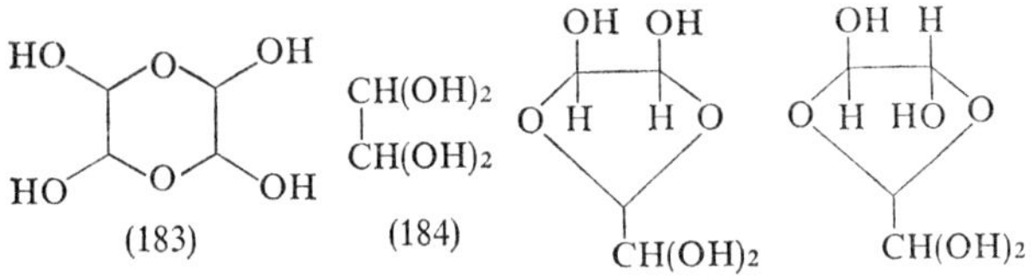

(183)   (184)   (185)   (186)

7 Amines

Preparation.—Alkylation of guanidine with alkyl halides followed by basic hydrolysis affords a new method for the preparation of pure primary amines; primary alkyl halides give the most satisfactory results.[379]

A method conceptually similar to the Gabriel synthesis is described for the preparation of primary amines; the reagent in this case[380] is the lithium salt of bisbenzenesulphenimide (187). An advantage over the Gabriel synthesis is the ease of hydrolysis of the N-alkyl product, permitting use of polyfunctional alkylating agents such as halogenoesters or halogenonitriles; the lithium salt can effect conjugate addition to αβ-unsaturated substrates, affording β-amino-derivatives. These transformations are outlined in Scheme 106.

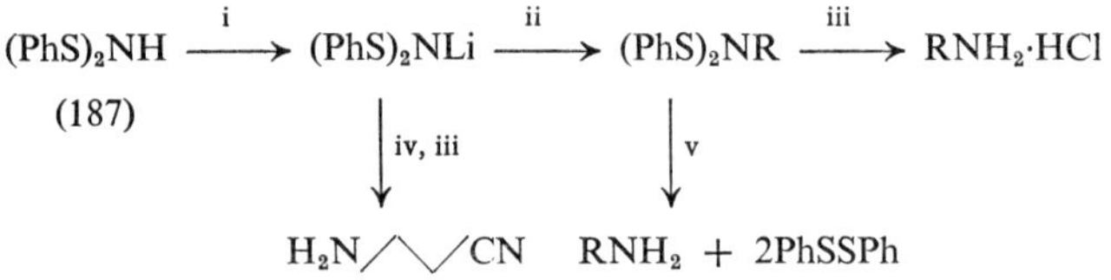

$$(PhS)_2NH \xrightarrow{i} (PhS)_2NLi \xrightarrow{ii} (PhS)_2NR \xrightarrow{iii} RNH_2 \cdot HCl$$
(187)

$$\downarrow iv, iii \qquad \downarrow v$$

$$H_2N\diagup\diagdown CN \qquad RNH_2 + 2PhSSPh$$

Reagents: i, Bu^{n}Li; ii, ROTos; iii, 3N-HCl; iv, H$_2$C=CHCN; v, 2PhSH

Scheme 106

[376] T. D. Inch, P. Watts, and N. Williams, *Chem. Comm.*, 1971, 174.
[377] G. C. S. Collins and W. O. George, *Chem. Comm.*, 1971, 501.
[378] E. B. Whipple, *J. Amer. Chem. Soc.*, 1970, **92**, 7183.
[379] P. Hebrard and M. Olmucki, *Bull. Soc. chim. France*, 1970, 1938.
[380] T. Mukaiyama and T. Taguchi, *Tetrahedron Letters*, 1970, 3411.

Functional Groups other than Acetylenes, Allenes, and Olefins

Triphenylphosphine imines, such as (188), are readily alkylated, the salt decomposing on hydrolysis to the secondary amine hydrochloride, as shown in Scheme 107; alkylating agents are limited to methyl and ethyl halides, as higher alkyl halides undergo preferential elimination.[381]

$$Ph_3\overset{+}{P}-\overset{-}{NR} \xrightarrow{R^1X} Ph_3\overset{+}{P}-NRR^1X^- \xrightarrow{H_2O} RR^1NH,HX + Ph_3PO$$
(188)

Scheme 107

The reduction of carboxylic acids by lithium in methylamine can be stopped at the imine stage;[210] the imines can be further reduced to secondary amines. Diamines are available from 2- or 3-bromoamines *via* the corresponding phosphoramides (Scheme 108);[382] the flexibility of this route has been enhanced by the demonstrated potential[383] of the intermediate β-aminophosphoramides to undergo alkylation.

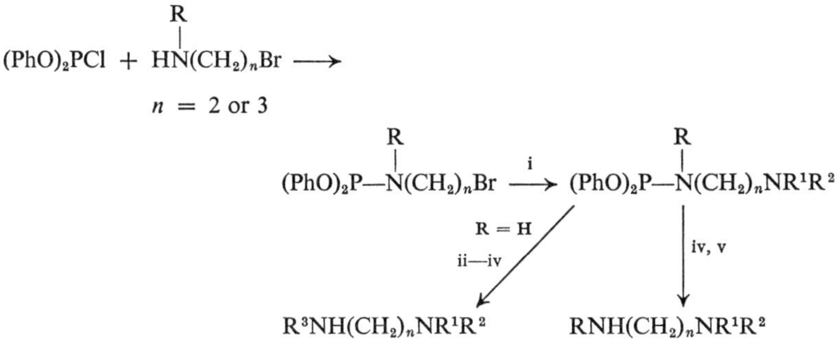

$$(PhO)_2PCl + H\overset{R}{\underset{|}{N}}(CH_2)_nBr \longrightarrow$$

$$n = 2 \text{ or } 3$$

$$(PhO)_2P-\overset{R}{\underset{|}{N}}(CH_2)_nBr \xrightarrow{i} (PhO)_2P-\overset{R}{\underset{|}{N}}(CH_2)_nNR^1R^2$$

$$\begin{array}{cc} R = H & \\ \text{ii—iv} \swarrow & \searrow \text{iv, v} \end{array}$$

$$R^3NH(CH_2)_nNR^1R^2 \qquad RNH(CH_2)_nNR^1R^2$$

Reagents: i, R^1R^2NH; ii, NaH; iii, R^3X; iv, H_3O^+; v, OH^-

Scheme 108

Sulphurated borohydrides[384] can reduce oximes to amines; the intermediate hydroxylamines are isolable. Sodium borohydride reduces nitriles to amines in high yield in the presence of Raney nickel as catalyst.[385] The titanocene-promoted fixation–reduction of molecular nitrogen[386] has been utilized in the conversion of ketones into amides and acid chlorides into nitriles[387] in an overall reductive deoxygenation, as exemplified in Scheme 109.

A French group[388] have reported that activation of alcohols as their alkoxy-trisdimethylaminophosphonium perchlorates permits subsequent

---

[381] H. Zimmer, M. Jayawant, and P. Gutsch, *J. Org. Chem.*, 1970, **35**, 2826.
[382] M. Dreux, P. Savignac, J. Chenault, and G. Plé, *Tetrahedron Letters*, 1971, 1557.
[383] M. Dreux, P. Savignac, and J. Chenault, *Tetrahedron Letters*, 1971, 4109.
[384] J. M. Lalancette and J. R. Brindle, *Canad. J. Chem.*, 1970, **48**, 735.
[385] R. A. Egli, *Helv. Chim. Acta*, 1970, **53**, 47.
[386] E. E. van Tamelen, D. Seeley, S. Schneller, H. Rudler, and W. Cretney, *J. Amer. Chem. Soc.*, 1970, **92**, 5253; E. E. van Tamelen, *Accounts Chem. Res.*, 1970, **3**, 361.
[387] E. E. van Tamelen and H. Rudler, *J. Amer. Chem. Soc.*, 1970, **92**, 5253.
[388] B. Castro and C. Selve, *Bull. Soc. chim. France*, 1971, 4368.

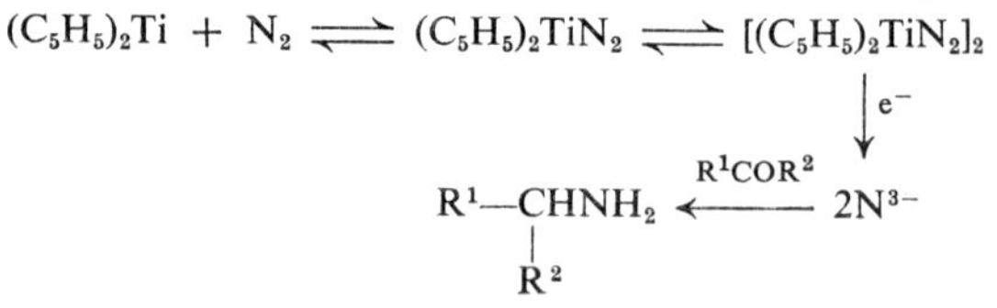

$$(C_5H_5)_2Ti + N_2 \rightleftharpoons (C_5H_5)_2TiN_2 \rightleftharpoons [(C_5H_5)_2TiN_2]_2$$

$$\downarrow e^-$$

$$R^1\text{—CHNH}_2 \xleftarrow{R^1COR^2} 2N^{3-}$$
$$\underset{R^2}{|}$$

Scheme 109

reaction with primary and secondary amines, giving products of exclusive monoalkylation; the intermediate perchlorate (189) is isolated prior to reaction with the amine in DMF.

$$\overset{+}{ROP}(NMe_2)_3 ClO_4^-$$
(189)

Organic azides react with unhindered organoboranes to afford secondary amines; nitrenes are not involved as intermediates, the postulated pathway[389] being as shown (Scheme 110).

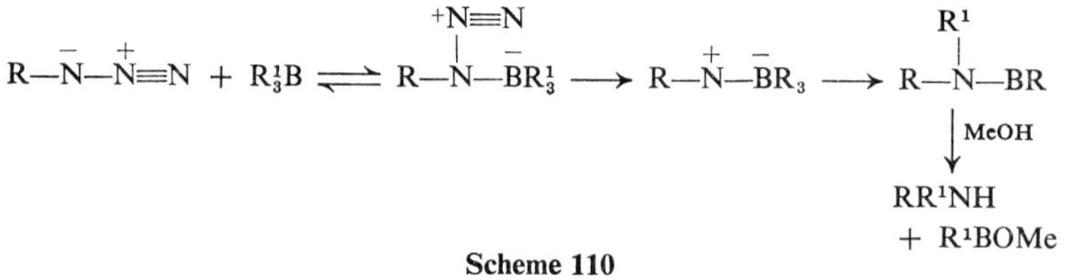

$$R\overset{-}{-}\overset{+}{N}\text{—}N{\equiv}N + R^1_3B \rightleftharpoons R\overset{+\,N{\equiv}N}{\underset{|}{-}N}\overset{-}{-}BR^1_3 \longrightarrow R\overset{+}{-}\overset{-}{N}\text{—}BR_3 \longrightarrow R\overset{R^1}{\underset{|}{-}N}\text{—}BR$$

$$\downarrow MeOH$$

$$RR^1NH + R^1BOMe$$

Scheme 110

Tertiary amines are degraded to secondary amines[390] by a combination of an organic hydroperoxide and nitrous acid; the reagent fulfilling both requirements, 2-nitroprop-2-yl hydroperoxide, is prepared by catalysed oxygenation of 2-nitropropane. Reaction with tertiary amines proceeds as shown in Scheme 111, the secondary amine produced being trapped as its N-nitroso-derivative by the generated nitrous acid.

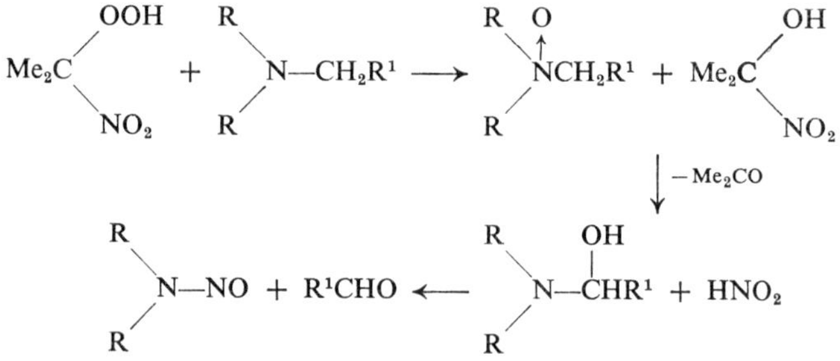

$$Me_2C\underset{NO_2}{\overset{OOH}{\diagup}} + \underset{R}{\overset{R}{\diagdown}}N\text{—}CH_2R^1 \longrightarrow \underset{R}{\overset{R}{\diagdown}}\overset{O\uparrow}{N}CH_2R^1 + Me_2C\underset{NO_2}{\overset{OH}{\diagup}}$$

$$\downarrow -Me_2CO$$

$$\underset{R}{\overset{R}{\diagdown}}N\text{—}NO + R^1CHO \longleftarrow \underset{R}{\overset{R}{\diagdown}}N\text{—}\overset{OH}{\underset{|}{C}}HR^1 + HNO_2$$

Scheme 111

[389] A. Suzuki, S. Sono, M. Itoh, H. C. Brown, and M. M. Midland, *J. Amer. Chem. Soc.*, 1971, 4329.
[390] B. Franck, J. Conrad, and P. Misbach, *Angew. Chem. Internat. Edn.*, 1970, **9**, 892.

Functional Groups other than Acetylenes, Allenes, and Olefins

In the mercuration-reductive demercuration of enamines, mercuration at carbon is the dominant reaction pathway when ionic mercury(II) salts are employed; this sequence[391] constitutes a general high-yield synthesis of tertiary amines (Scheme 112).

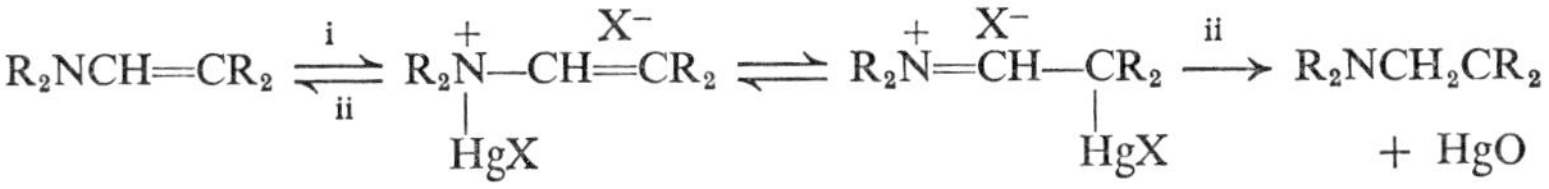

$$R_2NCH{=}CR_2 \underset{ii}{\overset{i}{\rightleftharpoons}} \underset{\underset{HgX}{|}}{R_2\overset{+}{N}{-}CH{=}CR_2}\;X^- \rightleftharpoons \underset{\underset{HgX}{|}}{R_2\overset{+}{N}{=}CH{-}CR_2}\;X^- \overset{ii}{\longrightarrow} R_2NCH_2CR_2 + HgO$$

Reagents: i, HgX_2; ii, $NaBH_4$

Scheme 112

Primary amines react with two equivalents of an alkyl-lithium to give lithio-imines, such as (190), which can be quenched to afford ketones or allowed to react with a further equivalent of alkyl-lithium to yield a more highly substituted primary amine (Scheme 113).[392]

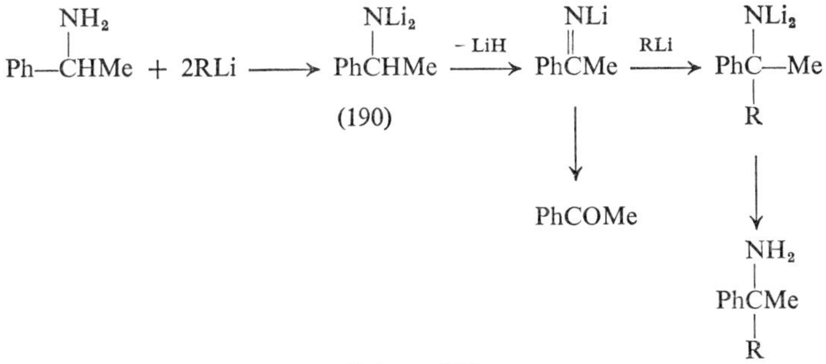

$$\underset{\underset{CHMe}{|}}{Ph}{-}NH_2 + 2RLi \longrightarrow \underset{\underset{(190)}{}}{PhCHMe\;|\;NLi_2} \overset{-LiH}{\longrightarrow} PhCMe\;\|\;NLi \overset{RLi}{\longrightarrow} \underset{\underset{R}{|}}{PhC{-}Me\;|\;NLi_2}$$

PhCOMe

$$\underset{\underset{R}{|}}{PhCMe\;|\;NH_2}$$

Scheme 113

Sommer[393] has described a general exhaustive alkylation procedure for the direct synthesis of quaternary ammonium compounds. As protonation of sterically hindered amines is only slightly affected by steric hindrance, whereas kinetic nucleophilicity is greatly decreased, the use of a sterically hindered base of greater base strength than the substrate amine will bind up the acid generated on alkylation; this system allows quaternization under mild conditions, with the minimum of manipulation.

The phenacylsulphonyl group affords[394] protection to amines; regeneration is effected by zinc in acetic acid: the protected amine can also be cleanly monoalkylated (Scheme 114).

$$RNHSO_2CH_2COPh \overset{i,\;ii}{\longrightarrow} RNHR^1$$

Reagents: i, $R^1X{-}K_2CO_3{-}Me_2CO$; ii, Zn–AcOH

Scheme 114

[391] R. D. Bach and D. K. Mitra, *Chem. Comm.*, 1971, 1433.
[392] H. G. Richey, jun., W. F. Erickson, and A. S. Heyn, *Tetrahedron Letters*, 1971, 2183, 2187.
[393] H. Z. Sommer, H. I. Lipp, and L. L. Jackson, *J. Org. Chem.*, 1971, 36, 824; H. Z. Sommer and L. L. Jackson, *J. Org. Chem.*, 1970, 35, 1558.
[394] J. B. Hendrickson and R. Bergeron, *Tetrahedron Letters*, 1970, 345.

A surprisingly high degree of asymmetric induction is observed[395] in the nitrogen-hydrogen insertion reaction of methyl alkoxycarbonyl carbene with optically active 1-phenylethylamine; after hydrogenolysis of the benzyl grouping and hydrolysis, alanine is obtained in good chemical yield with an optical purity of up to 20% (Scheme 115). The primary factor determining the configuration of the alanine produced is the chirality of the amine; the chirality of the ester seems to be of only secondary importance.

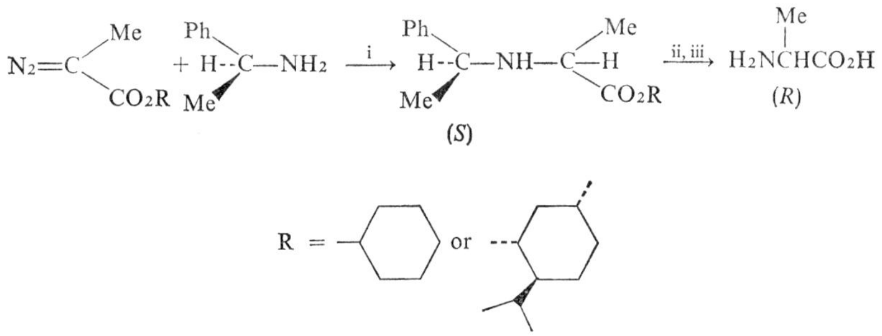

Scheme 115

Reagents: i, CuCN; ii, H_2–Pd; iii, OH^-

Full details have been published[327b] on the reduction of imines to optically active amines with chiral borohydride reagents. The homogeneous hydrogenation of azo-, imino-, and nitro-groups to amine derivatives is reported;[396] a rhodium salt–sodium borohydride system is used.

Tsuji[397] has described the preparation of long-chain amines by the palladium-catalysed telomerization of butadiene; with ammonia, tertiary amines are formed, whereas nitroalkanes undergo replacement of the α-hydrogen atoms with octa-2,7-dienyl groups. Nickel(0) complexes catalyse a similar telomerization.[398]

Olefins react with secondary amines, carbon monoxide, and water in the presence of rhodium and iron catalysts to give alkylamines by a process of aminomethylation (Scheme 116).[399]

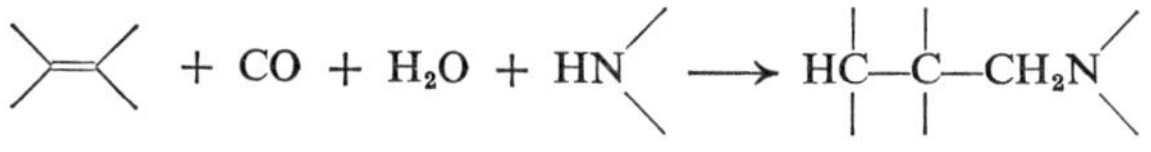

Scheme 116

[395] J.-F. Nicoud and H. B. Kagan, *Tetrahedron Letters*, 1971, 2065.
[396] I. Jardine and F. J. McQuillin, *Chem. Comm.*, 1970, 626.
[397] T. Mitsuyasu, M. Hara, and J. Tsuji, *Chem. Comm.*, 1971, 345.
[398] R. Baker, D. E. Halliday, and T. N. Smith, *Chem. Comm.*, 1971, 1583.
[399] A. F. M. Iqbal, *Helv. Chim. Acta*, 1971, **54**, 1440.

Functional Groups other than Acetylenes, Allenes, and Olefins

The oxidative addition of azide ion to olefins affords a simple route[400] to diamines (Scheme 117); styrene gives a dimeric product (191). When

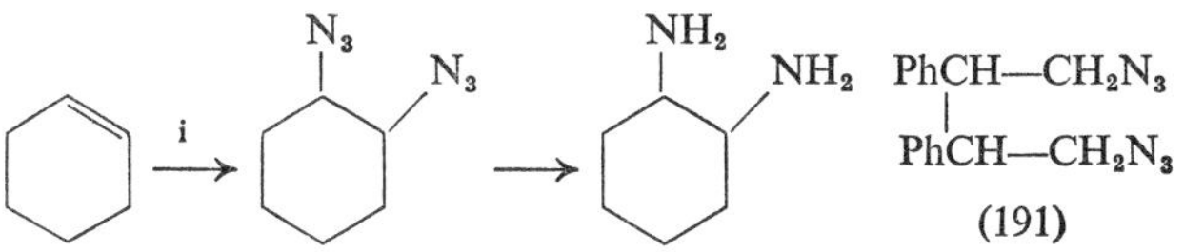

cyclohexene →i→ cyclohexane-N₃,N₃ → cyclohexane-NH₂,NH₂   PhCH—CH₂N₃
                                                                                                      PhCH—CH₂N₃
                                                                                                        (191)

Reagents: i, NaN₃–AcOH

Scheme 117

ozonides are catalytically reduced in the presence of amines, more highly substituted amines are formed by condensation with the intermediate aldehyde.[401]

Reactions.—An observation[402] that dinitrogen tetroxide oxidizes amines to the corresponding alkyl nitrates in low yield has been developed into a useful synthetic method.[403] Conversion of an amine into its trimethylsilyl derivative alters the course of the oxidation permitting isolation of the desired nitrate in high yield (Scheme 118); in a later communication, the use of THF as solvent is reported[404] to obviate the need for prior trimethylsilylation. This reaction seems to be general for primary and secondary amines and, although the mechanism of the reaction is unknown, retention of configuration at carbon is observed; the intermediacy of the solvated diazonium salt (192) is invoked when THF is solvent.

$$RNHSiMe_3 + N_2O_4 \longrightarrow RONO_2 + N_2 + (Me_3Si)_2O$$
$$R-\overset{+}{N}\equiv N \;\; \overset{-}{O}NO_2$$
$$(192)$$

Scheme 118

Carbonyl compounds are produced in the homogeneous oxidation of amines by complexed molybdenum salts.[405] Disubstituted cyanamides react with hydroxylamine to give, depending on the reaction conditions, hydroxyguanidines or amino-oxyformamidines.[406] Triethylamine is quaternized by dichloromethane to afford chloromethyltriethylammonium chloride in a state of very high purity.[407]

---

[400] H. Schäfer, *Angew. Chem. Internat. Edn.*, 1970, **9**, 158.
[401] R. W. White, S. W. King, and J. L. O'Brien, *Tetrahedron Letters*, 1971, 3591.
[402] E. H. White and W. R. Feldman, *J. Amer. Chem. Soc.*, 1957, **79**, 5832.
[403] F. Wudhl and T. B. K. Lee, *Chem. Comm.*, 1970, 490.
[404] F. Wudhl and T. B. K. Lee, *J. Amer. Chem. Soc.*, 1971, **93**, 271.
[405] F. W. S. Benfield and M. L. H. Green, *Chem. Comm.*, 1971, 1274.
[406] C. Bełżecki, B. Hintze, and S. Kwiatkowska, *Chem. Comm.*, 1970, 806.
[407] D. A. Wright and C. A. Wulff, *J. Org. Chem.*, 1970, **35**, 4252.

8 Alkyl Halides

Preparation.—Halogenonium ion oxidation of alcohol carbazates (193) affords a new method for the formation of alkyl halides from alcohols under neutral conditions;[408] this procedure (Scheme 119) provides an efficient means of conversion of bridgehead alcohols, such as adamantan-1-ol, into the corresponding halides.

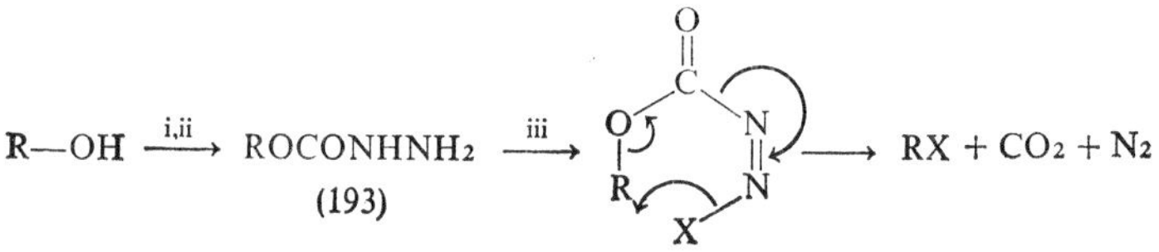

$$R-OH \xrightarrow{i,ii} ROCONHNH_2 \xrightarrow{iii} [\text{cyclic intermediate}] \longrightarrow RX + CO_2 + N_2$$
(193)

Reagents: i, $COCl_2$; ii, H_2NNH_2; iii, X^+

Scheme 119

The Syntex group[409] have given a full description of the efficacy of methyltriphenoxyphosphonium iodide for selective iodination of the primary alcohol group of pyrimidine nucleosides. Cyanuric chloride (194) is offered[410] as a laboratory reagent for the conversion of alcohols into halides; the reagent is most effective with primary alcohols, no structural isomerization being observed and no external base being necessary (Scheme 120).

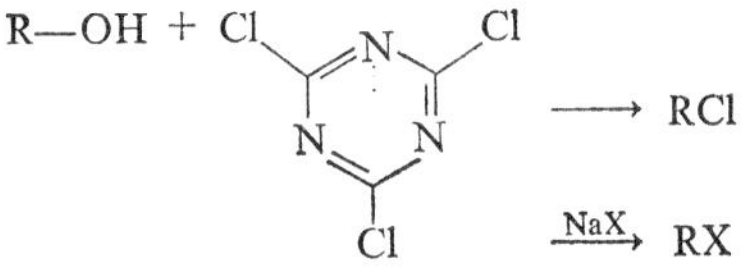

$$R-OH + \text{cyanuric chloride (194)} \longrightarrow RCl \xrightarrow{NaX} RX$$

Scheme 120

The free-radical oxidation of organoboranes with bromine in the absence of light[411] leads to the production of alkyl bromides (Scheme 121).

$$R_2B-CH_2R^1 + Br_2 \xrightarrow{fast} R_2B-\underset{\underset{Br}{|}}{C}HR^1 + HBr \xrightarrow{slow} R_2BBr + \underset{\underset{Br}{|}}{C}H_2-R^1$$

Scheme 121

Silver difluorochloroacetate[412] smoothly converts bromo-compounds into the chloro-analogues; for example, n-octyl bromide is converted into

---

[408] D. L. J. Clive and C. V. Denyer, *Chem. Comm.*, 1971, 1112.
[409] J. P. Verheyden and J. G. Moffat, *J. Org. Chem.*, 1970, **35**, 2319.
[410] S. R. Sandler, *J. Org. Chem.*, 1970, **35**, 3967; *Chem. and Ind.*, 1971, 1416.
[411] H. C. Brown and C. F. Lane, *J. Amer. Chem. Soc.*, 1970, **92**, 7212.
[412] J. A. Vida, *Tetrahedron Letters*, 1970, 3447.

Functional Groups other than Acetylenes, Allenes, and Olefins

n-octyl chloride in 95% yield: this reaction seems to be quite general, secondary and bridgehead tertiary bromides giving equally satisfactory results. Photolysis of lead(IV) carboxylates in carbon tetrachloride affords alkyl chlorides with one less carbon atom in synthetically acceptable yields.[413] Chloroiodomethane, useful in the Simmons–Smith methylene transfer reaction, can be prepared conveniently on a large scale by reaction of dichloromethane with sodium iodide in DMF.[414] Optically active 2-bromo-3-methylbutane has been obtained in a pure state for the first time by bromination of resolved 3-methylbutan-2-ol with triphenylphosphine dibromide.[415]

A further method[416] is described for the conversion of allylic alcohols into chlorides without rearrangement; methanesulphonyl chloride reacts with allylic alcohols at 0 °C in the presence of lithium chloride, DMF, and collidine to give pure, unrearranged allylic chlorides: non-allylic alcohols are inert to such conditions. This technique has been used by Meyers[262] in a synthesis of an insect pheromone.

Vicinal chloroiodoalkanes are readily obtained by reaction of olefins with copper(II) chloride and iodine;[417] conjugated dienes give dichlorides by halide exchange with the initially formed chloroiodides. The need for molecular halogen is obviated when strongly complexed copper(II) halides are used.[418] Triphenylphosphine dihalides convert epoxides into vicinal dihalides;[419] initial cleavage of the carbon–oxygen bond emanating from the more substituted carbon gives the halogenoalkoxyphosphonium salt (195), which undergoes a subsequent Arbusov reaction to give the dihalide (Scheme 122).

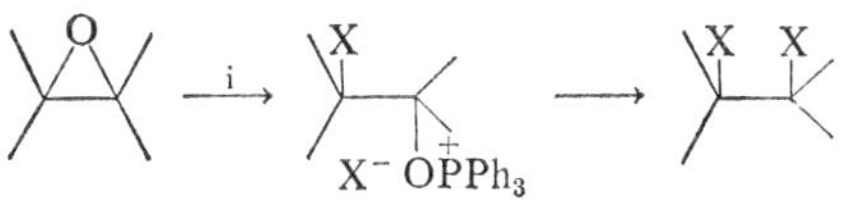

Reagent: i, Ph_3PX_2

Scheme 122

Reactions.—DMF is recommended[420] as solvent for the sodium borohydride reduction of primary and secondary alkyl halides to the corresponding hydrocarbons; this reaction, which proceeds at room temperature, shows the kinetic behaviour of an S_N2 process. The intermediacy of organoboranes in the analogous reduction of tertiary halides has been demonstrated;[421]

---

[413] V. Franzen and R. Edens, *Annalen*, 1970, **735**, 47.
[414] S. Miyano and H. Hashimoto, *Bull. Chem. Soc. Japan*, 1971, **44**, 2864.
[415] R. A. Arain and M. K. Hargreaves, *J. Chem. Soc.* (C), 1970, 67.
[416] E. W. Collington and A. I. Meyers, *J. Org. Chem.*, 1971, **36**, 3044; *cf.* G. Stork, M. Gregson, and P. A. Grieco, *Tetrahedron Letters*, 1969, 1393.
[417] W. C. Baird, jun., J. H. Surridge, and M. Buza, *J. Org. Chem.*, 1971, **36**, 2088.
[418] W. C. Baird, jun., J. H. Surridge, and M. Buza, *J. Org. Chem.*, 1971, **36**, 3324.
[419] A. N. Thakore, P. Pope, and A. C. Oehlschlager, *Tetrahedron*, 1971, **27**, 2617.
[420] M. Vol'pin, M. Dvolaitsky, and I. Levitin, *Bull. Soc. chim. France*, 1970, 1526.
[421] J. Jacobus, *Chem. Comm.*, 1970, 338.

attempted distillation of the crude product gave olefins, but prior addition of valeric acid allowed isolation of the desired hydrocarbons. An elimination–addition sequence is proposed for this reaction, which has been developed[422] as a convenient, one-step procedure for the direct reduction of tertiary halides; optically active substrates afford racemic products. Kochi[423] has investigated in detail the mechanism of the quantitative reduction of alkyl halides to hydrocarbons by chromium(II) complexes (Scheme 123); the intermediacy of a hydrolytically unstable alkyl ethylenediamine chromium(III) species is proposed.

$$R—X + 2Cr^{II}(en)_2^{2+} + H_2O \longrightarrow RH + Cr^{III}(en)_2(OH)^{2+} + Cr^{III}(en)_2(X)^{2+}$$

Scheme 123

The direct oxidation of alkyl halides to carbonyl compounds by DMSO occurs in good yield under mild conditions in the presence of silver ion,[424] which assists formation of the alkoxysulphonium intermediate; silver perchlorate is the salt of choice, other salts competing nucleophilically with DMSO for the substrate. Oxidation of primary halides stops cleanly at the aldehyde stage, except when an excess of silver salt is present. As has been already described,[211] sodium tetracarbonylferrate converts alkyl bromides into the homologous aldehydes; this conversion is most successful with primary halides, n-decanal being obtained in 77% yield from n-nonyl bromide.

Brown[425] has reported a facile coupling reaction of benzylic and allylic iodides. Trialkylboranes react with allylic or benzylic iodides in the presence of a limited amount of oxygen to give products of iodide transfer; when oxygen is present in excess, coupled products are observed (Scheme 124).

$$R_3B + R^1I \xrightarrow{\text{limited } O_2} R—I$$
$$Et_3B + R^1I \xrightarrow{\text{excess } O_2} R^1—R^1$$
$$R^1 = \text{allyl or benzyl}$$

Scheme 124

The coupling of Grignard reagents and alkyl halides is dramatically enhanced when the latter possesses proximate functionality capable of complexation, and hence of transition-state stabilization; whereas neither n-pentyl nor ethoxyethyl bromide react with iodomagnesium phenylacetylide, the bromide (196) does.[426]

Although vicinal dihalides normally react with triethyl phosphite in an Arbusov manner to give substituted phosphonates, dehydrohalogenation is

$$MeOCH_2CH_2OCH_2CH_2Br$$

(196)

---

[422] R. O. Hutchins, R. J. Bertsch, and D. Hoke, *J. Org. Chem.*, 1971, **36**, 1568.
[423] J. K. Kochi and J. W. Powers, *J. Amer. Chem. Soc.*, 1970, **92**, 137.
[424] W. W. Epstein and J. Ollinger, *Chem. Comm.*, 1970, 1338.
[425] A. Suzuki, S. Nozawa, M. Harada, M. Itoh, H. C. Brown, and M. M. Midland, *J. Amer. Chem. Soc.*, 1971, **93**, 1508.
[426] G. W. Cooper and R. P. Houghton, *Tetrahedron Letters*, 1970, 3915.

Functional Groups other than Acetylenes, Allenes, and Olefins

the principal reaction when electron-withdrawing groups are present, as in the conversion of 2,3-dibromopropionitrile into acrylonitrile.[427]

Primary alkyl halides are dehydrohalogenated in hexamethylphosphoramide at 200 °C; with some exceptions, unrearranged alk-1-enes are obtained in good yield.[428]

Sodium cyanoborohydride in hexamethylphosphoramide selectively reduces alkyl iodides, bromides, and toluene-p-sulphonates to the corresponding hydrocarbons; aldehydes, ketones, *etc.*, are not attacked. This allows the direct conversion of an alcohol into the corresponding hydrocarbon (Scheme 125) by hydride attack on the intermediate phosphite salt[324] (197).

$$RCH_2OH + (PhO)_3\overset{+}{P}\text{—MeI}^- \longrightarrow (PhO)_2\overset{+}{P}\text{—MeI}^- \overset{i}{\longrightarrow} RMe$$
$$\underset{(197)}{|\ OCH_2R}$$

Reagent: i, NaCNBH$_3$

Scheme 125

9 Alcohols

Preparation.—Variations continue to appear on the theme of alcohol production by hydroboration–oxidation of olefins. B-Methoxydialkylboranes[429] react with olefins in the presence of lithium aluminium hydride to afford a new route to trialkylboranes and thence, by carbonylation–oxidation, to trialkylcarbinols. Carbonylation with carbon monoxide is avoided in a new procedure;[225] in the presence of an excess of trifluoroacetic anhydride, trialkyl cyanoborates undergo a triple alkyl migration from boron to carbon to give, on oxidation, high yields of trialkylcarbinols (Scheme 126). Tri-

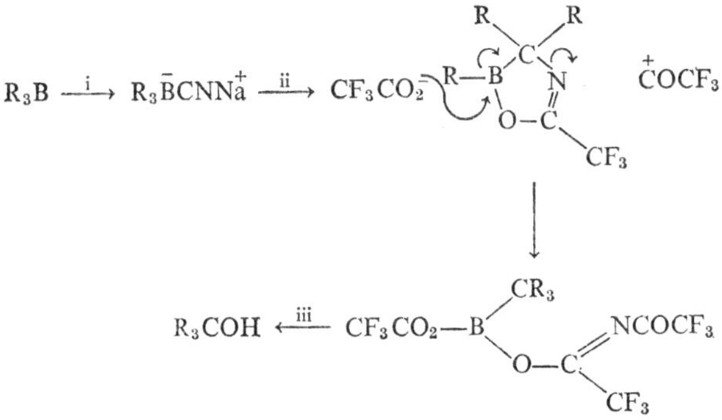

$$R_3B \overset{i}{\longrightarrow} R_3\bar{B}CNN\overset{+}{a} \overset{ii}{\longrightarrow} CF_3CO_2^- \cdots \overset{+}{COCF_3}$$

$$\downarrow$$

$$R_3COH \overset{iii}{\longleftarrow} CF_3CO_2\text{—B}\cdots$$

Reagents: i, NaCN; ii, (CF$_3$CO)$_2$O; iii, O

Scheme 126

[427] J. P. Schroeder, L. B. Tew, and V. M. Peters, *J. Org. Chem.*, 1970, **35**, 3181.
[428] R. S. Monson, *Chem. Comm.*, 1971, 113.
[429] H. C. Brown, E. Negishi, and S. K. Cupta, *J. Amer. Chem. Soc.*, 1970, **92**, 6648.

alkylboranes react with trihalogenomethanes[430] in the presence of alkoxide ions, affording a further entry into trialkylcarbinols (Scheme 127).

$$R_3B + HCClF_2 \xrightarrow{i, ii} R_3COH$$

Reagents: i, LiOCEt$_3$–THF; ii, O

Scheme 127

Highly substituted alcohols are formed in the light-induced reaction of trialkylboranes with bromine in the presence of water.[431] In common with the dark reaction,[411] the initial process is one of α-bromination; in the present case, migration from boron to carbon then occurs, both or only one of the alkyl groups migrating depending on whether the alkyl groups are primary or secondary, respectively (Scheme 128).

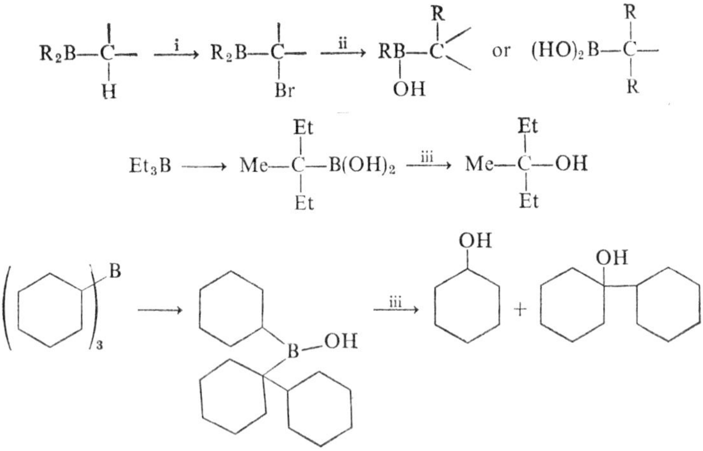

$$R_2B-\underset{H}{\overset{R}{C}}- \xrightarrow{i} R_2B-\underset{Br}{\overset{R}{C}}- \xrightarrow{ii} RB-\underset{OH}{\overset{R}{C}}\diagdown \text{ or } (HO)_2B-\underset{R}{\overset{R}{C}}-$$

$$Et_3B \longrightarrow Me-\underset{Et}{\overset{Et}{C}}-B(OH)_2 \xrightarrow{iii} Me-\underset{Et}{\overset{Et}{C}}-OH$$

Reagents: i, Br$_2$; ii, H$_2$O; iii, O

Scheme 128

A method of four-carbon homologation is described[432] in the oxygen-initiated free-radical reaction of trialkylboranes with butadiene monoxide; 4-alkylbut-2-en-1-ols are produced, the *trans*-olefin isomer predominating, as shown in Scheme 129.

$$R_3B + CH_2=CH-CH\underset{O}{\diagdown\diagup}CH_2 \xrightarrow{O_2-H_2O} RCH_2CH=CHCH_2OH + R_2BOH$$

Scheme 129

---

[430] H. C. Brown, B. A. Carlson, and R. H. Prager, *J. Amer. Chem. Soc.*, 1971, **93**, 2070.
[431] H. C. Brown and C. F. Lane, *J. Amer. Chem. Soc.*, 1971, **93**, 1025.
[432] A. Suzuki, N. Miyaura, M. Itoh, H. C. Brown, G. W. Holland, and E. Negishi, *J. Amer. Chem. Soc.*, 1971, **93**, 2792.

Functional Groups other than Acetylenes, Allenes, and Olefins

4,4-Dialkyl *cis*-but-2-en-1,4-diols are obtained in good yield by reaction of α-lithiofuran with trialkylboranes[433] (Scheme 130).

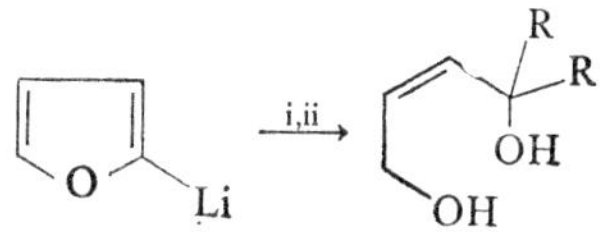

Reagents: i, R_3B; ii, O

Scheme 130

A very mild means of oxidation of trialkylboranes to the corresponding alcohols is provided in the stoicheiometrically controlled reaction with oxygen; at least a portion of the reaction is free-radical in nature, iodine acting as an inhibitor: this leads to some loss of stereospecificity in cases leading to chiral alcohols.[434] If the oxidation is performed at low temperature in dilute solution, alkyl hydroperoxides are obtained.[435]

The oxymercuration of olefins with chiral mercury(II) carboxylates affords alcohols with up to 15% optical purity; (+)-tartrate salts are more effective than the (+)-lactate analogues.[436]

A stereospecific synthesis of 1,2-diols is reported;[437] advantage is taken of the clean inversion occurring on acid-catalysed opening of an epoxide with DMSO. *o*-Sulphoperbenzoic acid (198), a water-soluble per-acid, converts olefins into *trans*-1,2-diols by spontaneous acid-catalysed cleavage of the intermediate epoxide; no esters are observed.[438] The epoxide can be isolated, if desired, by employment of a buffer. This per-acid also effects Baeyer–Villiger rearrangement and oxidizes amines to *N*-oxides.

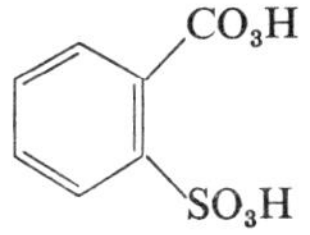

(198)

A novel method[439] for the degradation of primary alcohols to their next lower homologue is provided by acid-catalysed rearrangement of the derived hydroperoxide, *via* alkyl migration to electron-deficient oxygen. Tertiary alcohol hydroperoxides can be generated *in situ*, resulting in rearrangement to ω-hydroxyketones. Both of these transformations occur in synthetically

---
[433] A. Suzuki, N. Miyaura, and M. Itoh, *Tetrahedron*, 1971, **27**, 2775.
[434] H. C. Brown, M. M. Midland, and G. W. Kabalka, *J. Amer. Chem. Soc.*, 1971, **93**, 1024.
[435] H. C. Brown and M. M. Midland, *J. Amer. Chem. Soc.*, 1971, **93**, 4078.
[436] R. M. Carlson and A. H. Funk, *Tetrahedron Letters*, 1971, 3661.
[437] M. A. Khuddus and D. Swern, *Tetrahedron Letters*, 1971, 411.
[438] J. M. Bachhawat and N. K. Mathur, *Tetrahedron Letters*, 1971, 691.
[439] N. C. Deno, W. E. Billups, K. E. Kramer, and R. R. Lastomirsky, *J. Org. Chem.*, 1970, **35**, 3080.

useful yields. Secondary alcohols undergo cleavage. These conversions are presented in Scheme 131.

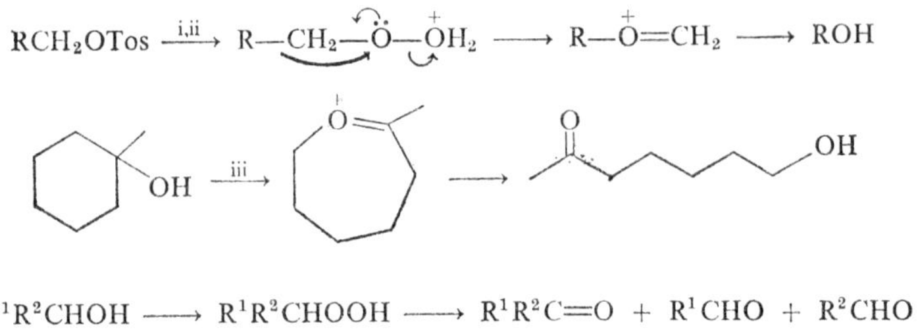

Reagents: i, H_2O_2–NaOH; ii, H_3O^+; iii, H_2SO_5–H_2SO_4

Scheme 131

The nucleophilic ring-opening of saturated and allylic epoxides by organometallic reagents has been studied in some considerable detail.[440] Dialkylcopperlithium reagents attack epoxides to give the expected, more highly substituted alcohol as major product;[441] no reaction was observed with cyclohexene oxide. Allylic epoxides are attacked by such reagents to give products of conjugate addition, an observation[442] extended into a useful method for direct stereoselective addition of an isoprenyl unit, to form a substituted allylic alcohol (Scheme 132).

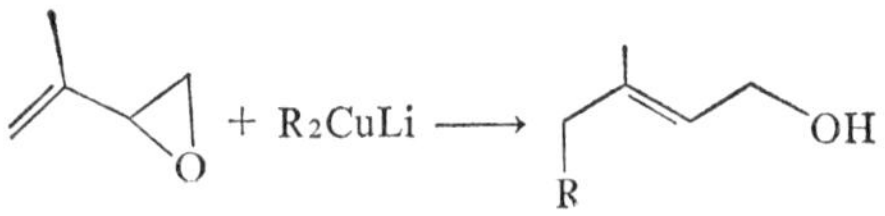

Scheme 132

Vinyl alanes react with carbonyl compounds to afford exclusively *trans*-allylic alcohols;[443] contrary to an earlier report,[444] prior conversion of the alanes into their 'ate' complexes is not a prerequisite for this reaction (Scheme 133). The deoxygenative propensity[445] of titanium(III) has been utilized further

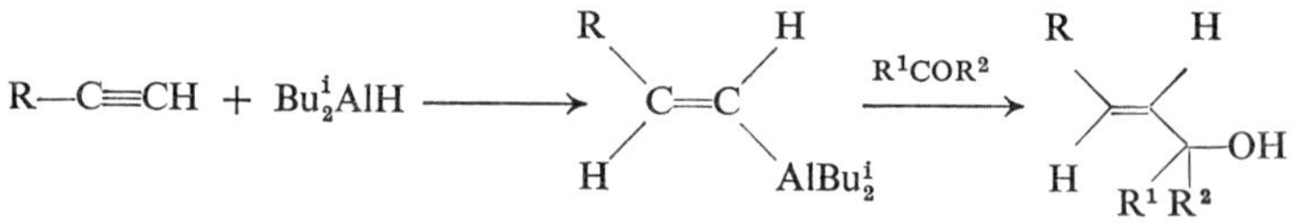

Scheme 133

[440] J. D. Morrison, R. L. Atkins, and J. E. Tomaszewski, *Tetrahedron Letters*, 1970, 4635; R. W. Herr and C. R. Johnson, *J. Amer. Chem. Soc.*, 1970, **92**, 4979.
[441] R. W. Herr, D. M. Wieland, and C. R. Johnson, *J. Amer. Chem. Soc.*, 1970, **92**, 3813.
[442] R. J. Anderson, *J. Amer. Chem. Soc.*, 1970, **92**, 4978.
[443] H. Newman, *Tetrahedron Letters*, 1971, 4571.
[444] G. Zweifel and R. B. Steele, *J. Amer. Chem. Soc.*, 1967, **89**, 2754.
[445] E. E. van Tamelen, B. Akermark, and K. B. Sharpless, *J. Amer. Chem. Soc.*, 1969, **91**, 1552.

Functional Groups other than Acetylenes, Allenes, and Olefins 163

in a synthesis of 1,2-diols from carboxylic acids;[446] in the presence of titanium(III) chloride, the doubly charged adduct of methyl-lithium and benzoic acid is deoxygenated to a radical-anion, which dimerizes (Scheme 134).

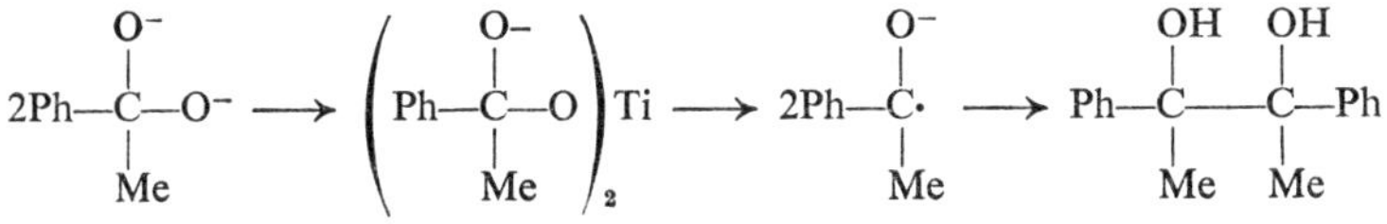

$$2\text{Ph}-\underset{\underset{\text{Me}}{|}}{\overset{\overset{\text{O}^-}{|}}{\text{C}}}-\text{O}^- \longrightarrow \left(\text{Ph}-\underset{\underset{\text{Me}}{|}}{\overset{\overset{\text{O}^-}{|}}{\text{C}}}-\text{O}\right)_2 \text{Ti} \longrightarrow 2\text{Ph}-\underset{\underset{\text{Me}}{|}}{\overset{\overset{\text{O}^-}{|}}{\text{C}}}\cdot \longrightarrow \text{Ph}-\underset{\underset{\text{Me}}{|}}{\overset{\overset{\text{OH}}{|}}{\text{C}}}-\underset{\underset{\text{Me}}{|}}{\overset{\overset{\text{OH}}{|}}{\text{C}}}-\text{Ph}$$

Scheme 134

Allylic alcohols can be obtained by decarboxylative elimination of $\beta\gamma$-epoxyacids.[447] Olefins are cleaved to alcohols by sequential ozonolysis and catalytic hydrogenation.[448] Oximes are reduced to alcohols[449] by aqueous alkaline sodium borohydride, α-oximinoketones giving the corresponding diols; selectivity is afforded by the failure of $\alpha\beta$-unsaturated oximes to react.

A synthon for the hypothetical vinyl anion (199) has been discovered by the demonstration of its synthetic equivalence[450] to the sulphoxide-stabilized anion (200). Sulphenate esters, such as (201), easily derivable from the corresponding alcohols, undergo a facile [2,3]sigmatropic rearrangement[451] to the allylic sulphoxide (202). It has now been shown that this rearrangement can be reversed by heating the sulphoxide (200) in the presence of a thiophile such as trimethyl phosphite; under such conditions, the isomeric sulphenate ester is efficiently trapped[452] to give the rearranged alcohol (203) in good yield. The synthetic utility of this rearrangement is exemplified in Scheme 135.

9-Anthryl ethers afford protection to alcohol functions;[453] treatment of the derivative with singlet oxygen, generated *via* the triphenyl phosphite–ozone adduct, followed by mild catalytic reduction, effects regeneration of the free alcohol (Scheme 136).

An alcohol protecting group cleavable by specific base catalysis is offered[454] as an alternative to the more common acid-labile functions. Tritylone ethers (204) are readily cleaved on Wolff–Kishner reduction, *via* the intermediate (205). *cis*-Diols can be protected as their acetals with DMF or dimethylbenzamide.[455]

---

[446] E. H. Axelrod, *Chem. Comm.*, 1970, 451.
[447] W. C. Still, jun., A. J. Lewis, and D. Goldsmith, *Tetrahedron Letters*, 1971, 1421.
[448] R. W. White, S. W. King, and J. L. O'Brien, *Tetrahedron Letters*, 1971, 3587.
[449] K. H. Bell, *Austral. J. Chem.*, 1970, **23**, 1415.
[450] D. A. Evans, G. C. Andrews, and C. L. Sims, *J. Amer. Chem. Soc.*, 1971, **93**, 4956.
[451] R. Tang and K. Mislow, *J. Amer. Chem. Soc.*, 1970, **92**, 2100.
[452] D. J. Abbott and C. J. M. Stirling, *J. Chem. Soc. (C)*, 1969, 818; R. D. G. Cooper and F. L. José, *J. Amer. Chem. Soc.*, 1970, **92**, 2575.
[453] W. E. Barnett and L. L. Needham, *Chem. Comm.*, 1970, 1383; *J. Org. Chem.*, 1971, **36**, 4134.
[454] W. E. Barnett and L. L. Needham, *Chem. Comm.*, 1971, 170.
[455] S. Hanessian and F. Moralioglu, *Tetrahedron Letters*, 1971, 813.

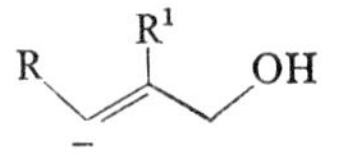

(199)

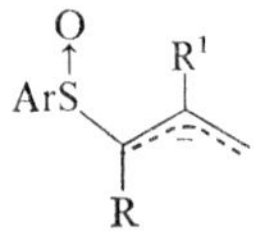

(200)

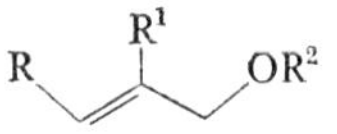

(201) R² = SAr
(203) R² = H

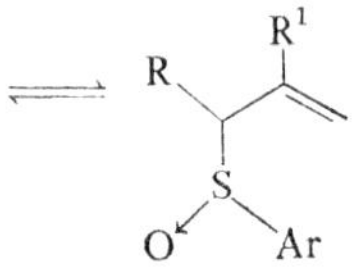

(202)

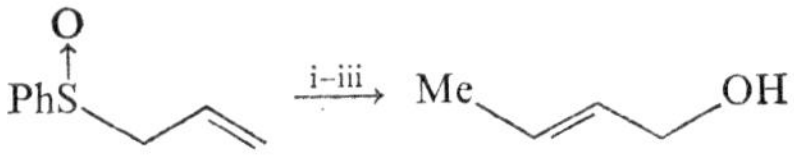

Reagents: i, Bu^{n}Li; ii, MeI; iii, (MeO)$_3$P

Scheme 135

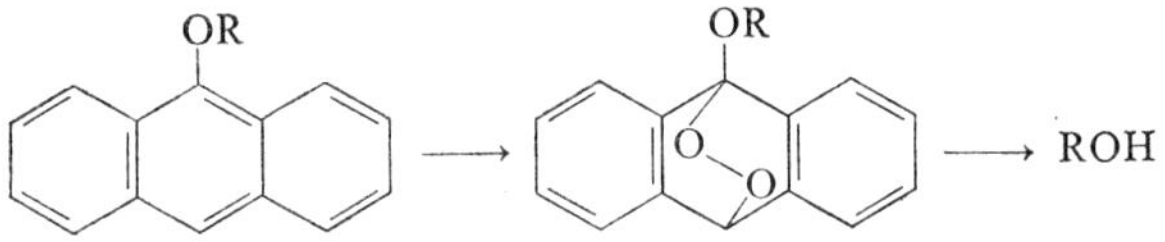

Scheme 136

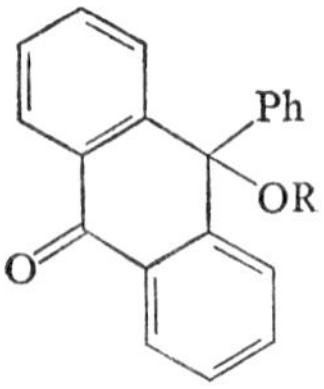

(204)

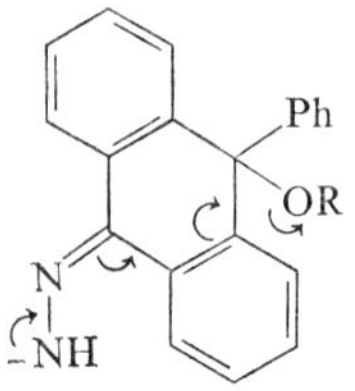

(205)

Whereas 3-phenyl allylic alcohols are reduced by lithium aluminium hydride with intramolecular hydride delivery to C-2,[456a] the 2-phenyl isomers[456b] undergo attack at C-3, with concomitant carbon–oxygen bond fission; for example, (206) is converted into (207).

The production of THF derivatives by oxidation of aliphatic unbranched

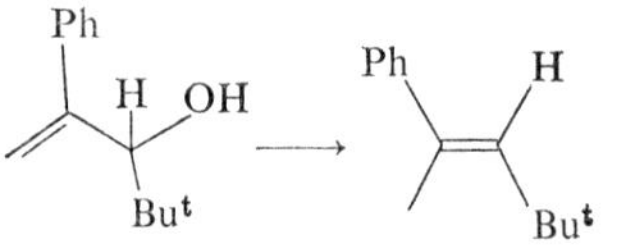

(206) (207)

[456] (a) W. T. Borden, *J. Amer. Chem. Soc.*, 1970, **92**, 4898; (b) W. T. Borden and M. Scott, *Chem. Comm.*, 1971, 381.

Functional Groups other than Acetylenes, Allenes, and Olefins

secondary alcohols with silver(I) compounds and bromine has been examined in detail;[457] variation in silver compound can lead to exclusive THF or ketone formation. Phenyl N-bromoketimine (208) is an oxidizing agent for alcohols with the same range of behaviour as N-bromosuccinimide.[458]

The Chugaev elimination of tertiary alcohols is improved by use of the potassium xanthate rather than the S-methyl xanthate; this method cannot be extended to primary or secondary alcohols.[459] Secondary alcohols are dehydrated on heating in hexamethylphosphoramide;[460] the E_2 transition state (209) is proposed.

$$Ph_2C=NBr \qquad (Me_2N)_3P \rightarrow O:$$

(208)

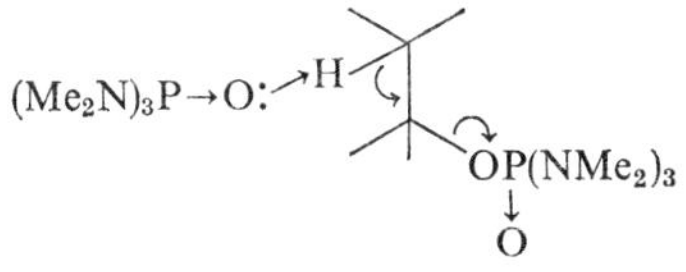

(209)

Hydrolysis of the borate ester (210) with deuterium oxide provides an improved procedure[461] for the preparation of t-butyl [^{2}H]alcohol (Scheme 137).

$$Bu^tOH + H_3BO_3 \longrightarrow (Bu^tO)_3B \xrightarrow{D_2O} 2Bu^tOD + \tfrac{1}{3}$$

(210)

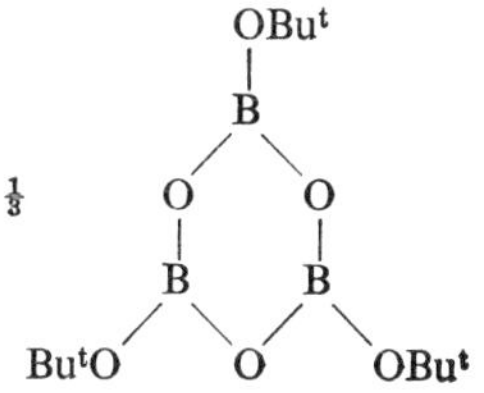

Scheme 137

The sulphene derived from methanesulphonyl chloride is involved in a high-yield synthesis of alcohol methanesulphonates;[462] the nucleophilicity and stability of the alcohol are unimportant, the only limiting factor being the stability of the product.

The C-4 carbon in phthiocerol A has been assigned the S-configuration.[463]

Acetals and Thioacetals.—A catalytic dehydrator for rapid acetal synthesis has been described; a combination of an acid ion-exchange resin and a drying agent is used.[464] Ingenious use is made of molecular sieves in a related

---

[457] A. Deluzarche, A. Maillard, P. Rimmelin, F. Schue, and J. M. Sommer, *Chem. Comm.*, 1970, 921; N. M. Roscher, *Chem. Comm.*, 1971, 474.
[458] C. G. McCarty and C. G. Leeper, *J. Org. Chem.*, 1970, **35**, 4245.
[459] K. G. Rutherford, R. M. Ottenbrite, and B. K. Tang, *J. Chem. Soc. (C)*, 1971, 582.
[460] R. S. Monson, *Tetrahedron Letters*, 1971, 567; R. S. Monson and D. N. Priest, *J. Org. Chem.*, 1971, **36**, 3826.
[461] A. T. Young and R. D. Guthrie, *J. Org. Chem.*, 1970, **35**, 853.
[462] R. K. Crossland and K. L. Servis, *J. Org. Chem.*, 1970, **35**, 3195.
[463] K. Maskens and N. Polgar, *Chem. Comm.*, 1970, 340.
[464] V. I. Stenberg, G. F. Vesley, and D. Kubik, *J. Org. Chem.*, 1971, **36**, 2550; see also ref. 49.

procedure;[465] a dual-function proton-exchanged sieve is employed, in which the outer surface catalyses the reaction and the inner surface selectively absorbs the water produced: alternatively, a simple sieve and toluene-p-sulphonic acid can be used. A simple one-step synthesis of dithiohemiacetals has been reported[466] and is shown in Scheme 138.

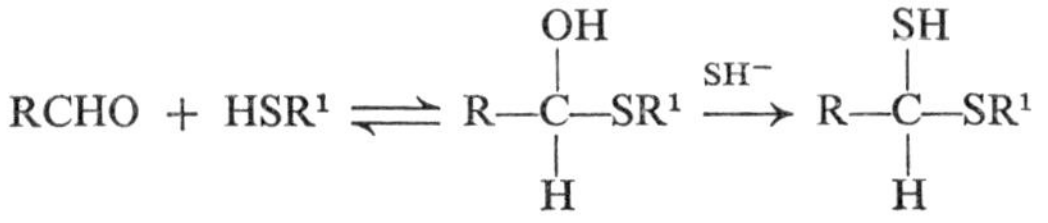

$$\text{RCHO} + \text{HSR}^1 \rightleftharpoons \text{R}\underset{\text{H}}{\overset{\text{OH}}{-\underset{|}{\overset{|}{\text{C}}}-}}\text{SR}^1 \xrightarrow{\text{SH}^-} \text{R}\underset{\text{H}}{\overset{\text{SH}}{-\underset{|}{\overset{|}{\text{C}}}-}}\text{SR}^1$$

Scheme 138

The ozonolysis of acetals and related ethers shows considerable promise as a synthetic tool.[467] Aldehyde acetals and alcohol tetrahydropyranyl ethers are smoothly oxidized to esters; in the latter case, production of δ-valerolactone by the alternate mode of cleavage is not observed. Benzylidene and ethylidene protecting groups are cleanly and specifically removed from sugar substrates; for such an oxidative cleavage, it is suggested that the acetal function must be capable of adopting a conformation in which both acetal oxygen atoms can have one of their lone pairs orientated *trans* to the acetal carbon–hydrogen bond. This criterion permits the specific oxidation of β-glycosides, the α-anomers being incapable of fulfilling such a requirement. These transformations are outlined in Scheme 139. Tetrahydropyranyl ethers are converted photochemically[468] into a mixture of δ-valerolactone and the linear ester (212), presumably by the alternate decomposition pathways of the intermediate (211), as shown in Scheme 140.

Steroidal alcohols, masked as their tetrahydropyranyl ethers, benzyl ethers or bismethylenedioxy-derivatives, can be deprotected by the trityl carbonium ion in a hydride-transfer process.[469] This method of regeneration should be equally applicable to other protected groups such as amines and carboxylic acids.

Certain sugar allylic acetals, such as the pyranoside (213), undergo an abnormal mode of reduction with lithium aluminium hydride to give vinyl ethers (Scheme 141); an intramolecular delivery of hydride ion is proposed,[470] by co-ordination of the reagent with the methoxy-oxygen atom.

Whereas lithium aluminium hydride–aluminium trichloride reduction of monothioacetals effects carbon–oxygen cleavage to give hydroxythioethers,

---

[465] D. P. Roelofsen, E. R. J. Wils, and H. van Bekkum, *Rec. Trav. chim.*, 1971, **90**, 1141; see also ref. 48.
[466] L. Schutte, *Tetrahedron Letters*, 1971, 2321.
[467] P. Deslongchamps and C. Moreau, *Canad. J. Chem.*, 1971, **49**, 2463.
[468] T. Yamagishi, T. Yoshimoto, and K. Minamini, *Tetrahedron Letters*, 1971, 2795.
[469] D. H. R. Barton, P. D. Magnus, G. Streckert, and D. Zurr, *Chem. Comm.*, 1971, 1109; see also ref. 306.
[470] B. Frazer-Reid and B. Radatus, *J. Amer. Chem. Soc.*, 1970, **92**, 6661.

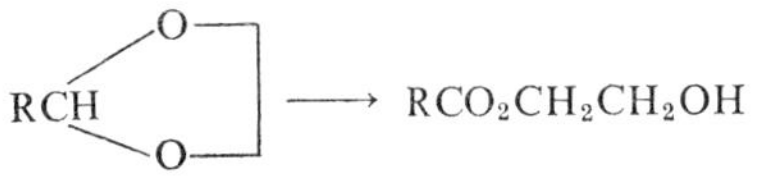

Functional Groups other than Acetylenes, Allenes, and Olefins

$$RCH\underset{O}{\overset{O}{<}}\!\!\!\!\!\!\!\!\!\!\!\!\!\!\!\!\!\!\! \longrightarrow RCO_2CH_2CH_2OH \qquad RCH(OR^1)_2 \longrightarrow RCO_2R^1$$

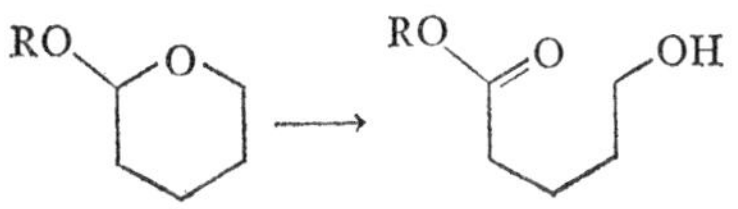

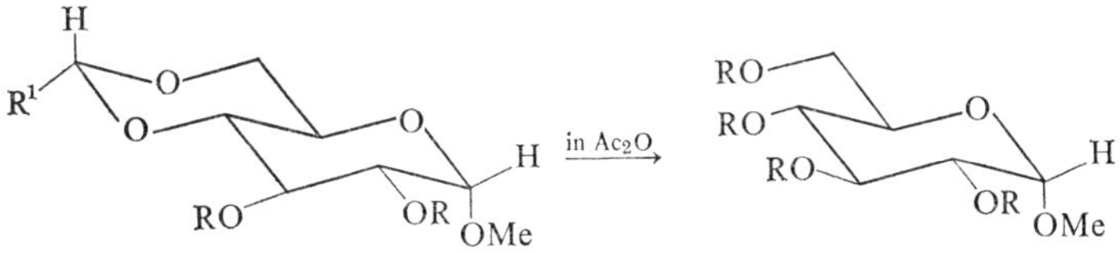

R^1 = Me or Ph
R = MeCO

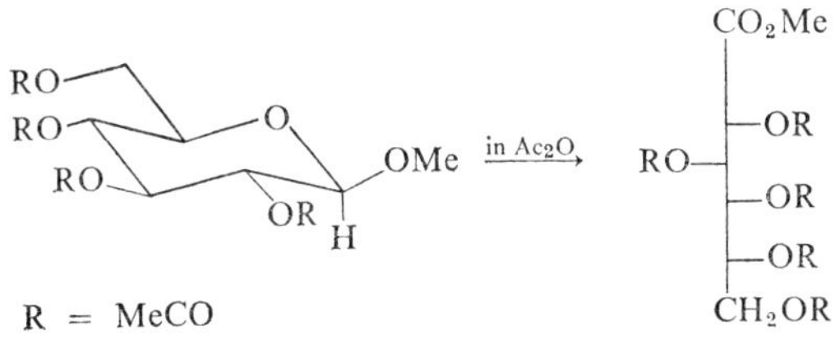

R = MeCO

Scheme 139

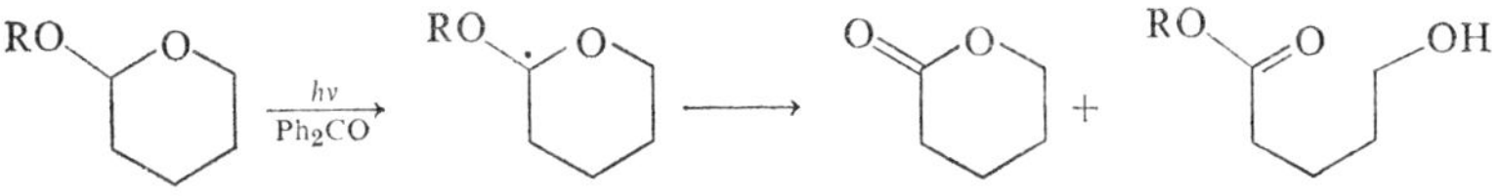

(211)   (212)

Scheme 140

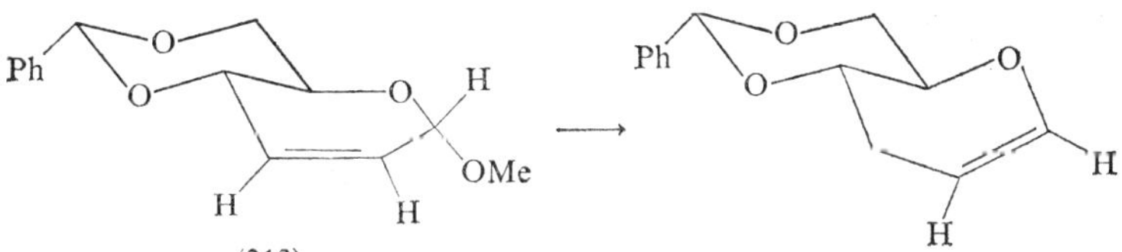

(213)

Scheme 141

alkali-metal–ammonia reduction[471] yields alkoxymercaptans by carbon–sulphur cleavage (Scheme 142).

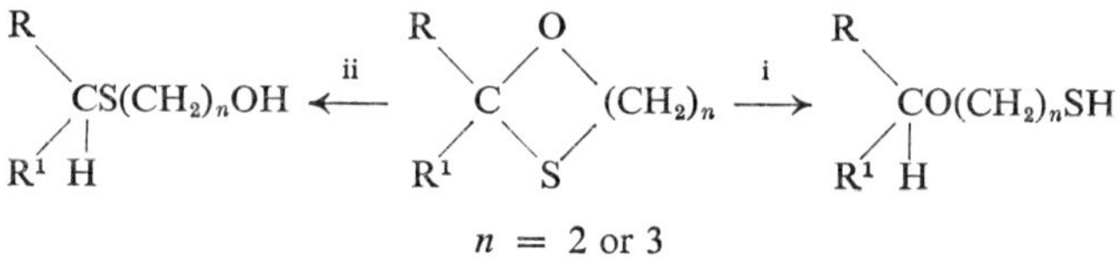

$$\underset{R^1\ H}{\overset{R}{\diagdown}}CS(CH_2)_nOH \xleftarrow{ii} \underset{R^1}{\overset{R}{\diagdown}}C\underset{S}{\overset{O}{\diagup}}(CH_2)_n \xrightarrow{i} \underset{R^1\ H}{\overset{R}{\diagdown}}CO(CH_2)_nSH$$

$n = 2$ or 3

Reagents: i, M–NH$_3$; ii, LiAlH$_4$–AlCl$_3$

Scheme 142

Dithians are smoothly cleaved by calcium in ammonia (Scheme 143).[472]

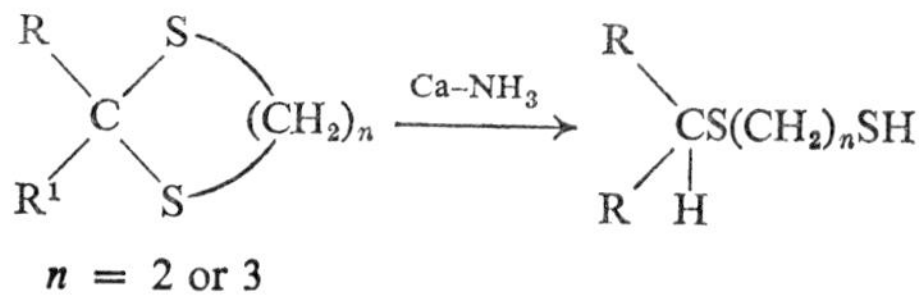

$$\underset{R^1}{\overset{R}{\diagdown}}C\underset{S}{\overset{S}{\diagup}}(CH_2)_n \xrightarrow{Ca-NH_3} \underset{R\ H}{\overset{R}{\diagdown}}CS(CH_2)_nSH$$

$n = 2$ or 3

Scheme 143

Full details have been published[291] for the preparation of the valuable carbonyl synthon, 1,3-dithian.

10 Sulphur Compounds

The synthetic potential of sulphur extrusion by tervalent phosphorus compounds continues to provide considerable stimulus. Of notable interest is Eschenmoser's application[473a] of such extrusion processes in a general method for the synthesis of secondary vinylogous amides and enolizable β-dicarbonyl compounds by sulphide contraction via alkylative coupling (as distinct from the previously reported[473b] oxidative coupling). α-Halogenocarbonyl compounds of type (214) react with secondary thioamides (or thiolactams) or thiocarboxylic acids to give products of S-alkylation; these intermediates can undergo intramolecular carbon–carbon condensation under the influence of base or acid, and, in the presence of a suitable tervalent phosphorus compound as thiophile, lead to the vinylogous amide or enolized β-diketone (Scheme 144). The required thioamides are readily prepared by reaction of the amides with phosphorus pentasulphide, or as their iminoesters with hydrogen sulphide. Such a general procedure has found considerable application in the total synthesis of Vitamin B$_{12}$.

---

[471] E. L. Eliel and T. W. Doyle, *J. Org. Chem.*, 1970, **35**, 2716.
[472] B. C. Newman and E. L. Eliel, *J. Org. Chem.*, 1970, **35**, 3641.
[473] (a) M. Roth, P. Dubs, E. Götschi, and A. Eschenmoser, *Helv. Chim. Acta*, 1971, **54**, 710; (b) A. Eschenmoser, *Quart. Rev.*, 1970, **24**, 710.

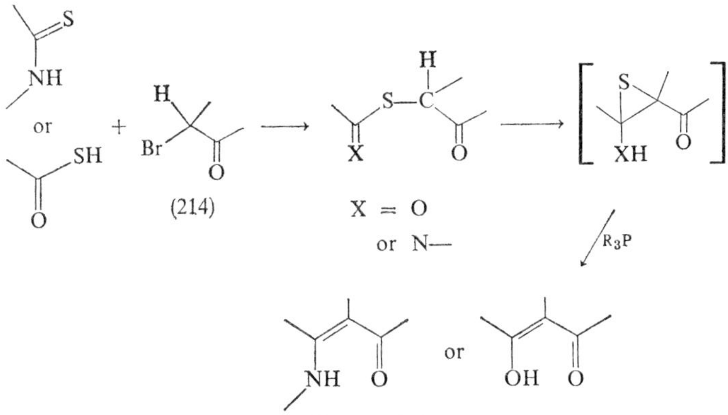

Scheme 144

A mechanistic dichotomy is noted[474] in the desulphurization of aliphatic trisulphides; triphenylphosphine removes the central sulphur atom, but tris(diethylamino)phosphine extrudes a sulphur atom originally bonded to carbon.

N-Thioalkylphthalimides (215) are converted into N-alkylphthalimides by tris(dimethylamino)phosphine, providing a method of transformation of thiols into amines.[475] Thiols react with such N-thioalkylphthalimides in a general procedure[476] for the preparation of unsymmetrical disulphides; a similar reaction with amines[477] gives N-thio-N'-diamides. These reactions are outlined in Scheme 145.

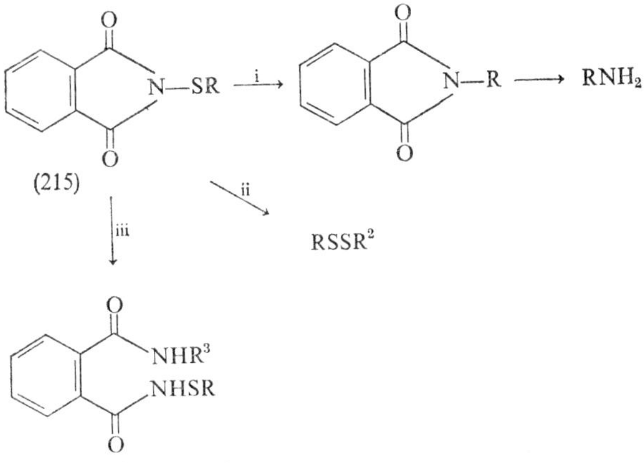

Reagents: i, R_3^1P; ii, R^2SH; iii, R^3NH_2

Scheme 145

[474] D. N. Harpp and D. K. Ash, *Chem. Comm.*, 1970, 811.
[475] D. N. Harpp and B. A. Orwig, *Tetrahedron Letters*, 1970, 2691.
[476] D. N. Harpp, D. K. Ash, T. G. Back, J. G. Gleason, B. A. Orwig, and W. F. Van Horn, *Tetrahedron Letters*, 1970, 3551; K. S. Boustany and A. B. Sullivan, *ibid.*, p.3547; K. S. Boustany, *Chimia (Switz.)*, 1970, **24**, 396.
[477] K. S. Boustany and J. P. Van der Kooi, *Tetrahedron Letters*, 1970, 4983.

Trialkylboranes react with elemental sulphur to give good yields of symmetrical disulphides;[478] oxygen-induced free-radical chain reaction with organic disulphides gives sulphides,[479] as shown in Scheme 146.

$$\text{RSMe} \xleftarrow{iii} R_3B \xrightarrow{i} R_2BSR \xrightarrow{ii} RSH \longrightarrow RSSR$$

Reagents: i, S; ii, H$_2$O; iii, MeSSMe-O$_2$

Scheme 146

The synthesis of β-ketosulphides from α-diketones is reported;[480] a key step is hydride-transfer reduction of an α-chloro-β-ketosulphide with triethylsilane (Scheme 147).

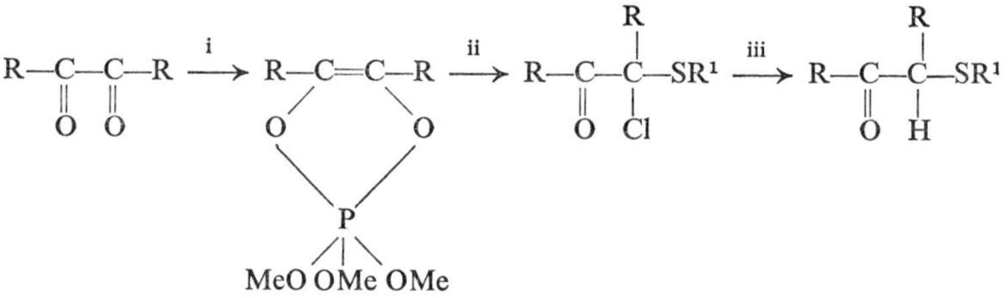

$$R-\underset{\underset{O}{\|}}{C}-\underset{\underset{O}{\|}}{C}-R \xrightarrow{i} R-\underset{O}{C}=\underset{O}{C}-R \xrightarrow{ii} R-\underset{\underset{O}{\|}}{C}-\underset{\underset{Cl}{|}}{\overset{R}{C}}-SR^1 \xrightarrow{iii} R-\underset{\underset{O}{\|}}{C}-\underset{\underset{H}{|}}{\overset{R}{C}}-SR^1$$

with P(OMe)(OMe)(OMe) bridge

Reagents: i, (MeO)$_3$P; ii, R^{1}SCl; iii, Et$_3$SiH–Bu$_3^n$N

Scheme 147

Sulphoxides are reduced quantitatively to sulphides by a sodium borohydride–cobalt chloride system; co-ordination of the transition-metal ion with the sulphoxide oxygen is suggested.[481]

A number of organic sulphur compounds have been isolated from the seaweed *Dictopteris*,[482] including the substances (216)—(219).

Woodward[483] has now reported reliable procedures for the preparation of trimethylene and ethylene dithiotoluene-*p*-sulphonates, reagents used in the formation of dithians and dithiolans from active methylene compounds, including those adjacent to a carbonyl group activated by prior hydroxymethylene or enamine formation. Applications include carbon–carbon bond cleavage[484] and ketone transposition[485] (Scheme 148).

---

[478] Z. Yoshida, T. Okushi, and O. Manabe, *Tetrahedron Letters*, 1970, 1641.
[479] H. C. Brown and M. M. Midland, *J. Amer. Chem. Soc.*, 1971, **93**, 3291.
[480] D. N. Harpp and P. Mathiaparanam, *J. Org. Chem.*, 1971, **36**, 2540.
[481] D. W. Chasar, *J. Org. Chem.*, 1971, **36**, 613.
[482] P. Roller, K. Au, and R. E. Moore, *Chem. Comm.*, 1971, 503; R. E. Moore, *ibid.*, p. 1168.
[483] R. B. Woodward, I. J. Pachter, and M. L. Scheinbaum, *J. Org. Chem.*, 1971, **36**, 1137.
[484] J. A. Marshall and H. Roebke, *Tetrahedron Letters*, 1970, 1555.
[485] J. A. Marshall and H. Roebke, *J. Org. Chem.*, 1969, **34**, 4188.

Functional Groups other than Acetylenes, Allenes, and Olefins

C_8H_{17} CO(CH$_2$)$_2$SCOMe C_8H_{17} CO(CH$_2$)$_2$S$_n$(CH$_2$)$_2$COC$_8$H$_{17}$
(216)   (217) n = 2, 3, or 4

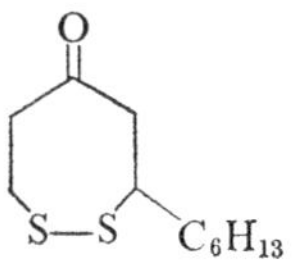

(218) C_6H_{13} CH$\overset{t}{=}$CHCO(CH$_2$)$_2$SCOMe
                         (219)

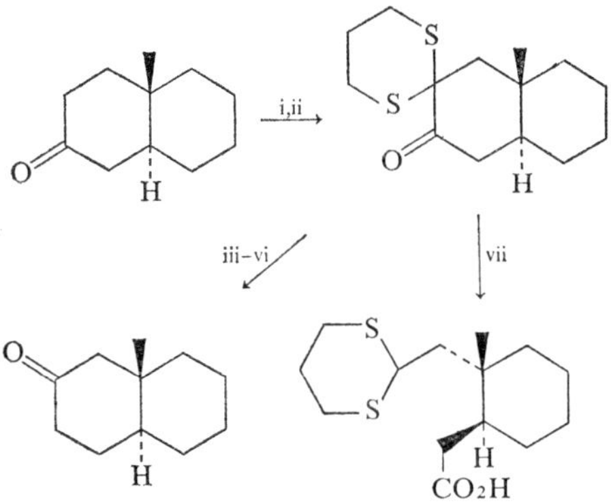

Reagents: i, HCO$_2$Et–NaH; ii, TosS(CH$_2$)$_3$STos–KOAc; iii, LiAlH$_4$; iv, Ac$_2$O;
v, HgCl$_2$–H$_2$O; vi, Ca–NH$_3$; vii, NaOMe–DMSO

Scheme 148

1,3-Dithianyltoluene-*p*-sulphonates, such as (220), are reported[486] to fragment to acyclic olefin keten thioacetals, which can be converted into ω-unsaturated carboxylic acids (Scheme 149).

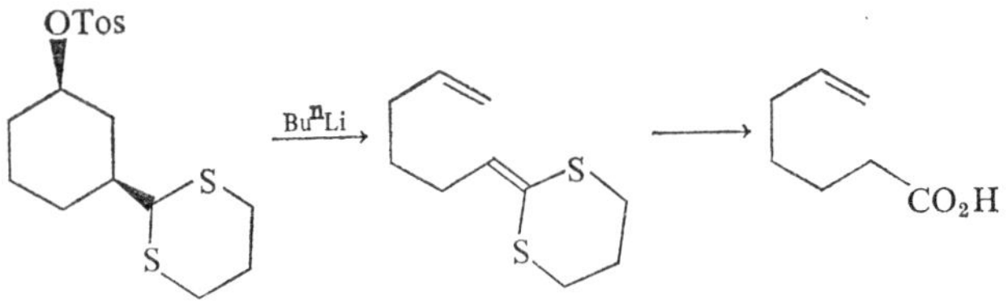

Scheme 149

α-Sulphinyl carbanions have been shown to react with ketones with retention of carbanion configuration; reaction of the lithio-carbanion from *S*-benzyl methyl sulphoxide with acetone gives a mixture of diastereoisomers in

[486] J. A. Marshall and J. L. Belletire, *Tetrahedron Letters*, 1971, 871.

the ratio 15:1, the major isomer having the *R* configuration at carbon.[487] This has been developed as a synthesis of chiral epoxides (Scheme 150).

Cyclodextrin is reported to be effective for the partial resolution of sulphoxides;[488] in most cases, the *R*-sulphoxide is preferentially included into β-cyclodextrin. Fenton's reagent is recommended[489] for the generation of alkyl radicals from sulphoxides.

The metal-ion–base-catalysed rearrangement of α-ketohemithioacetals

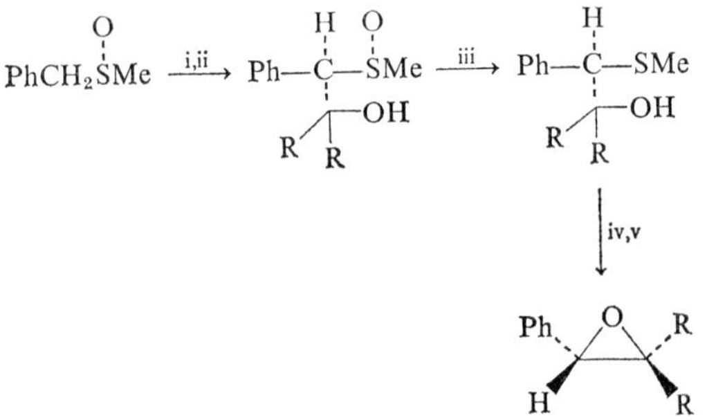

$$PhCH_2SMe \xrightarrow{i,ii} Ph-\underset{R}{\underset{|}{C}}(OH)(R)-SMe \xrightarrow{iii} Ph-\underset{R}{\underset{|}{C}}(OH)(R)-SMe$$

$$\downarrow iv,v$$

Ph, O, R / H, R (epoxide)

Reagents: i, MeLi; ii, RCOR; iii, MeCOCl–SnCl$_4$–DMF; iv, MeI–AgClO$_4$; v, 2% KOH

Scheme 150

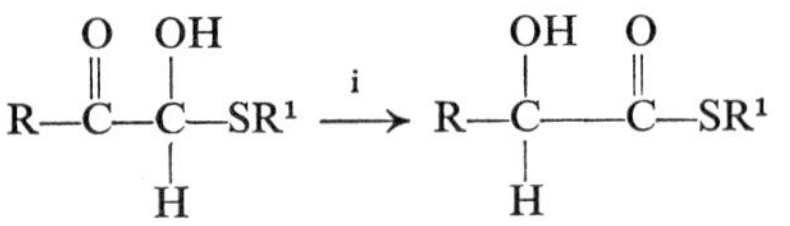

$$R-\underset{}{\overset{O}{\overset{\|}{C}}}-\underset{H}{\underset{|}{\overset{OH}{\overset{|}{C}}}}-SR^1 \xrightarrow{i} R-\underset{H}{\underset{|}{\overset{OH}{\overset{|}{C}}}}-\overset{O}{\overset{\|}{C}}-SR^1$$

Reagents: i, Mg(NO$_3$)$_2$–NaOAc–DMF

Scheme 151

to α-hydroxythiolesters has been described[490] (Scheme 151); this rearrangement is pertinent to a study of the glyoxalase-1 enzyme system. Triethyloxonium fluoroborate smoothly converts thionoesters into thiolesters[491] (Scheme 152).

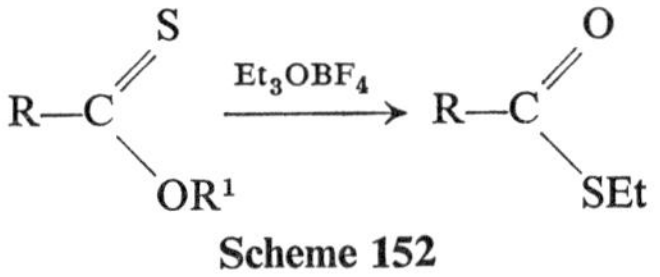

$$R-\overset{S}{\overset{\|}{C}}-OR^1 \xrightarrow{Et_3OBF_4} R-\overset{O}{\overset{\|}{C}}-SEt$$

Scheme 152

The arylsulphonate group is photosensitive, affording an easy method for the regeneration of protected aliphatic alcohols and amines from their

---

[487] T. Durst, R. Viau, R. van den Elzen, and C. H. Nguyen, *Chem. Comm.*, 1971, 1334.
[488] M. Mikolajczyk, J. Drabowicz, and F. Cramer, *Chem. Comm.*, 1971, 317.
[489] B.-M. Bertilsson, B. Gustafsson, I. Kühn, and K. Torssell, *Acta Chem. Scand.*, 1970, 24, 3590.
[490] S. S. Hall and A. Poet, *Tetrahedron Letters*, 1970, 2867.
[491] T. Oishi, M. Mori, and Y. Ban, *Tetrahedron Letters*, 1971, 1777.

sulphonates and sulphonamides, respectively.[492] The scope and mechanism of a patented reaction for the preparation of sulphones by addition of sulphur dioxide to terminal olefins has been explored.[493]

11 Miscellaneous Reactions

The intermediacy of azido-radicals in the cerium(IV) oxidation of metallic azides to molecular nitrogen has been demonstrated[494] by olefin trapping; α-azido-β-nitratoalkanes are formed (Scheme 153), with a regioselectivity consistent with initial addition of the azido-radical.

$$Ce^{IV} + N_3^- \longrightarrow Ce^{III} + N_3^{\cdot}$$

$$2N_3^{\cdot} \longrightarrow 3N_2$$

$$R_2C{=}CH_2 + NaN_3 + Ce^{IV} \longrightarrow R_2\underset{ONO_2}{C}{-}\underset{N_3}{CH_2}$$

Scheme 153

Although acetyl and other acyl nitrates are dangerously explosive, in solution they undergo smooth thermolysis to alkyl nitrates with one less carbon atom (Scheme 154); a number of methods are discussed for the preparation of acyl nitrates.[495] Thermolysis is conducted at high temperature to minimize the alternative mode of decomposition, production of carboxylic acid anhydrides and dinitrogen pentoxide.

$$RCO_2NO_2 \longrightarrow RNO_2 + CO_2$$

Scheme 154

Vicinal nitro-nitrates, readily available by reaction of olefins with nitrogen oxides and oxygen, are cleanly converted into terminal nitroalkanes by reduction with sodium borohydride in non-acidic conditions: the intermediate nitro-olefin is reduced very rapidly, precluding dimerization.[496]

Kornblum[497] has reported on the alkylation of some tertiary nitro-compounds with nitro-paraffin salts (Scheme 155); a radical-anion chain mechanism is proposed, with initial electron transfer from the nitro-paraffin anion to the substrate.

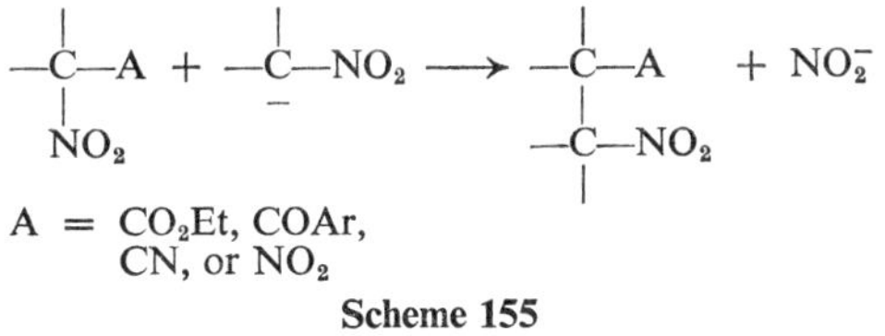

$$\underset{NO_2}{-\overset{|}{C}-A} + -\overset{|}{\underset{|}{C}}-NO_2 \longrightarrow -\overset{|}{\underset{-\overset{|}{C}-NO_2}{C}}-A + NO_2^-$$

A = CO₂Et, COAr, CN, or NO₂

Scheme 155

---

[492] A. Abad, D. Mellier, J. P. Pète, and C. Portella, *Tetrahedron Letters*, 1971, 4555.
[493] H. W. Gibson and D. A. McKenzie, *J. Org. Chem.*, 1970, **35**, 2994.
[494] W. S. Trahanovsky and M. D. Robbins, *J. Amer. Chem. Soc.*, 1971, **93**, 5256.
[495] G. B. Bachman and T. F. Biermann, *J. Org. Chem.*, 1970, **35**, 4229.
[496] J. M. Larkin and K. L. Kreuz, *J. Org. Chem.*, 1971, **36**, 2574.
[497] N. Kornblum, S. D. Boyd, and F. W. Stuchal, *J. Amer. Chem. Soc.*, 1970, **92**, 5783, 5785.

A facile method for the synthesis of azoalkanes has been reported;[498] direct oxidation of semicarbazides with copper(II) chloride complexes affords the base-labile *cis*-azoalkane–copper(I) chloride complexes and isocyanates (Scheme 156).

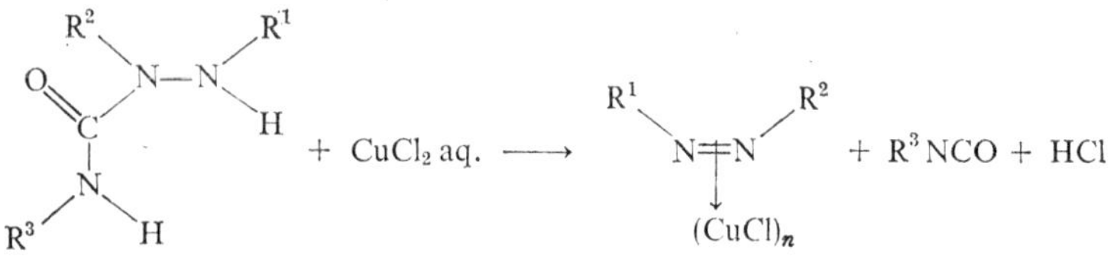

$$\underset{R^3}{\underset{|}{\overset{R^2}{\underset{|}{\overset{|}{\text{N}}}}}}\overset{O}{\underset{H}{\overset{\|}{\text{C}}}}\underset{H}{\overset{R^1}{\underset{|}{\text{N}-\text{N}}}} + \text{CuCl}_2 \text{ aq.} \longrightarrow \underset{(\text{CuCl})_n}{R^1{\text{N}{=}\text{N}}R^2} + R^3\text{NCO} + \text{HCl}$$

Scheme 156

Oximes are reduced in benzene solution to hydroxylamines by sodium borohydride adsorbed on silica gel.[499] In turn, silver carbonate on Celite cleanly oxidizes hydroxylamines to *C*-nitroso-compounds,[500] isolate as nitroso-dimers; no oximes are formed. Such dimers can be thermally dissociated to the more reactive monomers, which are tautomerized to oximes on thermal or red-light treatment.[501]

A new polarographic reduction wave has been observed for aqueous solutions of formaldehyde and secondary amines; this is ascribed[502] to the hitherto undetected Mannich intermediate, the aminomethyl carbonium ion; 1.9 electrons are utilized per molecule, in experimental accord with Scheme 157.

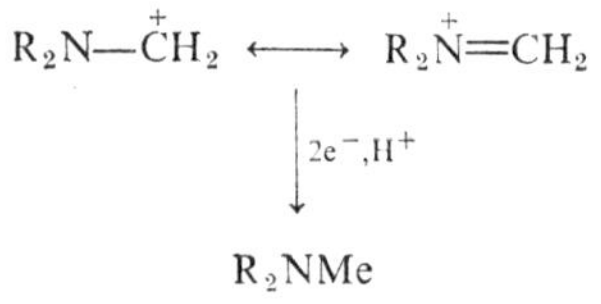

$$R_2N-\overset{+}{C}H_2 \longleftrightarrow R_2\overset{+}{N}{=}CH_2$$
$$\downarrow {\scriptstyle 2e^-, H^+}$$
$$R_2NMe$$

Scheme 157

A convenient synthesis of the spin-trap reagent, t-butyl nitroxide, has been reported;[503a] a cautionary note[503b] on the use of such traps has been sounded, nitroxyl radicals being produced by reaction of the reagents with anions.

Whereas mercuration of $\alpha\beta$-unsaturated carbonyl compounds normally affords products of α-mercuration, substrates alkylated solely in the α-position undergo β-mercuration (Scheme 158).[504]

---

[498] M. Heyman, V. T. Bandurco, and J. P. Snyder, *Chem. Comm.*, 1971, 297.
[499] F. Hodoşan and V. Ciurdaru, *Tetrahedron Letters*, 1971, 1997.
[500] J. A. Maasen and Th. J. de Boer, *Rec. Trav. chim.*, 1971, **90**, 373.
[501] A. Mackor and Th. J. de Boer, *Rec. Trav. chim.*, 1970, **89**, 164.
[502] M. Masui, K. Fujita, and H. Ohmori, *Chem. Comm.*, 1970, 182.
[503] (*a*) R. J. Holman and M. J. Perkins, *J. Chem. Soc.* (*C*), 1970, 2195; (*b*) A. R. Forrester and S. P. Hepburn, *J. Chem. Soc.* (*C*), 1971, 701.
[504] A. J. Bloodworth and R. J. Bunce, *Chem. Comm.*, 1970, 753.

Functional Groups other than Acetylenes, Allenes, and Olefins 175

$$\underset{\text{R}}{\text{CH}_2=\overset{|}{\text{C}}-\text{COX}} + \text{Hg(OAc)}_2 + \text{R}^1\text{OH} \longrightarrow \text{AcOHgCH}_2\underset{\text{R}}{\overset{\text{OR}^1}{\underset{|}{\text{C}}}}\text{COX}$$

Scheme 158

A new reactive leaving group, 2,2,2-trifluoroethanesulphonate (tresylate) has been described;[505] solvolysis rates for a range of substrates are intermediate between those of the corresponding toluene-*p*-sulphonate and trifluoromethanesulphonate esters.

Newman[506] has reported the synthesis of an epoxyketen acetal; this spiro-compound collapses on treatment with acid or base, as shown in Scheme 159. The structure of the antifungal antibiotic lipoxamycin (221) has been elucidated.[507]

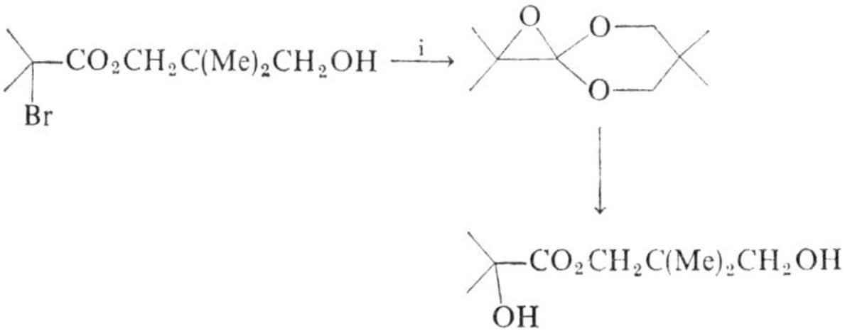

Reagent: i, NaH–glyme

Scheme 159

$$\text{Me}_2\text{CH(CH}_2)_3\text{CO(CH}_2)_6\text{COCH}_2\text{CH}_2\underset{|}{\text{N}}\text{COCHCH}_2\text{OH}$$
$$\phantom{\text{Me}_2\text{CH(CH}_2)_3\text{CO(CH}_2)_6\text{COCH}_2\text{CH}_2\text{N}}\overset{|}{\text{OH}}\ \overset{|}{\text{NH}_2}$$

(221)

12 Reviews

Critical reviews on ylides,[508] enamines,[509] and nitrenes[510] have been published. Brown has reported on the reactivity and selectivity of diborane[511] and disiamylborane[512] with a range of functional groups. The properties of hexamethylphosphoramide as a solvent and a reagent have been catalogued;[513]

[505] R. K. Crossland, W. E. Wells, and V. J. Shiner, jun., *J. Amer. Chem. Soc.*, 1971, **93**, 4217.
[506] M. S. Newman and E. Kilbourn, *J. Org. Chem.*, 1970, **35**, 3186.
[507] H. A. Whaley, *J. Amer. Chem. Soc.*, 1971, **93**, 3767.
[508] P. A. Lowe, *Chem. and Ind.*, 1970, 1070.
[509] P. W. Hickmott and H. Suschitzky, *Chem. and Ind.*, 1970; 1118; S. Hunig and H. Hoch, *Fortschr. Chem. Forsch.*, 1970, **14**, 236.
[510] R. K. Smalley and H. Suschitzky, *Chem. and Ind.*, 1970, 1338.
[511] H. C. Brown, P. Heim, and N. M. Yoon, *J. Amer. Chem. Soc.*, 1970, **92**, 1637.
[512] H. C. Brown, D. B. Bigley, S. K. Arora, and N. M. Yoon, *J. Amer. Chem. Soc.*, 1970, **92**, 7161.
[513] H. Normant, *Russ. Chem. Rev.*, 1970, 457.

its use as a solvent for electron-transfer reduction has been reviewed.[334b] Kuivila has assessed the utility of organotin hydrides as reducing agents.[514] Conditions affecting the regioselective alkylation of ambident anions have been discussed.[515] Bestmann[516] has published an account of his studies on the nucleophilicity of phosphoranes, and Olah[517] has described his accumulated results on protonated heteroaliphatic substrates such as acids, ethers, aldehydes, and ketones. The use of chlorotrimethylsilane as a trapping agent in the acyloin condensation has been reviewed.[518] Reviews have appeared on aliphatic diazo-compounds,[519] aliphatic N-nitrosamines,[520] trialkyloxonium fluoroborates,[521] nitrenium ions,[522] aliphatic deamination[523] with nitrous acid, migration reactions of alkoxycarbonyl groups,[524] and the wide synthetic utility of organothallium compounds.[216]

[514] H. Kuivila, *Synthesis*, 1970, 499.
[515] W. J. LeNoble, *Synthesis*, 1970, 1.
[516] H.-J. Bestmann, *Bull. Soc. chim. France*, 1971, 1619.
[517] G. A. Olah, A. M. White, and D. H. O'Brien, *Chem. Rev.*, 1970, 561.
[518] K. Rühlmann, *Synthesis*, 1971, 236.
[519] O. P. Studzinskii and I. K. Korobitsyna, *Russ. Chem. Rev.*, 1970, 834; G. W. Cowell and A. Ledwith, *Quart. Rev.*, 1970, **24**, 119.
[520] A. L. Fridman, F. M. Mukhametshin, and S. S. Novikov, *Russ. Chem. Rev.*, 1971, 34.
[521] V. G. Grank, B. M. Pyatin, and R. G. Glushkov, *Russ. Chem. Rev.*, 1971, 747.
[522] P. G. Gassman, *Accounts Chem. Res.*, 1970, 3, 26.
[523] C. J. Collins, *Accounts Chem. Res.*, 1971, 4, 315.
[524] R. M. Acheson, *Accounts Chem. Res.*, 1971, **4**, 177.

3
Fatty Acids and Related Compounds

BY F. D. GUNSTONE

1 Introduction

This review is concerned with developments in the chemistry of fatty acids and of some related compounds during 1970 and 1971.

An abbreviated form of nomenclature, widely used for long-chain acids, will be employed here. Numbers and letters are used to indicate chain-length, the number of unsaturated centres, the position(s) of unsaturation, and its nature (*a*, acetylenic; *c*, *cis*-olefinic; *e*, ethylenic; and *t*, *trans*-olefinic). For example, 18:2(9c12c) is the symbol for octadeca-*cis*-9,*cis*-12-dienoic acid (linoleic). Trivial names are either explained in the text or in the footnote* to this page.

Following IUPAC rules, fatty acids should always be numbered with the carboxy-group as C-1. In an alternative system, commonly employed for the designation of many methylene-interrupted polyethenoid acids, the end methyl group is designated as C-1 and since all the double bonds are methylene-interrupted it is sufficient to indicate only the number of double bonds and the position of the first. A IUPAC–IUB Commission has proposed that unsaturation should then be indicated by symbols such as $(n-6)$ rather than $\omega 6$, but the latter remains more popular. Linoleic acid can thus be represented as 18:2(9c12c) or 18:2(*n*-6) or 18:2($\omega 6$). IUPAC rules will be followed in this review except when there seems good reason for using the alternative ω system.

2 Natural Compounds

Long-chain Acids.—Though new long-chain acids continue to be isolated and identified, most contain only new arrangements of familiar structural features. Acids reported here are discussed in order of increasing chain-length. Reviews

---

* Trivial names for fatty acids used frequently in this article are (in alphabetical order): arachidonic(20:4 5c8c11c14c), crepenynic(18:2 9c12a), elaidic(18:1 9t), eleostearic(18:3 9c11t13t), helenynolic(9-OH 18:2 10t12a), kamlolenic(18-OH 18:3 9c11t13t), linoleic (18:2 9c12c), linolenic(18:3 9c12c15c), malvalic [see structure (20)], oleic(18:1 9c) ricinoleic(12-OH 18:1 9c), stearolic (18:1 9a), sterculic [see structure (21)].

have appeared dealing with epoxy-acids,[1] cyclopropane and cyclopropene acids,[153a] alkyl-branched acids,[154d] acids with conjugated unsaturation,[155b] acids produced from hydrocarbons by micro-organisms,[155d] and with the absolute configuration of optically active long-chain compounds.[153e]

The insect body fat of *Llaveia axin* is known as aje, and contains C_{12} and C_{14} acids with conjugated pentaene unsaturation.[2] The configuration of the unsaturated system has not been determined but the structures of the two acids (1) and (2) suggest that the C_{14} member is formed from the C_{12} acid by chain extension.

$$12:5(3,5,7,9,11) \xrightarrow{+C_2} 14:5(5,7,9,11,13)$$
$$(1) \qquad\qquad\qquad (2)$$

The allenic ester produced by the male Dried Bean beetle (*Acanthoscelides obtectus*), with a half-life of only 10 h at room temperature or 20 days at $-13\,°C$, has been identified as ($-$)-methyl tetradeca-*trans*-2,4,5-trienoate[3] and its ($\pm$)-isomer has been synthesized.[4]

The lipids of the sea anemone *Metridium dianthus* contain both the 16:1(6*t*) acid and alcohol.[5] In contrast to plants and animals, bacilli introduce *cis* double bonds at the Δ^5 and Δ^{10} positions. Reporting the presence of the 16:2(5c10c) acid, Fulco claims this is the first polyenoic acid of bacillic origin arising by *de novo* synthesis.[6] Two enzymes are responsible for these desaturation processes; that responsible for Δ^5 desaturation is effective at $20\,°C$ but not at $35\,°C$, whereas that for Δ^{10} desaturation is effective at both these temperatures. It has been shown that Δ^{10} desaturation in *B. cereus* occurs with C_{16} and C_{18} substrates but not with the C_{14} or C_{17} homologues.[7]

Long-chain acids containing 1,4-epoxides (2,5-disubstituted tetrahydrofurans) were unknown prior to their recent preparation from linoleic acid and other C_{18} acids (see Section 5, p. 190), but since the earlier publication a series of such acids have been discovered in wool fat[8] and some dihydrofurans have been reported as metabolic products of prostaglandins (see Section 6, p. 198).[9] A polar fraction from wool fat (1%) contains mainly the C_{16} tetrahydrofuran (3) (in 85% yield) along with minor amounts of similar C_{14}, C_{15}, C_{17}, and C_{18} acids.

The thermophilic bacterium *B. acidocalderius* produces, at low pH and high temperatures, appreciable proportions of the cyclic C_{17} (27%) and C_{19} (31%)

---

[1] F. R. Earle, *J. Amer. Oil Chemists Soc.*, 1970, **47**, 510.
[2] J. Cason, R. Davis, and M. H. Sheehan, *J. Org. Chem.*, 1971, **36**, 2621.
[3] D. F. Horler, *J. Chem. Soc.* (C), 1970, 859.
[4] P. D. Landor, S. R. Landor, and S. Mukasa, *Chem. Comm.*, 1971, 1638.
[5] S. N. Hooper and R. G. Ackman, *Lipids*, 1971, **6**, 341.
[6] A. J. Fulco, *J. Biol. Chem.*, 1970, **245**, 2985.
[7] R. K. Dart and T. Kaneda, *Biochim. Biophys. Acta*, 1970, **218**, 189.
[8] Shô Itô, K. Endo, and S. Inoue, *Tetrahedron Letters*, 1971, 4011.
[9] C. Pace-Asciak and L. S. Wolfe, *Chem. Comm.*, 1970, 1234, 1235; *Biochemistry*, 1971, **10**, 3657; C. Pace-Asciak, *ibid.*, p. 3664.

Fatty Acids and Related Compounds

$$CH_3(CH_2)_5 \overset{H}{\underset{OH}{\diagup}} \overset{H}{\underset{(CH_2)_4CO_2H}{O}}$$
(3)

acids (4).[10] The C_{17} member has previously been recognized as a minor component (0.01%) of butterfat, possibly from bovine rumen bacteria.

$$\text{Cyclohexyl}-(CH_2)_n CO_2 H$$

(4) $n = 10$ or 12

The component acids of *Monnina emarginata* seed oil contain 13(*S*)-hydroxyoctadeca-*cis*-9,*trans*-11-dienoic acid (~30%), enantiomeric with the known (*R*)-coriolic acid. Although mainly present as triglyceride, some of the acid (~4%) occurs as the lactone (a 14-membered ring).[11] More recently, the 12-membered ring lactone of 11-hydroxydodec-*trans*-8-enoic acid has been isolated from the fungus *Cephalosporium recifei*, grown on glucose solution.[12]

The presence of 5,8,12-trihydroxyoctadeca-*trans*-9-enoic acid in wheat bran has been demonstrated.[13] Three hitherto unknown oxygenated acids [(5)—(7)] have been found in *Stenachaenium macrocephalum* after storage for 2 years at 5 °C.[14] These are probably derived from the uncommon octadecatrienoic acid (3*t*9*c*12*c*) present in this seed oil, and these results are reminiscent of the isolation of unsaturated epoxy- and hydroxy-acids from sunflower seed oil after prolonged storage.[15]

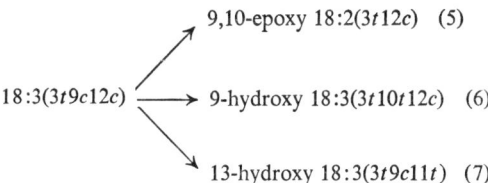

[10] M. de Rosa, A. Gambacorta, L. Minale, and J. D. Bu'Lock, *Chem. Comm.*, 1971, 1334.
[11] B. E. Phillips and C. R. Smith, jun., *Biochim. Biophys. Acta*, 1970, **218**, 71; B. E. Phillips, C. R. Smith, jun., and L. W. Tjarks, *ibid.*, 1970, **210**, 353; *J. Org. Chem.*, 1970, **35**, 1916.
[12] R. F. Vesonder, F. H. Stodola, L. J. Wickerham, J. J. Ellis, and W. K. Rohwedder, *Canad. J. Chem*, 1971, **49**, 2029.
[13] P. W. Albro and L. Fishbein, *Phytochemistry*, 1971, **10**, 631.
[14] R. Kleiman, G. F. Spencer, L. W. Tjarks, and F. R. Earle, *Lipids*, 1971, **6**, 617.
[15] K. L. Mikolajczak, R. M. Freidinger, C. R. Smith, jun., and I. A. Wolff, *Lipids*, 1968, **3**, 489; K. L. Mikolajczak, C. R. Smith, jun., and I. A. Wolff, *J. Amer. Oil Chemists' Soc.*, 1970, **47**, 24.

180   Aliphatic, Alicyclic, and Saturated Heterocyclic Chemistry

The acetylenic epoxy-acid (9), *cis*-9,10-epoxyoctadec-12-ynoic acid, discovered along with other epoxy-acids and other acetylenic acids in the seed oil of *Helichrysum bracteatum*,[16] may be an intermediate in the bioconversion of crepenynic acid (8) to helenynolic acid (10), both of which had been identified previously in this source.

$$CH_3(CH_2)_4C\overset{c}{\equiv}CCH_2CH\overset{}{=}CH(CH_2)_7CO_2H \quad (8)$$

$$CH_3(CH_2)_4C\equiv CCH_2\overset{O}{\overset{}{CH-CH}}(CH_2)_7CO_2H \quad (9)$$

$$CH_3(CH_2)_4C\equiv CCH\overset{t}{=}CHCH(OH)(CH_2)_7CO_2H \quad (10)$$

Two 2*t* acids (22:1 and 24:1) have been identified as natural products for the first time.[17] They occur in wheat leaf wax, esterified with C_9—C_{12} αω-diols.

Evidence has been adduced that lactones, present as minor components of animal and vegetable fats, are optically active. Both enantiomers are sometimes present but they are probably formed by different pathways.[18]

French workers have continued their study of mycolic acids.[19]

Other Long-chain Compounds.—The absolute configuration of phthiocerol A (11a), phthiotriol A (11b), and phthiodolone A (11c) has been established:[20]

$$CH_3\cdot(CH_2)_n\cdot\overset{R}{CH(OH)}\cdot CH_2\cdot\overset{R}{CH(OH)}\cdot(CH_2)_4\cdot\overset{R}{CHMe}\cdot X\cdot CH_2\cdot CH_3$$

$$n = 20 \text{ and } 22$$

[11a; X = (*S*)-CH(OMe)]   [11b; X = (*S*)-CH(OH)]   [11c; X = C=O]

Waxes.—The development of the newer spectroscopic and chromatographic techniques has led to a renewed interest in wax composition and structure. Among commercial waxes, carnauba wax[21] and beeswax[22] have been reexamined. Other reports detail the hydrocarbons present in the surface lipids of cockroaches,[23] the long-chain β-hydroxy-ketones in cabbage surface lipids,[24] and the diesters of diols in wheat leaf wax.[17] A review of plant waxes has also appeared (see ref. 155*f*).

---

[16] H. B. S. Conacher and F. D. Gunstone, *Lipids*, 1970, **5**, 137.
[17] A. P. Tulloch, *Lipids*, 1971, **6**, 641.
[18] B. van der Ven and K. de Jong, *J. Amer. Oil Chemists' Soc.*, 1970, **47**, 299.
[19] C. Asselineau, G. Tocanne, and J. F. Tocanne, *Bull. Soc. chim. France*, 1970, 1455; M. A. Laneélle and G. Laneélle, *European J. Biochem.*, 1970, **12**, 296; M. Welby-Gieusse, M. A. Laneélle, and J. Asselineau, *ibid.*, 1970, **13**, 164.
[20] M. Welby-Gieusse and J. F. Tocanne, *Tetrahedron*, 1970, **26**, 2875; K. Maskens and N. Polgar, *Chem. Comm.*, 1970, 673.
[21] L. E. Vandenburg and E. A. Wilder, *J. Amer. Oil Chemists' Soc.*, 1970, **47**, 514.
[22] A. P. Tulloch, *Chem. and Phys. Lipids*, 1971, **6**, 235.
[23] K. Tartivita and L. L. Jackson, *Lipids*, 1970, **5**, 35; L. L. Jackson, *ibid.*, p. 38.
[24] H. H. O. Schmid and P. C. Bandi, *J. Lipid Res.*, 1971, **12**, 198.

Fatty Acids and Related Compounds

New Lipids.—Studies, mainly by Mikolajczak and Smith,[25] have unravelled the structures of the unusual alcohols in the cyano-lipids (13—55%) which accompany the more conventional triglycerides in several sapindaceous seed oils and in one borage. The branched-chain C_5 cyano-alcohols (12)—(15) occur as mono- or di-acyl derivatives

$$\underset{(12)}{HOCH_2\overset{\overset{CH_2}{\|}}{C}CH(OH)CN} \qquad \underset{(13)}{CH_3\overset{\overset{CH_2}{\|}}{C}CH(OH)CN}$$

$$\underset{(14)}{HOCH_2\overset{\overset{CH_3}{|}}{C}{=}CHCN} \qquad \underset{(15)}{HOCH_2\overset{\overset{CH_2OH}{|}}{C}{=}CHCN}$$

Most phospholipids contain glycerol, phosphoric acid, a short-chain hydroxy-compound, and long-chain acids or alcohols. Minor variants of this pattern and new ways of associating these components continue to be identified, as for example the phosphatidyl-*N*-(2-hydroxyethyl)alanine (16) from rumen protozoa.[26]

$$\begin{array}{c} CH_2OCOR \\ | \\ RCOOCH \quad O \\ | \quad \| \\ CH_2OPOCH_2CH_2\overset{+}{N}H_2CHMeCO_2^- \\ | \\ O \\ (16) \end{array}$$

The so-called triglycerides of starfish contain ~35% of the monoacyl monoalkyl (or monoalk-1'-enyl) derivatives of ethanediol.[27] It is considered that the fatty tissues in the melon and jaw oils of dolphins, porpoises, and toothed whales may play a major role in the echolocation system of these animals. Litchfield *et al.*[28] have examined the melon oil and blubber of the Beluga whales, and report that the former remains clear at temperatures well below 0 °C, a property which is related to its unusual triglyceride composition.

When glycerol 1-phosphorylcholine is treated with an acyl halide, the expected phosphatidylcholine is accompanied by 'cyclic lysolecithin,'

---

[25] D. S. Seigler, K. L. Mikolajczak, C. R. Smith, jun., I. A. Wolff, and R. B. Bates, *Chem. and Phys. Lipids*, 1970, **4**, 147; K. L. Mikolajczak, C. R. Smith, jun., and L. W. Tjarks, *Biochim. Biophys. Acta*, 1970, **210**, 306; *Lipids*, 1970, **5**, 672, 812; K. Mikolajczak and C. R. Smith, jun., *ibid.*, 1971, **6**, 349; D. Seigler, F. Seamen, and T. J. Mabry, *Phytochemistry*, 1971, **10**, 485.
[26] P. Kemp and R. M. C. Dawson, *Biochem. J.*, 1969, **113**, 555.
[27] V. A. Vaver, N. A. Pisareva, B. V. Rozynov, and A. N. Ushakov, *Chem. and Phys. Lipids*, 1971, **7**, 75.
[28] C. Litchfield, R. G. Ackman, J. C. Sipos, and C. A. Eaton, *Lipids*, 1971, **6**, 674.

thought to have the structure (17a) or (17b). This has now been shown to be 1-chlorophosphoglyceride (18), with minor amounts of the 2-chloro-isomer.[29]

$$
\begin{array}{c}
\text{CH}_2\text{—O} \\
\text{RCO}_2\text{CH} \quad\quad \text{P} \\
\text{CH}_2\text{—O} \quad \text{OCH}_2\text{CH}_2\overset{+}{\text{N}}\text{Me}_3 \\
(17\text{a})
\end{array}
\qquad
\begin{array}{c}
\text{CH}_2\text{OCOR} \\
\text{CHO} \\
\text{CH}_2\text{O} \quad\quad \text{P} \\
\text{CH}_2\text{O} \quad \text{OCH}_2\text{CH}_2\overset{+}{\text{N}}\text{Me}_3 \\
(17\text{b})
\end{array}
$$

$$
\begin{array}{c}
\text{CH}_2\text{Cl} \\
\text{RCO}_2\text{CH} \\
\text{CH}_2\text{OPCH}_2\text{CH}_2\overset{+}{\text{N}}\text{Me}_3 \\
\text{O}^- \\
(18)
\end{array}
$$

The chemistry of the sulpholipids has been reviewed (see ref. 152*f*). The major sulpholipid from *Ochromonas danica* is a hexachloro-C_{22} disulphate (19).[30]

$$\text{CH}_3(\text{CH}_2)_5\text{CHClCHClCH(OSO}_3\text{H)CHClCH}_2\text{CHCl(CH}_2)_8\text{CCl}_2\text{CH}_2\text{OSO}_3\text{H}$$
(19)

3 Synthetic Compounds

Fatty Acids and Related Compounds.—Although C_{18} acids are still the most common — whether natural or synthetic — C_{14} compounds attract increasing interest, and in the period under review standard synthetic procedures involving acetylenic intermediates and/or the Wittig reaction have been used to prepare *cis*-tetradec-9-enyl acetate (a pheromone),[31] racemic methyl tetradeca-*trans*-2,4,5-trienoate (an allenic sex hormone produced by the male Dried Bean beetle),[4] and several geometrical isomers of tetradeca-3,5-dienoic acid (megatomoic acid, the sex attractant of the Black Carpet beetle).[32]

A C_{17} hydroxy-acid (12-hydroxyheptadeca-*trans*-8,*trans*-10-dienoic acid), a by-product of prostaglandin synthesis, has been prepared from short-chain starting materials.[33]

The synthesis of labelled and unlabelled methyl malvalate (20) and methyl sterculate (21) has attracted attention. In the best known route an acetylenic

$$
\begin{array}{c}
\quad\quad \text{CH}_2 \\
\text{CH}_3(\text{CH}_2)_7\text{C}=\text{C}(\text{CH}_2)_n\text{CO}_2\text{R}
\end{array}
\qquad
\begin{array}{l}
\text{methyl malvalate, } n = 6 \text{ (20; R = Me)} \\
\text{methyl sterculate, } n = 7 \text{ (21; R = Me)}
\end{array}
$$

[29] R. Aneja and J. S. Chadha, *Biochim. Biophys. Acta*, 1971, **239**, 84.
[30] J. Elovson and P. R. Vagelos, *Biochemistry*, 1970, **9**, 3110.
[31] H. J. Bestman, P. Range, and R. Kunstmann, *Chem. Ber.*, 1971, **104**, 65.
[32] J. O. Rodin, M. A. Leaffer, and R. M. Silverstein, *J. Org. Chem.*, 1970, **35**, 3152.
[33] E. Crundwell and A. L. Cripps, *Chem. and Ind.*, 1971, 767.

Fatty Acids and Related Compounds

compound is treated with ethyl diazoacetate and the resulting cyclopropene-carboxylic acid is decarboxylated as shown in Scheme 1.[34] It has been

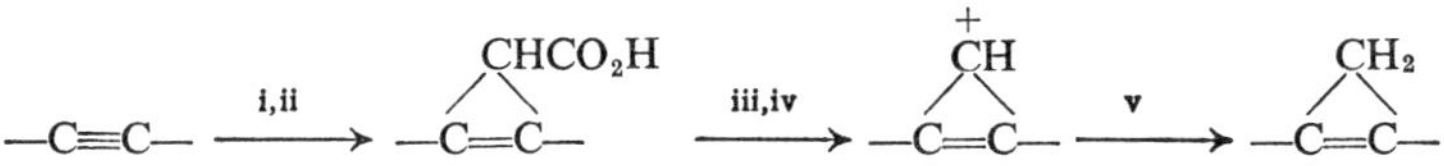

$$-C{\equiv}C- \xrightarrow{i, ii} -\overset{\overset{\displaystyle CHCO_2H}{\triangle}}{C{=}C}- \xrightarrow{iii, iv} -\overset{\overset{\displaystyle \overset{+}{C}H}{\triangle}}{C{=}C}- \xrightarrow{v} -\overset{\overset{\displaystyle CH_2}{\triangle}}{C{=}C}-$$

Reagents: i, N_2CHCO_2Et, copper bronze; ii, KOH; iii, $SOCl_2$ or $ClCOCOCl$; iv, $ZnCl_2$; v, $NaBH_4$.

Scheme 1

claimed that the overall yield (∼30%) of this process can be raised to 60—65% by the use of fluorosulphonic acid, as shown in the sequence of Scheme 2.[35]

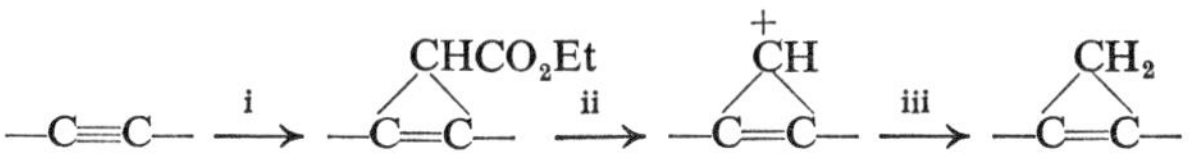

$$-C{\equiv}C- \xrightarrow{i} -\overset{\overset{\displaystyle CHCO_2Et}{\triangle}}{C{=}C}- \xrightarrow{ii} -\overset{\overset{\displaystyle \overset{+}{C}H}{\triangle}}{C{=}C}- \xrightarrow{iii} -\overset{\overset{\displaystyle CH_2}{\triangle}}{C{=}C}-$$

Reagents: i, N_2CHCO_2Et, copper bronze; ii, FSO_3H, CH_2Cl_2; iii, $NaBH_4$.

Scheme 2

Methyl sterculate (21) was also obtained from methyl stearolate (18:1 9a), albeit in lower yield (∼10%), by irradiation in the presence of diazomethane.[36]

Several methylene-interrupted polyunsaturated acids, including some labelled compounds, required usually for metabolic studies, have been prepared by standard procedures. The Oxford group[37] have prepared crepenynic acid (18:2 9c12a) and polyethenoid acids, including members of the ω3 series (19:5, 20:5, 22:5, 22:6, 24:6), the ω4 series (18:4), the ω6 series (19:4, 20:2, 20:3, 20:4, 22:4, 22:5, 24:5), the ω7 series (18:2, 18:3, 20:2, 20:3), and the ω8 series (22:4), along with the following non-methylene-interrupted acids: 20:3(8c12t14c), 20:4(4c8c11c14c and 8c11c14c18c), and 21:4(2t8c11c14c).[38] Kunau's[39] syntheses are based on improved preparations of the intermediate alkynes (22) and alkynoic acids (23), the former being obtained from the corresponding methyl ethers (22; X = OMe) by reaction with the appropriate acetyl halide.

$CH_3(CH_2)_m[C{\equiv}CCH_2]_nX$ $HC{\equiv}CCH_2C{\equiv}C(CH_2)_nCO_2H$
(22; n = 2—4, X = Br or I)   (23; n = 2—6)

[34] W. J. Gensler, K. B. Pober, D. M. Solomon, and M. B. Floyd, *J. Org. Chem.*, 1970, **35**, 2301; W. J. Gensler, M. B. Floyd, R. Yanase, and K. W. Pober, *J. Amer. Chem. Soc.*, 1970, **92**, 2472; W. J. Gensler, D. M. Solomon, R. Yanase, and K. W. Pober, *Chem. and Phys. Lipids*, 1971, **6**, 280.
[35] J. L. Williams and D. S. Sgoutas, *J. Org. Chem.*, 1971, **36**, 3064.
[36] E. R. Altenburger, J. W. Berry, and A. J. Deutschman, jun., *J. Amer. Oil Chemists' Soc.*, 1970, **47**, 77.
[37] R. W. Bradshaw, A. C. Day, E. R. H. Jones, C. B. Page, V. Thaller, and R. A. Vere Hodge, *J. Chem. Soc.* (*C*), 1971, 1156.
[38] (a) R. K. Beerthuis, D. H. Nugteren, H. J. J. Pabon, A. Steenhoek, and D. A. van Dorp, *Rec. Trav. chim.*, 1971, **90**, 943; (b) H. W. Sprecher, *Biochim. Biophys. Acta*, 1971, **231**, 122; J. Budny and H. W. Sprecher, *ibid.*, 1971, **239**, 190; W. H. Kunau, H. Lehmann, and R. Gross, *Z. physiol. Chem.*, 1971, **352**, 542.
[39] W. H. Kunau, *Chem. and Phys. Lipids*, 1971, **7**, 101, 108.

In continuation of a project in which a complete series of isomeric acids is synthesized in order to study their physical, chemical, and biological properties,[40] reports have now appeared on all the cis- and trans-octadecenoic acids, all the octadecynoic acids, on many octadecadiynoic and octadecadienoic acids, and on cyclopropane esters made from the mono- and dienoates.[41] Details of the chromatographic and spectroscopic behaviour of these isomeric esters are included. The cis- and trans-octadecenoic acids show an alternation in the melting points when related to the position of unsaturation: this is not apparent among isomeric octadecynoic acids.

It is sometimes possible to prepare a long-chain acid by modification of a more readily available acid of the same chain length. For example, a new octadecatrienoic acid (2t9c12c) has been prepared from linoleic acid, and it has been evaluated as a honeybee sex attractant.[42] Also, rac-helenynolic acid (10) has been prepared from natural crepenynic acid (8) by a route (Scheme 3) which may be of biosynthetic interest.[16]

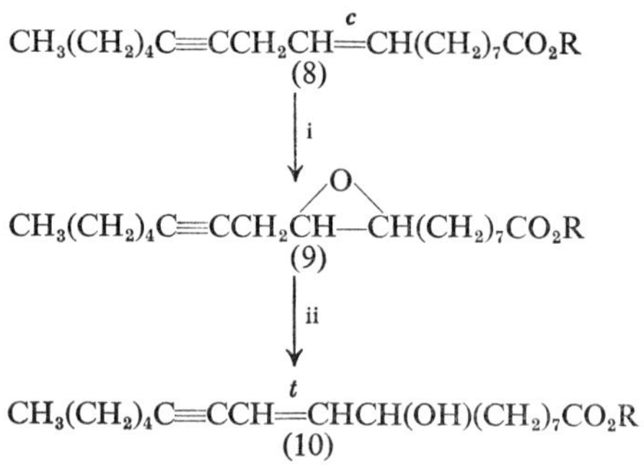

$$CH_3(CH_2)_4C{\equiv}CCH_2CH\overset{c}{=}CH(CH_2)_7CO_2R$$
(8)

↓ i

$$CH_3(CH_2)_4C{\equiv}CCH_2\overset{O}{\overset{\diagup\,\diagdown}{CH-CH}}(CH_2)_7CO_2R$$
(9)

↓ ii

$$CH_3(CH_2)_4C{\equiv}CCH\overset{t}{=}CHCH(OH)(CH_2)_7CO_2R$$
(10)

Reagents: i, ArCO$_3$H; ii, LiNEt$_2$

Scheme 3

Cyclic compounds continue to find use as chain extenders. 5-Methylcyclohexane-1,3-dione has been used for the synthesis of 3-methyl- and 3,6-dimethyl-tridecanoic ester,[43] and the enamine synthesis, usually utilizing cyclopentanone or cyclohexanone, has been extended to the use of cyclodecanone and cyclododecanone. Reaction occurs through a cyclobutanone

---

[40] E. C. Reitz, M. El-Sheikh, W. E. M. Lands, I. A. Ismail, and F. D. Gunstone, *Biochim. Biophys. Acta*, 1969, **176**, 480; H. M. Jenkin, L. E. Anderson, R. T. Holman, I. A. Ismail, and F. D. Gunstone, *J. Bacteriol.*, 1969, **98**, 1026; *Exp. Cell. Res.*, 1970, **59**, 1; H. Okuyama, W. E. M. Lands, W. W. Christie, and F. D. Gunstone, *J. Biol. Chem.*, 1969, **244**, 6514; H. J. Goller, D. S. Sgoutas, I. A. Ismail, and F. D. Gunstone, *Biochemistry*, 1970, **9**, 3072.

[41] F. D. Gunstone and M. Lie Ken Jie, *Chem. and Phys. Lipids*, 1970, **4**, 1, 131, 139; F. D. Gunstone, M. Lie Ken Jie, and R. T. Wall, *ibid.*, 1971, **6**, 147; J. A. Barve and F. D. Gunstone, *ibid.*, 1971, **7**, 311; S. G. Morris, *J. Amer. Oil Chemists' Soc.*, 1971, **48**, 376.

[42] A. N. Starratt and R. Boch, *Canad. J. Chem.*, 1971, **49**, 251.

[43] P. D. Grimwood, D. E. Minnikin, N. Polgar, and J. E. Walker, *J. Chem. Soc. (C)*, 1971, 870.

Fatty Acids and Related Compounds 185

intermediate (24), and acids formed from $R^{14}CO_2H$ are labelled at C-1 and at C-11 (or C-13).[44]

$$\underset{n\ =\ 8\ \text{or}\ 10}{\underset{(CH_2)_n}{\overset{}{\bigg(}}} \overset{}{\underset{}{N}}\overset{}{\underset{}{C}}{=}CH \xrightarrow{RCH_2{}^{14}COCl} \underset{(CH_2)_n}{\underset{}{\bigg(}}\overset{N}{\underset{}{\overset{|}{C}}}\overset{R}{\underset{CH}{\overset{|}{CH}}}{}^{14}C{=}O \quad (24)$$

$$(CH_2)_{n+1}\overset{CO}{\underset{{}^{14}CO}{\bigg\langle}}CHR \longrightarrow RCH_2{}^{14}CO(CH_2)_{n+1}CO_2H + RCH_2CO(CH_2)_{n+1}{}^{14}CO_2H$$

A new method of preparing 2-ynoic acids (and hence *cis*-2-enoic acids) from 3-oxo-esters is reported.[45] A series of branched-chain acids (25) was obtained by Wittig condensation of a phosphorane with methyl 10-oxo-hexadecanoate or methyl 12-oxo-octadecanoate.[46]

$$RCH{=}PPh_3 + CH_3(CH_2)_5CO(CH_2)_nCO_2Me \longrightarrow CH_3(CH_2)_5\underset{\underset{(25)}{CHR}}{\overset{\|}{C}}(CH_2)_nCO_2Me$$

(R varies from H to $C_{17}H_{35}$) $n = 8$ or 10

New methods of preparing alkenes have been outlined[47] but not yet applied to the synthesis of long-chain acids. Stereoselective and stereospecific olefin syntheses have been reviewed.[48]

Glycerides.—There have been no radical developments in the area of glyceride syntheses but standard methods have been reviewed (see ref. 155*a*) and continue to be used and improved, particularly for enantiomeric glycerides of known configuration.

Symmetrical 1,3-diacylglycerols have been made by acylation and reduction of dihydroxyacetone.[49] One French report describes several ways of making triacylglycerols having three different acyl groups[50] and another is concerned with the mono- and di-oleates of 2-methyl- and 2,2-dimethyl-butane-1,4-diol.[51]

1,2- and 2,3-Diacyl-*sn*-glycerols with mixed acyl groups have been prepared from D-mannitol[52] *via* the appropriate glycerol carbonates with the help of

---

[44] J. S. V. Hunter and R. J. Light, *Biochemistry*, 1970, **9**, 4283.
[45] A. Silveira, jun., T. J. Weslowski, T. A. Weil, V. Kumar, and J. P. Gillespie, *J. Amer. Oil Chemists' Soc.*, 1971, **48**, 661.
[46] D. G. Chasin and E. G. Perkins, *Chem. and Phys. Lipids*, 1971, **6**, 8.
[47] D. H. R. Barton and B. J. Willis, *Chem. Comm.*, 1970, 1225; D. H. R. Barton, B. J. Willis, and E. H. Smith, *ibid.*, p. 1226.
[48] J. Reucroft and P. G. Sammes, *Quart. Rev.*, 1971, **25**, 135.
[49] P. H. Bentley and W. McCrae, *J. Org. Chem.*, 1970, **35**, 2082.
[50] J. P. Careau, *Bull. Soc. chim. France*, 1970, 4104, 4107, 4111.
[51] M. Derbesy, M. Chicheportiche, and M. Naudet, *Bull. Soc. chim. France*, 1971, 1789
[52] D. Buchnea, *Lipids*, 1971, **6**, 734.

2,2,2-trichloroethyl chloroformate.[53] Certain advantages are claimed for the trichloroethoxycarbonyl protecting group.

Other Lipids.—Baer and his colleagues[54] have synthesized several phosphonic acid analogues [such as (26)] of the more familiar phosphoglycerides, and the phosphinic acid derivatives (27) have also been prepared.[55]

$$
\begin{array}{cc}
CH_2OCOR & CH_2OR^1 \\
| & | \\
RCO_2CH & CHOR^2 \\
| \quad O & | \quad O \\
| \quad \| \quad + & | \quad \| \quad + \\
CH_2OPCH_2CH_2NMe_3 & (CH_2)_nPCH_2CH_2NMe_3 \\
| & | \\
O^- & O^- \\
(26) & (27;\ n\ =\ 1\ \text{or}\ 2)
\end{array}
$$

Tri-isopropylbenzenesulphonyl chloride condenses alcohols and phosphatidic acids in 60—90% yield. Phospholipids produced by this method have a higher molar rotation than previously reported, an observation which raises doubts about the purity of previous preparations.[56] 1,2-Dipalmitoyl-*sn*-glycerol 3-(2-aminoethyl hydrogen phosphate) has been synthesized, labelled with ^{3}H, ^{14}C, and ^{32}P.[57]

$\alpha\beta$-Unsaturated ether derivatives (plasmalogens) of both ethanediol and glycerol have been synthesized.[58]

Continued study of sphingosine compounds has resulted in the synthesis of *cis*- and *trans*-sphingosine by Wittig condensation of $CH_3(CH_2)_{12}CH=PPh_3$ with an appropriate ribose derivative,[59] and the synthesis of D-dihydrosphingosine,[59] of all the racemic diastereoisomers of phytosphingosine,[60] and of D(+)-*ribo*-2-amino-1,3,4-trihydroxyeicosane (C_{20}-phytosphingosine).[61]

4 Physical Properties

Gas–Liquid Chromatography.—Much structural information can be derived from the chromatographic behaviour of long-chain esters, and this topic has been reviewed (see ref. 153c). Useful retention data are presented in several

---

[53] F. R. Pfeiffer, C. K. Miao, and J. A. Weisbach, *J. Org. Chem.*, 1970, **35**, 221.
[54] E. Baer and K. V. Jagannadha Rao, *Canad. J. Biochem.*, 1970, **48**, 184; E. Baer and S. K. Pavanaram, *ibid.*, pp. 221, 979, 988.
[55] A. F. Rosenthal and S. V. Chodsky, *J. Lipid Res.*, 1971, **12**, 277.
[56] R. Aneja, J. S. Chadha, and A. P. Davies, *Biochim. Biophys. Acta*, 1970, **218**, 102; R. Aneja and J. S. Chadha, *Chem. and Phys. Lipids*, 1970, **4**, 60; *Biochim. Biophys. Acta*, 1971, **248**, 455.
[57] J. S. Owen, G. H. Scott, M. S. Harvey, and J. D. Billimoria, *Chem. and Ind.*, 1971, 727.
[58] J. K. G. Kramer and H. K. Mangold, *Chem. and Phys. Lipids*, 1970, **4**, 332; I. B. Vtorov, G. A. Serebrennikovo, and R. P. Evstigneeva, *Tetrahedron Letters*, 1971, 4605.
[59] E. J. Reist and P. H. Christie, *J. Org. Chem.*, 1970, **35**, 3521, 4127.
[60] K. Sisido, N. Hirowatari, H. Tamura, H. Kobata, H. Takagisi, and T. Isida, *J. Org. Chem.*, 1970, **35**, 350.
[61] A. Kisić, N. Ž. Stanaćev, L. Gospoćić, and M. Prostenik, *Chem. and Phys. Lipids* 1970, **7**, 135.

Fatty Acids and Related Compounds

papers[62,63] and a further report on the computerized treatment of g.l.c. data has appeared.[64]

Polymorphism.—The polymorphism of monoacid triglycerides from C_8 to C_{22} has been examined.[65] Both the odd and even series show three distinct melting levels and there is alternation of melting points between the two series. A French group have examined binary mixtures of the six possible glycerides containing palmitic and/or stearic acid by differential thermal analysis, and found not more than two forms in any of the mixtures.[65]

Infrared and Raman Spectra.—The solid-state i.r. spectra of many deuteriated long-chain acids and esters have been examined by Dinh-Nguyên and Fischmeister, who find the spectra of the isomeric *vic*-dideuteriostearates to be characteristic and therefore useful for identification.[66] The i.r. spectra of over one hundred long-chain vinyl derivatives and the application of laser Raman spectra to drying oils and alkyd resins have been discussed.[67]

Nuclear Magnetic Resonance Spectroscopy.—Long-chain acids and their derivatives contain many protons which cannot be distinguished by n.m.r. spectroscopy, and this has limited the value of this technique in the study of these compounds. This deficiency may be partly remedied by the use of more sophisticated instruments and by the application of chemical shift reagents. The brief paper by Swern and Wineburg[68] on this latter topic must surely be the forerunner of many more. Using the transition-metal complex Eu(fod)$_3$,* these authors distinguished separate peaks for all the protons designated *a—j* in the methyl oleate structure:

$$\overset{a}{CH_3}(CH_2)_6\overset{b}{CH_2}\overset{c}{CH}=\overset{c}{CH}\overset{d}{CH_2}\overset{d}{CH_2}\overset{e}{CH_2}\overset{f}{CH_2}\overset{h}{CH_2}\overset{i}{CH_2}\overset{j}{CH_2}CO_2CH_3$$

Frost and Barzilay,[69] using a 220 MHz spectrometer, claim that significant long-range deshielding effects may operate up to a distance of six carbon atoms. They report, for example, that all the *cis*-octadecenoates can be distinguished except for the Δ^{10}- and Δ^{11}-isomers, and that it is possible to characterize double bonds in polyunsaturated acids.

The earlier review by Hopkins[70a] has been updated by a list of references to

---

* fod is 1,1,1,2,2,3,3-heptafluoro-7,7-dimethyloctane-4,6-dione.

[62] C. R. Scholfield and H. J. Dutton, *J. Amer. Oil Chemists' Soc.*, 1970, **47**, 1.
[63] F. D. Gunstone and M. Lie Ken Jie, *Chem. and Phys. Lipids*, 1970, **4**, 131, 139; F. D. Gunstone, M. Lie Ken Jie, and R. T. Wall, *ibid.*, 1971, **6**, 147.
[64] A. W. Boyne and W. R. H. Duncan, *J. Lipid Res.*, 1970, **11**, 293.
[65] E. S. Lutton and A. J. Fehl, *Lipids*, 1970, **5**, 90; R. Perron, J. Petit, and A. Mathieu, *Chem. and Phys. Lipids*, 1971, **6**, 58.
[66] N. Dinh-Nguyên and I. Fischmeister, *Arkiv Kemi.*, 1970, **32**, 181, 205.
[67] G. E. McManis and L. E. Gast, *J. Amer. Oil Chemists' Soc.*, 1971, **48**, 668; L. A. O'Neill and N. A. R. Falla, *Chem. and Ind.*, 1971, 1349.
[68] D. Swern and J. P. Wineburg, *J. Amer. Oil Chemists' Soc.*, 1971, **48**, 371.
[69] D. J. Frost and J. Barzilay, *Analyt. Chem.*, 1971, **43**, 1316; *Rec. Trav. chim.*, 1971, **90**, 705.
[70] (a) C. Y. Hopkins, *Progr. Chem. Fats and Lipids* (ed. R. T. Holman), 1966, **8**, 213; (b) O. Suzuki, T. Hashimoto, K. Hayamizu, and O. Yamamoto, *Lipids*, 1970, **5**, 457; F. D. Gunstone, M. Lie Ken Jie, and R. T. Wall, *Chem. and Phys. Lipids*, 1969, **3**, 297; *ibid.*, 1971, **6**, 147.

the n.m.r. spectra of long-chain acids and their derivatives (see ref. 154*f*). The n.m.r. spectra of four conjugated C_{18} dienoates and trienoates, and of some methyl octadecadiynoates and octadecadienoates, and of the biscyclopropane esters derived from the dienoates, are detailed.[70b] The ratio of *cis*- and *trans*-isomers in a mixture of monoenoates can be determined by n.m.r. examination of the methoxymercuriacetates; in particular, the signals due to the O$\underline{C}$H$_3$ group in the *threo*- and *erythro*-adducts[71] (see also ref. 121). This spectral procedure has been applied to the study of internal and external membrane surfaces using paramagnetic Mn^{2+} and Eu^{3+} ions.[72]

Mass Spectrometry.—Mass spectrometry is widely employed by lipid chemists and its application to long-chain compounds has been reviewed.[153f] In addition to the routine use of this technique in the identification of unknown compounds (see, for example, refs. 100—102), it has been applied systematically to the study of many classes of lipids, including: perdeuteriated fatty acid esters,[73] the branched-chain esters (25),[46] alkyl (and alk-1-enyl) ether esters of diols,[74] mono- and di-alkyl ethers of diols,[75] deuteriated glycerol 1,3-distearates,[76] deuterium-labelled triglycerides and triglyceride mixtures,[77] glyceryl ether diesters,[78], wax esters,[79] phosphatidylcholines,[80] various glycerophospholipids,[81] alkyl 2-diethylphosphonoalkanoates,[82] synthetic and natural ceramides,[83] and neutral mono- and di-sialoglycosphingolipids.[84]

The application of negative-ion mass spectrometry to the identification of long-chain aldehydes and alcohols has been discussed.[85]

5 Chemical Properties

In addition to the carboxy-group, most natural long-chain acids contain one or more olefinic and/or acetylenic centres and, occasionally, an additional oxygenated function (hydroxy or epoxy). The reactions and interactions of these polyfunctional systems continue to be actively explored.

[71] K. Schaumburg, *Lipids*, 1970, **5**, 505.
[72] V. F. Bystrov, N. I. Dubrovina, L. I. Barsukov, and L. D. Bergelson, *Chem. and Phys. Lipids*, 1971, **6**, 343.
[73] G. Wendt and J. A. McCloskey, *Biochemistry*, 1970, **9**, 4854.
[74] J. K. G. Kramer, W. J. Baumann, and R. T. Holman, *Lipids*, 1971, **6**, 492.
[75] J. K. G. Kramer, R. T. Holman, and W. J. Baumann, *Lipids*, 1971, **6**, 727.
[76] A. Morrison, M. D. Barratt, and R. Aneja, *Chem. and Phys. Lipids*, 1970, **4**, 47.
[77] W. M. Lauer, A. J. Aasen, G. Graff, and R. T. Holman, *Lipids*, 1970, **5**, 861; A. J. Aasen, W. M. Lauer, and R. T. Holman, *ibid.*, p. 869; R. A. Hites, *Analyt. Chem.*, 1970, **42**, 1736.
[78] P. R. Le Tellier and W. W. Nawar, *J. Agric. Food Chem.*, 1971, **19**, 196.
[79] A. J. Aasen, H. H. Hofstetter, B. T. R. Iyengar, and R. T. Holman, *Lipids*, 1971, **6**, 502.
[80] R. A. Klein, *J. Lipid Res.*, 1971, **12**, 123, 628.
[81] J. H. Duncan, W. J. Lennarz, and C. C. Fenselau, *Biochemistry*, 1971, **10**, 927.
[82] D. G. Chasin and E. G. Perkins, *Chem. and Phys. Lipids*, 1971, **6**, 311.
[83] K. Samuelsson and B. Samuelsson, *Chem. and Phys. Lipids*, 1970, **5**, 44; K. Samuelsson, B. Samuelsson, and S. Hammarström, *J. Lipid Res.*, 1970, **11**, 150; S. Hammarström, *ibid.*, p. 175.
[84] G. Dawson and C. C. Sweeley, *J. Lipid Res.*, 1971, **12**, 56.
[85] P. C. Rankin, *Lipids*, 1970, **5**, 825.

Fatty Acids and Related Compounds

Double-bond Migration and Cyclization Reactions.—Linolenic acid (18:3 9c12c15c) is converted by potassium hydroxide at 180 °C or by sodium hydroxide at 220 °C, first to a conjugated trienoic acid, and then to a di-substituted cyclohexadiene (13—21%), which is identified after aromatization with palladium–charcoal. A series of octadecatrienoic acids, when submitted to this reaction, gave the product(s) indicated in parentheses: 9c12c15c (28; $n = 3$), 9t12t15t (28; $n = 2, 3$, and 4), 9c11t13t (28; $n = 3, 4$, and 5),

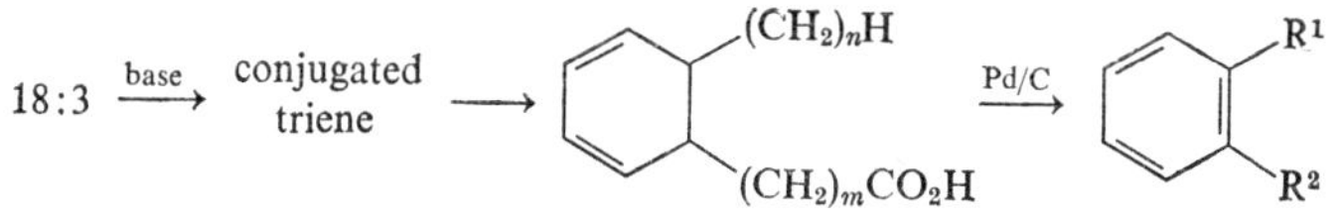

$$18:3 \xrightarrow{\text{base}} \text{conjugated triene} \longrightarrow \underset{(28;\ m+n=11)}{\text{[cyclohexadiene with }(CH_2)_nH\text{ and }(CH_2)_mCO_2H]} \xrightarrow{Pd/C} \text{[arene with }R^1, R^2]$$

and 9t11t13t (28; $n = 3, 4$, and 5).[86] The base-catalysed isomerization of acetylenes has been reviewed.[87]

An Indian group[88] report that the thermal cyclization of methyl α-eleo-stearate (18:3 9c11t13t) and of its 18-hydroxy-derivative (kamlolenate) in the presence of sulphur furnishes disubstituted cyclohexadienes [such as (29)], with the double-bond positions varying somewhat with the reaction temperature. The diene (29) obtained at 160 °C undergoes an interesting reaction with dimethyl but-2-ynedioate (acetylenedicarboxylic ester) since the Diels–Alder adduct is decomposed to methyl phthalate and methyl tetradec-cis-9-enoate. The results are summarized in Scheme 4.

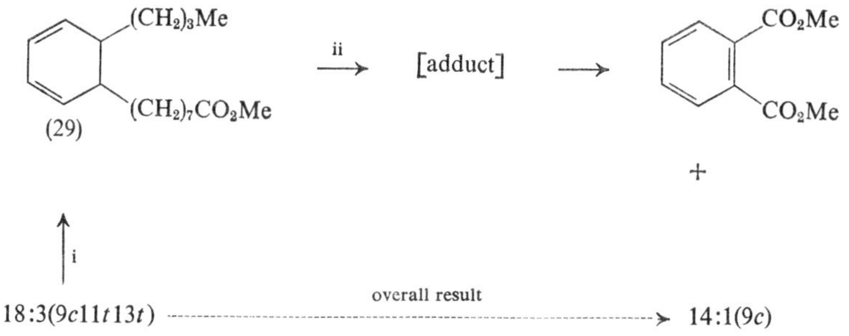

$$(29)\ [\text{cyclohexadiene with }(CH_2)_3Me, (CH_2)_7CO_2Me] \xrightarrow{ii} [\text{adduct}] \longrightarrow [\text{benzene-1,2-dicarboxylate, }CO_2Me, CO_2Me] \ +$$

$$\uparrow i$$

$$18:3(9c11t13t) \xrightarrow{\text{overall result}} 14:1(9c)$$

Reagents: i, S, 160 °C; ii, MeO₂CC≡CCO₂Me, 200 °C, 2 h

Scheme 4

Diene addition of ethylene to alkali-isomerized linoleic acid furnishes a C_{20} acid containing a cyclohexene group. Attempts have been made to

---

[86] A. N. Sagredos, *Fette, Seifen, Anstrichm.*, 1970, **72**, 871; A. N. Sagredos, J. D. von Mikusch, and V. Wolff, *Annalen*, 1971, **745**, 169.
[87] R. J. Bushby, *Quart. Rev.*, 1970, **24**, 585.
[88] U. R. Nayak, A. H. Kapadi, and Sukh Dev. *Tetrahedron*, 1970, **26**, 5071, 5083.

improve the usefulness of this product by peracid oxidation and by ozonolysis of the cyclohexene double bond.[89] The major routes to the monomeric cyclic acids have been reviewed.[90]

Methyl linoleate and methyl linolenate are isomerized to conjugated isomers with potassium t-butoxide at 60 °C and with tris(triphenylphosphine)-chlororhodium at 65 °C.[91] Linoleyl alcohol and its methyl ether react with butyl-lithium and carbon dioxide to give mainly the 9-carboxy-$\Delta^{10,12}$- and 13-carboxy-$\Delta^{11,13}$-derivatives:

$$\overset{13}{-}CH=CHCH_2\overset{9}{CH}=CH- \xrightarrow[CO_2]{BuLi} \begin{array}{l} \overset{13}{-}CH(CO_2H)CH=CHCH=CH- \\ + \\ -CH=CHCH=\overset{9}{C}HCH(CO_2H)- \end{array}$$

linoleyl alcohol

Reaction at C-13 seems to be slightly favoured. It is reported that 'the most probable explanation of the position discrimination is that the oxygen atom of the functional group helps, after coiling the chain, to solvate the lithium cation in the ion pair with the carbanion, and keeps it in such a position that most of the charge is located at C-13'.[92]

Ansell and Weedon have presented further reports on their studies of the alkali fusion of unsaturated acids, and of oxo-, hydroxy-, epoxy-, and alkoxy-acids.[93] The fission of 9(10)-hydroxy-10(9)-oxo-octadecanoic acid by alcoholic alkali at room temperature has been confirmed and shown to be an autoxidation reaction.[94]

Dimerization.—Unsaturated acids, whether monoenoic or polyenoic, furnish dimers which are in demand because of the valuable surface-active properties of their various derivatives. Methods of dimerization have therefore been extensively examined, but understanding of the reaction and the structural identification of the products have lagged behind. Dimerization is effected in several ways but clay catalysts are the most widely employed, and it is now recognized that such catalysts operate in several ways. They may promote modification of monoenoic and dienoic acids to more reactive monomers in addition to assisting both the dimerization process and the subsequent changes in the dimer molecules. In particular, hydrogen transfer seems to be important: monoenoic acids are thereby converted to more reactive dienoic acids and the dimer (probably a cyclohexene derivative resulting from Diels–Alder reaction) is converted to a substituted aromatic compound.[95,96]

[89] E. J. Dufek, L. E. Gast, and J. P. Friedrich, *J. Amer. Oil Chemists' Soc.*, 1970, **47**, 47; E. J. Dufek, J. C. Cowan, and J. P. Friedrich, *ibid.*, p. 51.
[90] J. D. von Mikusch and A. N. Sagredos, *Fette, Seifen, Anstrichm.*, 1971, **73**, 384.
[91] T. L. Mounts, H. J. Dutton, and D. Glover, *Lipids*, 1970, **5**, 997; W. J. De Jarlais and L. E. Gast, *J. Amer. Oil Chemists' Soc.*, 1971, **48**, 21.
[92] J. Klein, S. Glily, and D. Kost, *J. Org. Chem.*, 1970, **35**, 1281.
[93] M. F. Ansell, B. C. L. Weedon, and I. S. Shepherd, *J. Chem. Soc.*, (C), 1971, 1840; M. F. Ansell, B. C. L. Weedon, and D. J. Redshaw, *ibid.*, p. 1846; M. F. Ansell, B. C. L. Weedon, and A. N. Radziwill, *ibid.*, p. 1851.
[94] M. F. Ansell, I. S. Shepherd, and B. C. L. Weedon, *J. Chem. Soc.* (C), 1971, 1857.
[95] D. H. Wheeler, A. Milun, and F. Linn, *J. Amer. Oil Chemists' Soc.*, 1970, **47**, 242.
[96] M. J. A. M. den Otter, *Fette, Seifen, Anstrichm.*, 1970, **72**, 667, 875, 1056.

Fatty Acids and Related Compounds

Cyclopropane and Cyclopropene Acids.—The undesirable physiological properties of cyclopropene acids such as malvalic acid (20; R = H) and sterculic acid (21; R = H) and their presence, albeit in small amount, in cottonseed oil has aroused interest in these compounds. The content of cyclopropene acids is reduced by heating with palladium catalysts, though not with nickel or platinum catalysts.[97] Palladium treatment cleaves the cyclopropene ring and produces methyl- and methylene-substituted acids. Hydrogenolysis of cyclopropane acids is reported to give methyl-substituted fatty acids which can be identified by their g.l.c. behaviour on capillary columns.[98] The chemical reactions of cyclopropene acids (malvalic and sterculic) shown in Scheme 5 have been reported.[99]

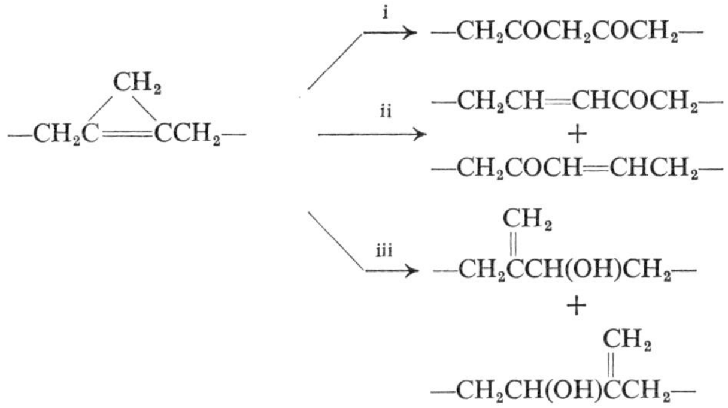

$$
\begin{array}{c}
 \xrightarrow{i} -CH_2COCH_2COCH_2- \\
 \\
-CH_2\overset{\overset{\displaystyle CH_2}{\diagup\diagdown}}{C}=CCH_2- \xrightarrow{ii} \begin{array}{c} -CH_2CH=CHCOCH_2- \\ + \\ -CH_2COCH=CHCH_2- \end{array} \\
 \\
 \xrightarrow{iii} \begin{array}{c} \overset{\displaystyle CH_2}{\underset{\|}{}} \\ -CH_2CCH(OH)CH_2- \\ + \\ \overset{\displaystyle CH_2}{\underset{\|}{}} \\ -CH_2CH(OH)CCH_2- \end{array}
\end{array}
$$

Reagents: i, alkaline $KMnO_4$; ii, $MeCO_3H$; iii, $MeCO_2H$, alkaline hydrolysis.

Scheme 5

Two research groups report the formation of appreciable quantities of cyclopropane compounds by reaction of appropriate derivatives of methyl ricinoleate, which contains a homoallylic system. One group employed the crystallized tosyl ester of the *trans*-isomer, the other used methyl 12-mesyloxy-oleate.[100] Hydroxy-, methoxy-, acetoxy-, and chloro-cyclopropane esters (30) were formed (30—60% yield) under controlled pH conditions; in acidic media the more stable homoallylic system is regenerated.

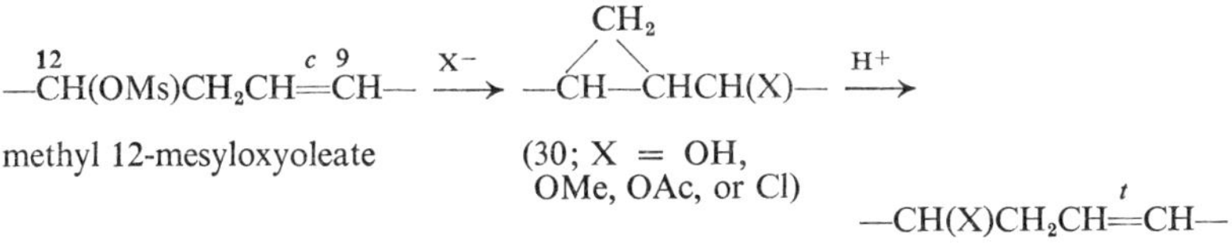

$$-\overset{12}{C}H(OMs)CH_2\overset{c\ 9}{CH}=CH- \xrightarrow{X^-} -\overset{\overset{\displaystyle CH_2}{\diagup\diagdown}}{CH}-CHCH(X)- \xrightarrow{H^+}$$

methyl 12-mesyloxyoleate (30; X = OH, OMe, OAc, or Cl)

$$-CH(X)CH_2\overset{t}{CH}=CH-$$

[97] Z. M. Zarins, R. K. Willich, and R. O. Feuge, *J. Amer. Oil Chemists' Soc.*, 1970, **47**, 215.
[98] R. G. Ackman and S. N. Hooper, *J. Amer. Oil Chemists' Soc.*, 1970, **47**, 525.
[99] M. W. Roomi and C. Y. Hopkins, *Canad. J. Biochem.*, 1970, **48**, 759.
[100] E. Ucciani, A. Vantillard, and M. Naudet, *Chem. and Phys. Lipids*, 1970, **4**, 225; F. D. Gunstone and A. I. Said, *ibid.*, 1971, **7**, 121.

1,4-Epoxides.—When heated with toluene-p-sulphonic acid in the presence of methanol or dioxan, methyl linoleate (and certain related esters) furnishes isomeric methyl octadecadienoates and hitherto unknown 1,4-epoxides (tetrahydrofurans) such as methyl 9,12-epoxystearate.[101] Such compounds (formed in ~45% yield by this procedure) are produced in higher yields by the acid-catalysed cyclization of di- and tri-hydroxy-acids, the rearrangement of hydroxy-epoxides, and by oxidative cyclization (with lead tetra-acetate or with halogens and metal salts) of saturated hydroxy-esters.[102] 1,4-Epoxides are also formed during some oxymercuration reactions (see below), and natural long-chain acids containing this heterocyclic system have recently been reported.[8,9]

Oxymercuration Reactions.—Two research groups have drawn attention to the value of the oxymercuration–demercuration reaction when applied to unsaturated long-chain compounds:

$$-CH=CH- \xrightarrow[\text{ii, NaBH}_4]{\text{i, Hg(OAc)}_2,\text{MeOH;}} -CH(OMe)CH_2- \; + \; -CH_2CH(OMe)-$$

The simple reaction with mercuric acetate and methanol, followed by reduction with sodium borohydride, converts methyl oleate, for example, into a mixture of methyl 9- and 10-methoxystearates. Other mono- and polyenoic esters behave in a similar way and it is possible by g.c.—m.s. to identify these reaction products and so determine the position of unsaturation in the original ester.[103]

The methanol can be replaced by other nucleophilic solvents such as ethanol, acetic acid, or water (in tetrahydrofuran as cosolvent) to give the corresponding ethoxy-, acetoxy-, or hydroxy-esters. Cyclic ethers (substituted tetrahydrofurans and tetrahydropyrans) are formed by intramolecular reaction when the unsaturated ester also contains an appropriately placed hydroxy-group, even in the presence of a reactive solvent.[104] This has been developed into a procedure for the identification, analysis, and isolation of long-chain alcohols and acids having alkene unsaturation in positions 3 (*trans* only), 4 (*cis* or *trans*), or 5 (*cis* or *trans*).[105] Such acids (or natural mixtures in which they are present) are reduced to alcohols and subjected to oxymercuration (in DMF as a non-participating solvent) and demercuration. Cyclic ethers are formed only when there is unsaturation at positions 3, 4, or 5; other double bonds are unaffected. For example, methyl arachidonate

---

[101] G. G. Abbot, F. D. Gunstone, and S. D. Hoyes, *Chem. and Phys. Lipids*, 1970, **4**, 351.
[102] G. G. Abbot and F. D. Gunstone, *Chem. and Phys. Lipids*, 1971, **7**, 279, 290, 303.
[103] P. Abley, F. J. McQuillin, D. E. Minnikin, K. Kusamran, K. Maskens, and N. Polgar, *Chem. Comm.*, 1970, 348.
[104] F. D. Gunstone and R. P. Inglis, *Chem. Comm.*, 1970, 877.
[105] F. D. Gunstone and R. P. Inglis, *Chem. Comm.*, 1972, 12.

Fatty Acids and Related Compounds

gives an unsaturated C_{20} tetrahydropyran (31) thus:

$CH_3(CH_2)_4(CH=CHCH_2)_3CH=CH(CH_2)_3CO_2Me$

$\xrightarrow{i, LiAlH_4; \; ii, Hg(OAc)_2, DMF; \; iii, NaBH_4}$

$CH_3(CH_2)_4(CH=CHCH_2)_3CH_2$—[tetrahydropyran ring]

(31)

Cyclic ethers are readily separated from unreacted alcohols and from one another by t.l.c. and g.l.c. If necessary, unreacted alcohol can be separated from the mercury derivative, from which the unsaturated alcohol is easily regenerated.

Catalytic Hydrogenation.—The search for a catalyst which selectively reduces linolenate (18:3 9c12c15c) in preference to linoleate (18:2 9c12c) has now focussed on copper, which is the most effective heterogeneous catalyst for this purpose. Hydrogenation is accompanied or preceded by extensive double-bond migration and stereomutation, and there is good evidence that reduction only occurs after conjugation of polyene systems to 1,3-dienes. Copper does not catalyse the hydrogenation of monoene esters, and diene esters remaining after prolonged hydrogenation are mainly those with widely separated double bonds.[106]

Reduction with homogeneous catalysts has also attracted attention. Chromium carbonyl complex catalysts, for example, hydrogenate polyunsaturated esters to give products with a high content of *cis*-monoene esters. In the absence of hydrogen these same catalysts give high yields of conjugated polyunsaturated esters.[107]

Hydrogenations with homogeneous and heterogeneous catalysts have been reviewed.[153d]

Other Addition Reactions of Unsaturated Acids or Esters.—During the period under review the reactions of methyl oleate or other unsaturated esters with a range of reagents have been reported. These include the following (the products being indicated in parentheses): hydrogen sulphide (mercaptans),[108] *NN*-dichlorourethane (aziridines),[109] *NN*-dibromobenzenesulphonamide (β-bromosulphonamides, convertible to *N*-sulphonylaziridines),[110] benzene

---

[106] E. Kirschner and E. R. Lowrey, *J. Amer. Oil Chemists' Soc.*, 1970, **47**, 237, 467; S. Koritala, *ibid.*, pp. 106, 269, 463; S. Koritala and C. R. Scholfield, *ibid.*, p. 262; S. Koritala, R. O. Butterfield, and H. J. Dutton, *ibid.*, p. 266; S. Koritala and E. Selke, *ibid.*, 1971, **48**, 222; J. C. Cowan, C. D. Evans, H. A. Moser, G. R. List, S. Koritala, K. J. Moulton, and H. J. Dutton, *ibid.*, 1970, **47**, 470.

[107] E. N. Frankel, *J. Amer. Oil Chemists' Soc.*, 1970, **47**, 11, 33; E. N. Frankel, F. L. Thomas, and J. C. Cowan, *ibid.*, p. 497; J. C. Bailar, jun., *ibid.*, p 475; R. L. Augustine and J. F. van Peppen, *ibid.*, p. 478.

[108] A. W. Schwab and L. E. Gast, *J. Amer. Oil Chemists' Soc.*, 1970, **47**, 371.

[109] T. A. Foglia, G. Maerker, and G. R. Smith, *J. Amer. Oil Chemists' Soc.*, 1970, **47**, 384.

[110] T. A. Foglia, E. T. Haeberer, and G. Maerker, *J. Amer. Oil Chemists' Soc.*, 1970, **47**, 27.

194   *Aliphatic, Alicyclic, and Saturated Heterocyclic Chemistry*

(phenylalkanoic acids and cyclized derivatives),[111] diary lnitrones (isoxazolidones, 1,3-amino-alcohols, and alcohols),[112] and hydrogen and carbon monoxide in the presence of a rhodium catalyst and triphenylphosphine (formyl-substituted esters).[113] There are new reviews on ozonolysis,[154a] allylic halogenation and oxidation of unsaturated esters,[154b] nitrogen and sulphur analogues of epoxy- and hydroxy-acids,[154c] and olefin reactions catalysed by transition-metal catalysts.[154]

Other Reactions.—An Indian group[114] have investigated three means of effecting hydroxylation of unsaturated acids: sulphation and hydrolysis, autoxidation and reduction, and allylic bromination (*N*-bromosuccinimide) and hydrolysis. French chemists[115] have prepared and examined some mono- and di-oxo-enoic acids. They find that unsaturated 1,4-dioxo-acids rearrange spontaneously to 1,2-dioxo-isomers:

$$-COCH=CHCO- \rightarrow -COCOCH=CH-$$

12-Oxostearic acid has been converted to 12-aminostearic acid by reductive amination.[116]

It has been demonstrated that oxygen, excited to its singlet state by photosensitization, plays the important role of forming the original hydroperoxides whose presence is necessary before the normal radical autoxidation process can begin.[117] A study of the decomposition of linoleate hydroperoxide in the presence of unsaturated acid indicates that dimer is formed from one molecule of hydroperoxide and one of fatty ester.[117] New products are continually identified in thermally oxidized glycerides, and a recent study has led to the recognition of pentatriacontan-18-one, methyl hendecanedioate, and isomeric methyl oxo-octadecenoates, methyl oxo-octadecanoates, and aromatic C_{18} esters from glycerol 1-linoleate 2,3-distearate.[118].

The possibility that fatty acids should react preferentially at the ω-position when the acid molecules are suitably oriented in a packed layer has been examined with chlorine atoms and with ethyl radicals, with partial success. For example, chlorination of octanoic acid usually gives all possible monochloro-octanoic acids but in the presence of alumina, on which the fatty acid is adsorbed and aligned, the content of the 2- to 5-chloro-octanoic acids

---

[111] M. F. Ansell and G. F. Whitfield, *J. Chem. Soc. (C)*, 1971, 1098.
[112] H. Basu and H. Schlenk, *Chem. and Phys. Lipids*, 1971, **6**, 266.
[113] E. N. Frankel, *J. Amer. Oil Chemists' Soc.*, 1971, **48**, 248.
[114] G. V. Rao and K. T. Achaya, *J. Amer. Oil Chemists' Soc.*, 1970, **47**, 286, 289; G. V. Rao, K. T. Achaya, and R. Subbarao, *ibid.*, p. 292.
[115] A. Tubul, J. Arnoux, E. Ucciani, and M. Naudet, *Chem. and Phys. Lipids*, 1970, **4**, 1208; M. Naudet, J. Arnoux, and A. Tubul, *ibid.*, p. 217.
[116] B. Freedman, *J. Amer. Oil Chemists' Soc.*, 1970, **47**, 305; B. Freedman and G. Fuller, *ibid.*, p. 311.
[117] H. Rawls and P. J. van Senten, *J. Amer. Oil Chemists' Soc.*, 1970, **47**, 121; T. L. Mounts, D. J. McWeeny, C. D. Evans, and H. J. Dutton, *Chem. and Phys. Lipids*, 1970, **4**, 197.
[118] L. R. Wantland and E. G. Perkins, *Lipids*, 1970, **5**, 187, 191.

Fatty Acids and Related Compounds

falls from 41 to 17% and the content of the 7- and 8-chloro-octanoic acids rises from 41 to 63%.[119] The photobromination of aliphatic acids has also been examined.[120]

The g.l.c. separation of *cis*- and *trans*-epoxy-esters forms the basis of a new procedure for determining the proportion of *cis*- and *trans*-isomers in a mixture.[121] The report that *m*-chloroperbenzoic acid can be used at 90 °C without decomposition in the presence of suitable radical inhibitors may be of value when epoxidizing alkenes of reduced reactivity.[122] Several fluorostearic acids have been prepared by interaction of mesyloxy-acids with tetrabutyl-ammonium fluoride in acetonitrile.[123] A procedure for methylation (by diazomethane in the presence of boron trifluoride) and demethylation (by reaction with sodium borohydride and iodine, followed by methanol), with retention of optical activity, could prove useful in the synthesis of compounds of known configuration.[124]

Isopropenyl esters, conveniently prepared from carboxylic acids and propyne in the presence of the zinc salt, readily effect the acylation of isethionic acid and of *N*-methyltaurine.[125]

$$RCO_2H \xrightarrow{CH_3C\equiv CH} RCO_2CMe=CH_2 \begin{array}{c} \xrightarrow{HOCH_2CH_2SO_3Na} RCO_2CH_2CH_2SO_3Na \\ \xrightarrow{MeNHCH_2CH_2SO_3Na} RCON(Me)CH_2CH_2SO_3Na \end{array}$$

Dutton and his colleagues have described the microreactor that they have built which permits reactions to be carried out in an extension of the g.l.c. apparatus to which it is attached. The reactions so performed include: bromination, silylation, homogeneous and heterogeneous catalytic reduction, and esterification and transesterification by several procedures.[126]

The formation of cyclic acetals of glycerol has received further attention[127] and methods of preparing alkyl esters from fatty acids and lipids have been reviewed.[155]

---

[119] N. C. Deno, R. Fishbein, and C. Pearson, *J. Amer. Chem. Soc.*, 1970, **92**, 1451; C. B. Johnson and A. T. Wilson, *Lipids*, 1971, **6**, 181, 186.
[120] E. Ucciani, F. Pierri, and M. Naudet, *Bull. Soc. chim. France*, 1970, 791.
[121] E. A. Emken, *Lipids*, 1971, **6**, 686.
[122] Y. Kishi, M. Aratani, H. Tanino, F. Fukuyama, T. Goto, S. Inoue, S. Sugiura, and H. Kakoi, *Chem. Comm.*, 1972, 64.
[123] N. J. M. Birdsall, *Tetrahedron Letters*, 1971, 2675.
[124] G. Odham and B. Samuelson, *Acta Chem. Scand.*, 1970, **24**, 468.
[125] E. S. Rothman and S. Serota, *J. Amer. Oil Chemists' Soc.*, 1971, **48**, 373; R. G. Bistline jun., E. S. Rothman, S. Serota, A. J. Stirton, and A. N. Wrigley, *ibid.*, p. 657.
[126] E. D. Bitner, A. C. Lanser, and H. J. Dutton, *Lipids*, 1970, **5**, 707; T. L. Mounts, R. O. Butterfield, C. R. Scholfield, and H. J. Dutton, *J. Amer. Oil Chemists' Soc.*, 1970, **47**, 79.
[127] J. Gelas, *Bull. Soc. chim. France*, 1970, 2341, 2349, 3721, 4041, 4046, 4465; W. J. Baumann, *J. Org. Chem.*, 1971, **36**, 2743.

6 Biological Reactions

In the general area of the biological reactions of long-chain acids, interest is mainly in settling the details of biosynthesis and metabolism, in the prostaglandins (covered only briefly in this review), and in the selective modification of fatty acids by enzymes. The specificity of this last group of reactions makes them of particular interest to the chemist.

Hydrogenation.—Micro-organisms in the rumen have long been known to hydrogenate dietary unsaturated acids (mainly linoleic and linolenic) to stearic acid and to unsaturated C_{18} acids having double bonds in unusual positions and mainly with *trans* configuration. One species of rumen bacteria (*Butyrivibrio fibrosolvens*) rapidly converts linoleic acid to conjugated diene and then more slowly to *trans*-monoene:

$$18:2(9c12c) \rightarrow 18:2(9c11t) \rightarrow 18:1(11t)$$

The isomerase requires specifically a carboxy-group, a Δ^{9c12c}-diene system, and an $\omega 6$ double bond; these observations have been interpreted in terms of an enzyme with three active sites. It is of interest that the hydrogen atom removed from C-11 is thought to be accepted by the carboxylate anion of the same molecule.[128]

Another rumen bacterium which has been isolated promotes the reduction of oleic acid mainly to stearic acid, of linoleic acid mainly to stearic acid, and of linolenic acid mainly to octadec-*cis*-15-enoic acid.[129]

Rumen hydrogenation, which has been reviewed by Viviani,[150] effects almost complete removal of dietary polyunsaturated acids from the depot and milk fats of ruminants. This may be undesirable from a human viewpoint because of increased interest in dietary polyunsaturated fat, and an Australian group have suggested how this depletion can be avoided. Linseed or safflower or sunflower oils are fed to the animals in capsules of formaldehyde-treated resins which resist breakdown in the rumen but are hydrolysed under the more acidic conditions of the abomasum. By this means the content of polyunsaturated fatty acids in milk and depot fats can be raised from the 2—5% level to around 25—30%. This change requires only 24—48 hours for the milk fat but rather longer for the depot fat.[130]

Hydration.—*Pseudomonas* species of bacteria effect the stereospecific hydration of unsaturated acids, acting on the following substrate acids to give, in 20—40% yield, the product indicated in parentheses: oleic (10D-hydroxystearic acid), linoleic (10D-hydroxyoctadec-*cis*-12-enoic acid), linolenic

---

[128] C. R. Kepler, W. P. Tucker, and S. B. Tove, *J. Biol. Chem.*, 1970, **245**, 3612; 1971, **246**, 2765.
[129] R. W. White, P. Kemp, and R. M. C. Dawson, *Biochem. J.*, 1970, **116**, 767.
[130] T. W. Scott, L. J. Cook, and S. C. Mills, *J. Amer. Oil Chemists' Soc.*, 1971, **48**, 358.

Fatty Acids and Related Compounds 197

(10D-hydroxyoctadeca-*cis*-12,*cis*-15-dienoic acid), and ricinoleic (10D, 12D-dihydroxystearic acid). There is no reaction with dec-9-enoic acid, 12,13-epoxyoleic acid, 12-oxo-oleic acid, or sterculic acid.[131]

Another hydratase converts *cis*- and *trans*-9,10-epoxystearic acids to *threo*- and *erythro*-dihydroxy-acids, respectively. Only one enantiomeric epoxy-acid is hydrolysed and only one enantiomeric dihydroxy-acid is formed. The reaction involves water, and the oxygen from this source appears at C(10) only.[132]

Hydroperoxidation.—Lipoxygenases promote the hydroperoxidation of linoleic acid but enzymes from different sources may furnish different products. Lipoxygenase from corn germ (*Zea mays*) gives mainly the 9-hydroperoxide, that from soya bean and from flax seed gives mainly the 13-hydroperoxide, and that from alfalfa gives a 1:1 mixture of the two hydroperoxides. Corn and soya bean lipoxygenase also differ in substrate specificity. The latter oxidizes linoleic (18:2 ω6), γ-linolenic (18:3 ω6), arachidonic (20:4 ω6), and linolenic acids (18:3 ω3) but corn lipoxygenase is effective only with linoleic and linolenic acids. Biologically produced hydroperoxides, unlike those resulting from autoxidation, are single enantiomers.[133,134]

$$18:2(9c12c) \xrightarrow[\text{lipoxygenase}]{O_2} 18:2 \text{ 9-OOH } 10t12c \text{ and/or } 18:2 \text{ 13-OOH } 9c11t$$

The analytical investigation of these products, usually after reduction to hydroxy-acids, has always presented some difficulty, and Hamberg's procedure[134] represents a significant advance. As shown in Scheme 6, the hydroperoxides are reduced to hydroxy-esters, acylated to menthoxycarbonyl esters, and oxidized to the menthyl esters of methyl 2-hydroxysebacate [(32), from the 9-hydroperoxide] or of methyl 2-hydroxyheptanoate [(33), from the 13-hydroperoxide]. Since the D- and L-derivatives also separate on g.l.c., up to four products, derived from the two enantiomers from each hydroperoxide, may be present.

The fate of the unsaturated hydroperoxides has also been studied. Gardner[135] reports that 9-hydroperoxide results only when corn germ lipoxygenase is partially purified: crude enzyme preparations convert the hydroperoxide to the three products (34)—(36). The same enzyme preparation converts the 13-hydroperoxide to analogous compounds (37) and (38). A similar isomerase

[131] W. G. Niehaus, jun., A. Kisic, A. Torkelson, D. J. Bednarczyk, and G. J. Schroepfer, jun., *J. Biol. Chem.*, 1970, **245**, 3790; G. J. Schroepfer, jun., W. G. Niehaus, jun., and J. A. McCloskey, *ibid.*, p. 3798; L. L. Wallen, E. N. Davies, Y. V. Wu, and W. K. Rohwedder, *Lipids*, 1971, **6**, 745.
[132] W. G. Niehaus, jun., A. Kisic, A. Torkelson, D. J. Bednarczyk, and G. J. Schroepfer, jun., *J. Biol. Chem.*, 1970, **245**, 3802.
[133] (*a*) G. A. Veldink, J. F. G. Vliegenthart, and J. Boldingh, *Biochem. J.*, 1970, **120**, 55; (*b*) D. C. Zimmerman and B. A. Vick, *Lipids*, 1970, **5**, 392; H. W. Gardner and D. Weisleder, *ibid.*, p. 678; C. C. Chang, W. J. Esselman, and C. O. Clagett, *ibid.*, 1971, **6**, 100.
[134] M. Hamberg, *Analyt. Biochem.*, 1971, **43**, 515.
[135] H. W. Gardner, *J. Lipid Res.*, 1970, **11**, 311.

—CH=CHCH=CHCH(OOH)— $\xrightarrow{i}$ —CH=CHCH=CHCH(OH)— $\xrightarrow{ii}$
9-hydroperoxide

—CH=CHCH=CHCH(OR)— $\xrightarrow{iii}$ MeO$_2$CCH(OR)(CH$_2$)$_7$CO$_2$Me
(R = menthyl)   (32)

—CH(OOH)CH=CHCH=CH— $\xrightarrow{i}$ —CH(OH)CH=CHCH=CH— $\xrightarrow{ii}$
13-hydroperoxide

—CH(OR)CH=CHCH=CH— $\xrightarrow{iii}$ CH$_3$(CH$_2$)$_4$CH(OR)CO$_2$Me
(R = menthyl)   (33)

Reagents: i, reduction; ii, (−)-menthyl chloroformate; iii, oxidative ozonolysis and methylation

Scheme 6

from flax seed has been shown to convert the 13-hydroperoxide to (37) but to have no effect on the 9-hydroperoxide.[133a]

—CH=CHCH=^{t}CHCH(OOH)—
9-hydroperoxide

↗ —CH=^{c}CHCH$_2$COCH(OR)—
    {R = H   (34)
    {R = C$_{17}$H$_{31}$CO (35)
↘ —CH(OH)CH=^{t}CHCOCH$_2$— (36)

—CH(OOH)CH=^{t}CHCH=^{c}CH—
13-hydroperoxide

↗ —CH(OH)COCH$_2$CH=^{c}CH— (37)
↘ —CH$_2$COCH=^{t}CHCH(OH)— (38)

In the anaerobic reaction between linoleic acid and its 13-hydroperoxide in the presence of lipoxygenase, C_5 and C_{13} degradation products accompany the oxo-octadecadienoate:[136]

—CH(OOH)CH=^{t}CHCH=^{c}CH—
13-hydroperoxide

↗ CH$_3$(CH$_2$)$_4$COCH=^{t}CHCH=^{c}CH(CH$_2$)$_7$CO$_2$H

↘ CH$_3$(CH$_2$)$_4$H + OHCCH=^{t}CHCH=$^{c/t}$CH(CH$_2$)$_7$CO$_2$H

Graveland[137] has shown that in the enzymic oxidation of linoleic acid in doughs and in flour–water suspensions the products include two isomeric hydroperoxy-acids, two isomeric hydroxy-acids, two isomeric hydroxy-epoxy-octadecenoic acids, and some trihydroxy-octadecenoic acids.

Biosynthesis and Metabolism.—Although the main features of the biosynthetic pathways to both saturated and unsaturated acids are well known, many

[136] G. J. Garssen, J. F. G. Vleigenthart, and J. Boldingh, *Biochem. J.*, 1971, **122**, 327.
[137] A. Graveland, *J. Amer. Oil Chemists' Soc.*, 1970, **47**, 352.

Fatty Acids and Related Compounds 199

details remain to be settled. Reviews deal with the biosynthesis of cyclopropane rings,[138] the chemistry and metabolism of fatty aldehydes,[152b] the origin of hydrogen in fatty acid synthesis,[151b] fatty acid biosynthesis in aorta and heart,[151c] and mechanism and stereochemistry in fatty acid metabolism.[139]

In the reductive chain-extension process, Barron and Mooney[140] found no evidence of an intermediate oxo-acid in mitochondrial systems. Their results support a reaction mechanism whereby the unfavourable energetics of the C–C fusion are overcome by rapid reduction by DNPH, producing the 3-hydroxy-acid as the first intermediate.

$RCO_2H \rightarrow RCOCH_2CO_2H \rightarrow RCH(OH)CH_2CO_2H \rightarrow$

$RCH=CHCO_2H \rightarrow R(CH_2)_2CO_2H$

The desaturases from hen liver (converting 18:0 to 18:1) and from safflower (converting 18:1 to 18:2) have been shown to operate on CoA derivatives and not on the phosphatidylcholines. Safflower desaturase, however, contains an active acyltransferase, so that both oleic and linoleic CoA esters are readily converted to phosphatidylcholines.[141] From a large amount of data on the desaturation step in the production of polyunsaturated acids by animals, Brenner[142] describes a hypothetical model of a 6-olefinase.

Gurr[143] has reviewed recent studies in the biosynthesis of polyunsaturated acids in plants. Other workers from the same laboratory conclude that there are two routes for the α-oxidation of fatty acids in leaves:[144] the quicker and major pathway proceeds through the 2L-hydroxy-acid; the slower and minor pathway proceeds through the 2D-enantiomer.

Sgoutas[145] has examined the formation and hydrolysis of those sterol esters in which saturated acyl groups differ in chain-length and unsaturated acyl groups differ in the position and configuration of the olefinic centre.

Prostaglandins.—Chemical activity in the prostaglandin field is apparent in the synthesis of prostaglandins (reviewed in Part II, Ch. 6), in the identification of prostaglandin metabolites, and in the synthesis and screening of polyunsaturated acids for prostaglandin-like activity.

Several papers have appeared on the metabolism of various E and F prostaglandins by guinea pig liver, and guinea pig lung, in the rat, and in male and female human subjects. Metabolism seems to occur by several

---

[138] J. H. Law, *Accounts Chem. Res.*, 1971, **4**, 199.
[139] L. J. Morris, *Biochem. J.*, 1970, **118**, 681.
[140] E. J. Barron and L. A. Mooney, *Biochemistry*, 1970, **9**, 2143.
[141] I. K. Vijay and P. K. Stumpf, *J. Biol. Chem.*, 1971, **246**, 2910.
[142] R. R. Brenner, *Lipids*, 1971, **6**, 567.
[143] M. I. Gurr, *Lipids*, 1971, **6**, 266.
[144] C. H. S. Hitchcock and L. J. Morris, *European J. Biochem.*, 1970, **17**, 39.
[145] D. S. Sgoutas, *Biochemistry*, 1970, 9, 1826; *Biochim. Biophys. Acta*, 1971, **239**, 469; H. J. Goller, D. S. Sgoutas, I. A. Ismail, and F. D. Gunstone, *Biochemistry*, 1970, **9**, 3072.

pathways and to lead to a large number of metabolites differing in the degree of oxidation and in the extent of chain-shortening.[146]

After examining many synthetic unsaturated acids it has been concluded that an unsaturated acid will be converted to a prostaglandin type of molecule if it is a C_{19}, C_{20}, or C_{21} acid with *cis* unsaturated centres at positions 8, 11, and 14 or 5, 8, 11, and 14. Activity is retained by acids having an additional double bond closer to the carboxy-group (Δ^2, Δ^3, or Δ^4) but is lost when there is an additional double bond nearer to the ω-methyl group.[38a]

Rat stomach homogenates and sheep seminal vesicles have been shown to produce new C_{20} derivatives in addition to the familiar prostaglandins (PGE_2 and $PGF_{2\alpha}$). With rat stomach homogenates, compounds (39) (major) and (40) (minor) were formed from arachidonic acid, whilst with sheep seminal vesicles arachidonic acid gave (39) and (41) and eicosa-8,11,14-trienoic acid produced the 5,6-dihydro-derivative of (41).[9]

(39) (40)

(41)

7 Reviews

During the period under review, books have been published by Wakil,[147] James and Gurr,[148] and Hitchcock and Nichols.[149] Reviews have been published on the following topics: cholesterol turnover in man,[150a] arterial composition and metabolism,[150b] essential fatty acids,[150c] lipids in membrane

[146] K. Gréen, *Biochemistry*, 1971, **10**, 1072; *Biochim. Biophys. Acta*, 1971, **231**, 419; M. Hamberg and U. Israelsson, *J. Biol. Chem.*, 1970, **245**, 5107; M. Hamberg and B. Samuelsson, *ibid.*, 1971, **246**, 6713; E. Granström and B. Samuelsson, *ibid.*, 1971, **246**, 5254; E. Granström, *European J. Biochem.*, 1971, **20**, 451.
[147] S. J. Wakil, 'Lipid Metabolism', Academic Press, New York, 1970.
[148] M. I. Gurr and A. T. James, 'Lipid Biochemistry: An Introduction', Chapman and Hall, London, 1971.
[149] C. Hitchcock and B. W. Nichols, 'Plant Lipid Biochemistry', Academic Press, New York, 1971.
[150] *Adv. Lipid Res.* (ed. R. Paoletti and D. Kritchevsky), 1970, Volume 8: (*a*) P. J. Nestel, p. 1; (*b*) O. W. Portman, p. 41; (*c*) M. Guarneri and R. M. Johnson, p. 115; (*d*) G. S. Getz, p. 175; (*e*) M. Kates, p. 225; (*f*) R. Viviani, p. 268; (*g*) D. O. Shah, p. 348.

Fatty Acids and Related Compounds

development,[150d] plant phospholipids and glycolipids,[150e] metabolism of long-chain fatty acids in the rumen,[150f] surface chemistry of lipids,[150g] light- and electron-microscopic radiography of lipids,[151a] the origin of hydrogen in fatty acid synthesis,[151b] fatty acid biosynthesis in aorta and heart,[151c] structures of membranes and role of lipids therein,[151d] glycosphingolipids,[151e] biosynthesis of pregnane derivatives,[151f] lipid composition of vegetable oils,[151g] phospholipids, liquid crystals, and cell membranes,[152a] chemistry and metabolism of fatty aldehydes,[152b] occurrence of unusual fatty acids in plants,[152c] insect lipids,[152d] pesticide residues in fats and other lipids[152e] chemistry of the sulpholipids,[152f] lipolytic enzymes,[152g] cyclopropane and cyclopropene fatty acids,[153a] milk lipids,[153b] structure determination of fatty acids by g.l.c.,[153c] hydrogenation with homogeneous and heterogeneous catalysts,[153d] optically active long-chain compounds and their absolute configurations,[153e] mass spectrometry of fatty acid derivatives,[153f] ozonolysis,[154a] allylic halogenation and oxidation of unsaturated esters,[154b] nitrogen and sulphur analogues of epoxy- and hydroxy-acids,[154c] natural alkyl-branched long-chain acids,[154d] olefin reactions catalysed by transition-metal compounds,[154e] n.m.r. spectra of fatty acids and related compounds,[154f] synthetic glycerides,[155a] fatty acids with conjugated unsaturation,[155b] glyceride chirality,[155c] the production of fatty acids from hydrocarbons by microorganisms,[155d] the preparation of alkyl esters from fatty acids and lipids,[155e] plant waxes,[155f] base-catalysed isomerization of acetylenes,[87] stereoselective and stereospecific olefin synthesis,[48] biosynthesis of cyclopropane rings,[138] the biology of milk ketones,[156] sphingolipid long-chain bases,[157] recent developments in the chemistry of sphingolipids,[158] chemistry of phospholipids in relation to biological membranes,[159] epoxy oils from plant seeds,[1] g.c. and m.s. studies of ceramides,[160] and mechanism and stereochemistry in fatty acid metabolism.[139]

---

[151] *Adv. Lipid Res.* (ed. R. Paoletti and D. Kritchevsky), 1971, Volume 9: (a) O. Stein and Y. Stein, p. 1; (b) S. Rous, p. 73; (c) A. F. Whereat, p. 119; (d) F. A. Vandenheuvel, p. 161; (e) H. Wiegandt, p. 249; (f) S. Burstein and M. Gut, p. 291; (g) E. Fedeli and G. Jacini, p. 335.
[152] *Progr. Chem. Fats and Lipids* (ed. R. T. Holman), 1970—1971, Volume 11: (a) R. M. Williams and D. Chapman, p. 1; (b), V. Mahadevan, p. 81; (c) C. R. Smith, jun., p. 137; (d) P. G. Fast, p. 179; (e) A. M. Parsons, p. 243; (f) T. H. Haines, p. 297; (g) R. G. Jensen, p. 347.
[153] *Topics Lipid Chem.* (ed. F. D. Gunstone), 1970, Volume 1: (a) W. W. Christie, p. 1; (b) W. R. Morrison, p. 51; (c) G. R. Jamieson, p. 107; (d) E. N. Frankel and H. J. Dutton, p. 161; (e) C. R. Smith, jun., p. 277; (f) J. A. McCloskey, p. 369.
[154] *Topics Lipid Chem.* (ed. F. D. Gunstone), 1971, Volume 2: (a) E. H. Pryde and J. C. Cowan, p. 1; (b) M. Naudet and E. Ucciani, p. 99; (c) G. Maerker, p. 159; (d) N. Polgar, p. 207; (e) C. W. Bird, p. 247; (f) F. D. Gunstone and R. P. Inglis, p. 287.
[155] *Topics Lipid Chem.* (ed. F. D. Gunstone), 1972, Volume 3: (a) R. G. Jensen, p. 1; (b) C. Y. Hopkins, p. 37; (c) C. R. Smith, jun., p. 89; (d) C. W. Bird and P. M. Molton, p. 125; (e) W. W. Christie, p. 171; (f) S. Hamilton and R. J. Hamilton, p. 199.
[156] F. W. Forney and A. J. Markovetz, *J. Lipid Res.*, 1971, **12**, 383.
[157] K. A. Karlsson, *Lipids*, 1970, **5**, 879; *Chem. and Phys. Lipids*, 1970, **5**, 6.
[158] A. J. Slotboom and P. P. M. Bonsen, *Chem. and Phys. Lipids*, 1970, **5**, 301.
[159] L. L. M. van Deenen, *Pure Appl. Chem.*, 1971, **25**, 25.
[160] K. Samuelsson and B. Samuelsson, *Chem. and Phys. Lipids*, 1970, **5**, 44.

Author Index

Aasen, A. J., 188
Abad, A., 173
Abbott, D. J., 163
Abbot, G. G., 192
Abell, P. I., 45
Abley, P., 71, 192
Abramovitch, R. A., 139
Achaya, K. T., 194
Acheson, R. M., 13, 176
Achiwa, K., 53
Ackman, R. G., 178, 181, 191
Acton, E. M., 13
Adams, T., 125
Adembri, G., 58
Adickes, H. W., 126
Adkins, J. D., 117
Aguiar, A. M., 134
Ainsworth, C., 95
Ajo, M. A., 22
Akermark, B., 162
Alberola, A., 69
Albertson, D. A., 136
Albonico, S. M., 139
Albro, P. W., 179
Alford, J. A., 75
Allen, L. E., 148
Altenburger, E. R., 183
Altland, H. W., 104
Altman, L., 127
Amouroux, R., 144
Anatol, J., 115
Anderson, H. W., 50
Anderson, L. E., 184
Anderson, R. J., 54, 162
Ando, T., 6
Ando, W., 102
Andrews, G. C., 163
Andrews, L. J., 56
Andrist, A. H., 65
Andrzejewski, D., 60
Aneja, R., 182, 186, 188
Angelo, B., 95
Anner, G., 4
Ansell, M. F., 190, 194
Anselme, J.-P., 105, 112
Antonucci, F. R., 114
Appel, R., 109, 111
Arain, R. A., 157
Arapakos, P. G., 114
Arase, A., 128

Aratani, M., 195
Arbuzov, B. A., 14
Archer, J. F., 139
Arhart, R. J., 57
Arnold, Z., 120
Arnoux, J., 194
Arora, S. K., 175
Arzoumanian, H., 21
Ash, D. K., 169
Ashby, E. C., 140
Asinger, F., 79
Asselineau, G., 180
Asselineau, J., 180
Atkins, P. R., 92
Atkins, R. L., 162
Au, K., 170
Audette, R. J., 125
Augustine, R. L., 193
Aversa, M. C., 43
Awang, D. V. C., 77
Axelrod, E. H., 60, 163
Ayres, R. L., 73
Azoo, J. A., 123

Babler, J. H., 59
Bach, R. D., 39, 41, 46, 50, 60, 153
Bachhawat, J. M., 161
Bachman, G. B., 89, 173
Back, T. G., 169
Bacon, R. G., 123
Baer, E., 186
Bahn, C. A., 78
Bailar, J. C., jun., 193
Bailey, D. M., 91
Baird, W. C., jun., 157
Baker, L. M., 118
Baker, R., 154
Balabanova, F. B., 14
Baldwin, J. E., 13, 42, 65
Baldwin, S. W., 80
Ban, Y., 172
Bandi, P. C., 180
Bandurco, V. T., 174
Banucci, E., 43
Barnett, W. E., 91, 163
Baron, W. J., 30
Barratt, M. D., 188
Barrett, G. C., 145
Barron, E. J., 199
Barstow, L. E., 106

Barsukov, L. I., 188
Bartlett, P. A., 103
Bartlett, P. D., 41, 48, 72, 142
Barton, D. H. R., 54, 110, 135, 166, 185
Barton, T. J., 22
Barve, J. A., 184
Barzilay, J., 187
Basu, H., 194
Bates, R. B., 181
Battioni, P., 44
Batty, J. W., 50
Baumann, W. J., 188, 195
Bayer, H. O., 15
Bayne, W. F., 43
Beacham, L. M., 105
Beard, C. D., 36
Beasley, G. H., 78
Becke, F., 114
Becker, J., 90
Bedenbaugh, A. O., 117
Bedenbaugh, J. H., 117
Bednarczyk, D. J., 197
Beerthuis, R. K., 183
Behare, E. S., 91
Bell, K. H., 153
Belletire, J. L., 171
Belzecki, C., 155
Bender, C. O., 13
Benes, M. J., 3
Benfield, F. W. S., 155
Bentley, P. H., 185
Berenguer, M. J., 144
Bergelson, L. D., 188
Bergeron, R., 153
Bergin, W. A., 117
Berlin, K. D., 68
Bernstein, M. D., 139
Berry, J. W., 183
Bertilsson, B.-M., 172
Bertini, F., 56
Bertram, J., 81
Bertrand, M., 39, 48, 49
Bertsch, R. J., 158
Bestmann, H.-J., 85, 100, 103, 176, 182
Bhalerao, U. T., 74
Bianchi, G., 7, 149
Bianchini, J.-P., 45
Bickelhaupt, F., 71, 116

Author Index

Biermann, T. F., 173
Bigley, D. B., 80, 175
Billimoria, J. D., 186
Billups, W. E., 87, 103, 161
Binks, J. R., 143
Bird, C. W., 201
Birdsall, N. J. M., 195
Birkhäuser, A., 128
Birr, C., 107
Bishop, C. E., 90
Bistline, R. G., jun., 195
Bitner, E. D., 195
Bjorklund, C., 114
Black, W. A., 41
Blagoev, B., 104
Bloodworth, A. J., 174
Bloomer, J. L., 86
Bloomfield, J. J., 95
Blum, J., 140
Boch, R., 184
Bocher, S., 78
Böhme, P., 132
Boerwinkel, F. P., 68
Bogatskii, A. V., 147
Bohlmann, F., 32, 35
Boldingh, J., 197, 198
Bolourtchian, M., 77
Bonnet, P. H., 32
Bonnett, R., 13
Bonsen, P. P. M., 201
Bopp, R. J., 22
Borch, R. F., 139
Borcic, S., 62
Borden, W. T., 164
Borowitz, I. J., 137
Boswell, C. J., 47
Bottini, A. T., 46
Bourgain, M., 5, 18
Boussu, M., 116
Boustany, K. S., 169
Bowers, K. W., 141
Boyd, D. R., 139
Boyd, S. D., 55, 173
Boyer, J. H., 115
Boyne, A. W., 187
Bradshaw, R. W., 183
Brady, D. G., 90
Brady, T. W., 63
Breitmaier, E., 120
Bremner, J. B., 94
Brenner, R. R., 199
Brenner, S., 35
Breslow, R., 80
Brettle, R., 146
Brewer, J. D., 13
Bridson, J. N., 13
Brieger, G., 102
Brindle, J. R., 112, 151
Brown, C., 62
Brown, D. W., 31
Brown, H. C., 26, 57, 69, 70, 104, 113, 125, 128,

152, 156, 158, 159, 160, 161, 170, 175
Brummel, R. N., 50
Bryson, T. A., 24
Bubnov, Yu. N., 31
Buchnea, D., 185
Budny, J., 183
Bu'Lock, J. D., 179
Bunce, N. J., 87
Bunce, R. J., 174
Buncel, E., 110
Burgess, G. M., 57
Burgess, J. R., 75
Burkett, A. R., 95
Burkoth, T. L., 138
Burstein, S., 201
Burt, D. W., 29
Busch, P., 90
Bushby, R. J., 3, 189
Butterfield, R. O., 193, 195
Buza, M., 157
Byrd, L. R., 44
Byrn, S. R., 41
Bystrov, V. F., 188

Cabaret, D., 115
Cadiot, P., 4
Cainelli, G., 56
Calas, R., 77
Calvin, M., 79
Campbell, R. A., 32
Camps, F., 99
Cane, D. E., 108, 148
Canfield, N. D., 59
Cannon, J. G., 108
Cannon, J. R., 9
Canonica, L., 149
Capillon, J., 98
Careau, J. P., 185
Carey, F. A., 132
Carlson, B. A., 160
Carlson, R. M., 70, 161
Carpino, L., 105
Carrick, W. L., 118
Carver, J. R., 74
Caserio, M. C., 44, 45, 46
Cason, J., 178
Casper, E. W. R., 137
Castells, J., 99, 144
Castro, B., 106, 151
Chadha, J. S., 182, 186
Chamberlain, P., 31
Chambers, J. Q., 59
Chan, T. H., 56, 122
Chang, C. C., 49, 197
Chang, E., 56, 122
Chapman, D., 201
Chapman, O. L., 92
Chaser, D. W., 170
Chasin, D. G., 185, 188
Chastrette, M., 144
Chatta, M. S., 134

Chen, A., 44
Chen, F., 58, 95
Chenault, J., 151
Chérest, M., 143
Chicheportiche, M., 185
Chiusoli, G. P., 103
Chodkiewicz, W., 4
Chodsky, S. V., 186
Chow, P. W., 9
Christie, P. H., 186
Christie, W. W., 184, 201
Christmann, K. F., 54
Chumachenko, T. K., 147
Chun, M. C., 16
Chwang, T. L., 35
Ciabattoni, J., 32
Citron, J. D., 116
Ciurdaru, V., 174
Clagett, C. O., 197
Clardy, J. C., 92
Clark, G. M., 20, 21
Clark, G. R., 62
Clark, S. D., 50
Clarke, B. C., jun., 90
Clarke, T. G., 123
Claus, K., 15
Clifford, P. R., 70
Clive, D. L. J., 111, 156
Closson, W. R., 48
Coates, R. M., 103, 136
Cocordano, M., 45
Coffen, D. L., 13, 59
Cogdell, T. J., 48
Cohen, L. A., 93
Collington, E. W., 126, 157
Collins, C. J., 176
Collins, G. C. S., 150
Collins, J. C., 124
Colvin, E., 122
Colvin, E. W., 85
Combret, J. C., 56, 94
Conacher, H. B. S., 180
Concannon, P. W., 32
Conia, J.-M., 62
Conrad, J., 152
Cook, L. J., 196
Cooke, M. P., jun., 117
Cooper, G. W., 158
Cooper, R. D. G., 163
Cope, A. C., 39
Corey, E. J., 3, 32, 53, 54, 65, 79, 108, 120, 121, 129, 131, 133, 135, 139, 146, 148
Cornforth, J. W., 69
Corson, F. P., 46
Cowan, J. C., 190, 193, 201
Cowell, G. W., 176
Cox, D. A., 78
Cramer, F., 172
Crandall, J. K., 38, 73
Creger, P. L., 95, 97

Author Index

Cretney, W., 151
Cripps, A. L., 182
Crossland, R. K., 165, 175
Crouch, R. K., 137
Crowley, J. I., 100
Crump, D. R., 142
Crundwell, E., 182
Cum, G., 43
Cure, J. M., 93
Curtis, R. F., 5
Cuvigny, Th., 137

Dabek, H., 115
Dahmen, A., 65
Dai, S. H., 41, 43
Dale, J. A., 85
Dang, T. P., 87
Danieli, B., 149
Danishefsky, S., 138
Dart, R. K., 178
Dauben, W. G., 78
Daviaud, C., 98
Davidson, W. J., 13
Davies, A. P., 186
Davies, E. N., 197
Davis, B. R., 142
Davis, C. B., 109
Davis, R., 178
Dawson, G., 188
Dawson, M. I., 3
Dawson, R. M. C., 181, 196
Day, A. C., 13, 183
Day, M. J., 110
Day, R. A., jun., 145
de Bie, M. J. A., 50
Deboer, C. D., 62
de Boer, Th. J., 174
De Graaf, J. W. M., 89
de Haan, J. W., 50
Dehmlow, E. V., 36
Deitch, J., 97
De Jarlais, W. J., 190
de Jong, K., 180
de la Mare, P. B. D., 68
Delavarenne, S.-Y., 8, 28
Deluzarche, A., 165
Dennis, W. E., 111
Deno, N. C., 87, 161, 195
den Otter, M. J. A. M., 190
Denson, D. D., 90
Dent, S. P., 116
Denyer, C. V., 156
de Radzitzky, P., 69
Derbesy, M., 185
Derevitskaya, V. A., 88
de Rosa, M., 179
De Sio, F., 58
Deslongchamps, P., 166
Dessau, R. M., 72, 147

Deutschman, A. J., jun., 183
Devaprabhakara, D., 46
De Ville, T. E., 52
Devlin, J. P., 68
Dickinson, R. G., 149
Dimsdale, M. J., 94
Diner, U. E., 68, 125
Dines, M., 110
Dinh-Nguyên, N., 187
Dodd, J. R., 35
Dolbier, W. R., 41, 43
Dombrovskii, A. M., 113
Dormoy, J. R., 106
Dornauer, H., 103
Doumaux, A., jun., 108
Doyle, T. W., 13, 168
Drabowicz, J., 172
Dreiding, A. S., 91
Drenth, W., 35, 50
Dreux, M., 151
Dubois, J. E., 58, 116
Dubrovina, N. I., 188
Dubs, P., 168
Dueber, T. E., 48
Dufek, E. J., 190
Dulcere, J. P., 39
Duncan, J. H., 188
Duncan, W. R. H., 187
Dunogues, J., 77
Durst, H. D., 139
Durst, T., 172
Dutton, H. J., 187, 190, 193, 194, 201
Dvolaitsky, M., 157
Dyke, S. F., 31
Dzhemilev, U. M., 125

Eaborn, C., 116
Earle, F. R., 178, 179
Easton, D. B. J., 32
Eastwood, F. W., 58
Eaton, C. A., 181
Edens, R., 157
Edwards, J. A., 100
Effenberger, F., 74
Egli, R. A., 114, 151
Eglinton, G., 4
Eguchi, M., 88
Ehmann, W. J., 76
Eidus, Ya. T., 3
Ei. Negishi, 57
Eisch, J. J., 19
Eliel, E. L., 168
Elix, J. A., 13
Ellis, J. J., 179
Elovson, J., 182
El-Sheikh, M., 184
Emerson, D. W., 135
Emken, E. A., 195
Endo, K., 178
Endo, T., 89

Epstein, W. W., 158
Erickson, B. W., 3, 131, 135
Erickson, K. C., 88
Erickson, W. F., 19, 153
Eschenmoser, A., 4, 168
Esselman, W. J., 197
Evans, C. D., 193, 194
Evans, D. A., 163
Evstigneeva, R. P., 186
Ezimora, G. C., 36

Fagone, F. A., 83
Fahey, R. C., 67
Fajkos, J., 100
Falla, N. A. R., 187
Farine, J.-C., 90
Fast, P. G., 201
Fatiadi, A. J., 149
Faulkner, J. D., 52
Favorskaya, T. A., 32
Fedeli, E., 201
Fehl, A. J., 187
Feiner, S., 62
Feldman, W. R., 155
Felix, D., 4
Felkin, H., 143
Fenselau, C. C., 188
Fernández, J., 144
Ferrando, M. J., 99
Ferris, J. D., 114
Feuge, R. O., 191
Fetizon, M., 78
Ficini, J., 38, 50
Figeys, H. P., 8
Finding, R., 7
Findlay, M. C., 45
Fink, W., 77
Fischer, R. P., 21
Fischmeister, I., 187
Fishbein, L., 179
Fishbein, R., 87, 195
Fitzgerald, P. H., 77
Fitzgerald, R., 46
Fleig, H., 114
Fleischman, M., 81
Fleming, I., 8
Floyd, M. B., 183
Foglia, T. A., 193
Folli, U., 86
Folsom, T. K., 88
Font, J., 99
Forney, F. W., 201
Forrester, A. R., 174
Forte, P. A., 111
Foucaud, A., 116
Fouquey, C., 104
Frajerman, C., 143
Franck, B., 152
Fleming, I., 8
Franck, R. W., 15
Frank, F. J., 124
Frankel, E. N., 193, 194, 210

Author Index

Frankenfeld, J. W., 141
Franzen, V., 157
Fraser-Reid, B., 62, 166
Frasnelli, H., 98
Freedman, B., 194
Freidinger, R. M., 179
Frey, T. G., 17
Fridman, A. L., 176
Fried, F., 10
Fried, J. H., 100
Friedrich, E. C., 61
Friedrich, J. P., 190
Fries, R. W., 148
Frischleder, H., 37
Frolov, S. I., 31
Frost, D. J., 187
Frost, K. A., 46
Fryberg, E. C., 139
Frye, C. L., 72
Fuchs, P. L., 56, 134
Fueno, T., 45
Fujii, A., 27
Fujita, K., 174
Fujita, T., 95
Fujiyoshi, K., 95
Fuks, R., 14
Fukuyama, F., 195
Fulco, A. J., 178
Fuller, G., 194
Fuller, M. W., 9
Funamizu, M., 13
Funk, A. H., 70, 161
Funke, B., 89
Furukawa, J., 61
Furusato, M., 79

Gais, H. J., 14
Gajewski, J. J., 41
Gal, A., 48
Galard, R. M., 144
Gall, M., 136
Gambacorta, A., 179
Gandolfi, R., 7, 149
Ganem, B. E., 32, 108
Garcea, R. L., 94
Gardner, H. W., 197
Garg, C. P., 125
Gariano, P., 88
Garrard, T. F., 31
Garrett, P. E., 59
Garssen, G. J., 198
Gassenmann, S., 120
Gassman, P. G., 27, 176
Gast, L. E., 187, 190, 193
Gaudemar, M., 93
Gaudemar-Bardone, F., 93
Gaudry, M., 144
Gault, Y., 143
Gautschi, F. G., 4
Geiger, R., 107
Gelas, J., 195
Gelbcke, M., 8

Gensler, W. J., 183
Geohegan, P. J., 70
George, W. O., 150
Gerlach, O., 74
Getz, G. S., 200
Ghera, E., 123
Ghosez, L., 63
Ghosh, C. K., 140
Giacobbe, T. J., 92
Giacomelli, G. P., 139
Gibbs, D. E., 68
Gibson, H. W., 173
Giese, R. W., 141
Gillespie, J. P., 185
Gilman, N. W., 3, 32, 108
Glasgow, L. R., 136
Glass, T. E., 122
Gleason, J. G., 86, 169
Glily, S., 190
Glover, D., 190
Glushkov, R. G., 176
Götschi, E., 168
Goldsmith, D., 163
Goller, H. J., 184, 199
Gompper, R., 13, 42, 65
Gontarz, J. A., 71
Gopal, H., 32, 118, 125
Gordon, A. J., 32, 118
Gore, J., 38
Gospočić, L., 186
Goto, K., 105
Goto, T., 195
Gotthardt, H., 15
Graff, G., 100, 188
Grank, V. G., 176
Granström, E., 200
Grasselli, P., 56
Graveland, A., 198
Gravestock, M. B., 24
Greaves, P. M., 38
Gréen, K., 200
Green, M. L. H., 155
Gregson, M., 157
Grieco, P. A., 157
Griesbaum, K., 23
Grigat, E., 108
Grimaldi, J., 49
Grimshaw, J., 141
Grimwood, P. D., 184
Gross, M. A., 86
Gross, R., 183
Grubbs, R. H., 76
Gruber, P., 83
Grünanger, P., 7, 149
Grundon, M. F., 139
Guarneri, M., 200
Guetté, J.-P., 98
Guetté, M., 98
Gunn, D. M., 120
Gunstone, F. D., 180, 184, 187, 191, 192, 199, 201
Gupta, K. K., 123

Gupta, S. K., 159
Gurfinkel, E., 6
Gurr, M. I., 199, 200
Gustafsson, B., 172
Gut, M., 201
Guthrie, R. D., 165
Gutsch, P., 151

Hach, V., 139
Haeberer, E. T., 193
Hafner, K., 14
Hagendoorn, J. A., 89
Haines, T. H., 201
Hall, D. M., 145
Hall, J. H., 65
Hall, S. S., 172
Halliday, D. E., 154
Halpern, Y., 80
Hamberg, M., 197, 200
Hamberger, H., 65
Hamilton, G. A., 81
Hamilton, R. J., 201
Hamilton, S., 201
Hamlet, Z., 7, 149
Hammann, W. C., 79
Hammargren, D. D., 138
Hammarström, S., 188
Hammond, G. S., 62
Hampson, N. A., 123
Han, J., 79
Hanack, M., 23, 47, 78
Hanessian, S., 163
Hanna, I., 78
Hara, M., 154
Harada, M., 158
Harbert, C. A., 115
Harding, C. E., 23, 78
Hargreaves, M. K., 145, 157
Hargrove, R. J., 48
Harpp, D. N., 86, 169, 170
Harrington, K. J., 58
Harris, R. F., 81
Hartman, M. E., 101
Hartung, H., 100
Harvey, M. S., 186
Hase, T., 149
Hashimoto, H., 157
Hashimoto, T., 187
Hassner, A., 53, 68, 69
Hauptmann, S., 37
Hauser, C. R., 104
Hayamizu, K., 187
Hayasi, Y., 13
Haynes, P., 109
Heasley, G. E., 72
Heasley, V. L., 72
Heathcock, C., 69
Heaton, P. R., 135
Hebrard, P., 150
Heck, R. F., 52
Hederich, V., 135

Heiba, E. I., 72, 147
Heim, P., 175
Heindel, N. D., 16
Heinz, G., 61
Heiszwolf, G. J., 137
Helder, R., 13
Hendrick, C. A., 54
Hendrick, M. E., 30
Hendrickson, J. B., 84, 153
Hepburn, S. P., 174
Herr, R. W., 162
Herrmann, J. L., 124
Herron, D. K., 53, 54
Hess, W. W., 124
Hesse, R. H., 110
Heyman, M., 174
Heyn, A. S., 19, 153
Hickmott, P. W., 175
Hightower, L. E., 136
Hilfiker, F. R., 123
Hill, J. B., 36
Hintze, B., 155
Hirai, H., 130
Hirowatari, N., 186
Hiskey, R. G., 105
Hitchcock, C. H. S., 199, 200
Hites, R. A., 188
Hixon, S. H., 47
Ho, A. J., 113
Hoblitt, R. P., 69
Hoch, H., 175
Hodder, O. J. R., 13
Hodoşan, F., 174
Hodson, D., 120
Hörnfeldt, A.-B., 122
Hoffmann, R. W., 42
Hofstetter, H. H., 188
Hogeveen, H., 23
Hogg, D. R., 53
Hoke, D., 158
Holland, G. W., 26, 57 128, 160
Holman, R. J., 174
Holman, R. T., 184, 188
Holt, G., 120
Honma, S., 128
Hooper, S. N., 178, 191
Hooz, J., 120, 128
Hopkins, C. Y., 187, 191, 201
Hora, J., 52
Hordis, C. K., 19
Horeau, A., 98
Horler, D. F., 178
Horner, L., 117
Hornig, A., 39
Houghton, R. P., 158
House, H. O., 136, 141
Hovey, M. M., 69
Howes, P. D., 50
Hoyes, S. D., 192

Hruby, V. J., 106
Hsieh, C.-C., 13
Huang, H.-Y., 13
Huber, F. E., jun., 114
Huckin, S. N., 102
Huisgen, R., 15, 65
Hummel, K., 78
Humphrey, S. A., 124
Humski, K., 62
Hunig, S., 175
Hunt, D. F., 118
Hunt, J. D., 73, 118, 129
Hunter, D. H., 60
Hunter, J. S. V., 185
Hursthouse, M. B., 52
Hutchings, M. G., 85, 119
Hutchins, R. O., 139, 158
Huurdeman, W. F. J. 135

Iarossi, D., 86
Ichikawa, K., 138
Ikeda, T., 108
Ikenaga, S., 89
Imoto, E., 89
Inch, T. D., 150
Inglis, R. P., 192, 201
Inoue, S., 178, 195
Iorio, E. J., 122
Iqbal, A. F. M., 154
Ireland, R. E., 3
Isaacs, N. S., 63
Isele, G. L., 109
Isida, T., 186
Ismail, I. A., 184, 199
Israelsson, U., 200
Ito, Y., 146
Itoh, M., 26, 57, 128, 152, 158, 160, 161
Iyengar, B. T. R., 188
Izawa, K., 45

Jacini, G., 201
Jackson, L. L., 153, 180
Jackson, W. R., 139
Jacobs, T. L., 41
Jacobsen, N. W., 149
Jacobus, J., 157
Jacques, J., 104
Jagannadha Rao, K. V., 186
Jagar, V., 13
Jagt, J. C., 114
Jallabert, C., 116
James, A. T., 200
Jamieson, G. R., 201
Janic, M., 3
Janoušek, Z., 120
Jansing, J., 15
Jardine, I., 154
Jautelat, M., 33
Javaid, K. A., 74
Jayawant, M., 151

Jenkin, H. M., 184
Jensen, H., 15
Jensen, R. G., 201
Jira, R., 128
Johnson, C. B., 195
Johnson, C. R., 162
Johnson, F., 91
Johnson, M. R., 139
Johnson, R. E., 91
Johnson, R. M., 200
Johnson, W. S., 24, 103
Johnston, B. E., 62
Jończyk, A., 138
Jones, (Sir) E. R. H., 183
Jones, F. N., 33
Jones, J. B., 55
Jones, M., 30
Jones, R. A., 7
Jones, W. M., 37
Josan, J. S., 58
José, F. L., 163
Joshi, G. C., 46
Jurjevich, A. F., 126

Kabalka, G. W., 26, 128, 161
Kadunce, W. M., 139
Kagan, H. B., 87, 154
Kagan, J., 94
Kaiser, E. M., 88
Kakoi, H., 195
Kalli, M., 38
Kalvoda, J., 4
Kampeier, J. A., 26
Kanadka, Y., 111
Kandall, C., 84
Kaneda, T., 178
Kanematsu, K., 15
Kantlehner, W., 89
Kapadi, A. H., 189
Kapchan, S. M., 56
Kaplan, J. P., 141
Kaplan, M., 109
Kappler, F. E., 86
Karlsson, K. A., 201
Kataoka, M., 6
Kates, M., 200
Katner, A. S., 13
Katzenellenbogen, J. A., 3
Kaufmann, H., 4
Kawabata, N., 61
Kawatani, H., 105
Kay, I. T., 92
Keana, J. F. W., 144
Kehoe, L. J., 103
Keller, K., 65
Keller, L. S., 42
Kelly, J. F., 15
Kemp, P., 181, 196
Kennedy, J. P., 79
Kepler, C. R., 196
Kermer, W. D., 115

Author Index

Kesmarky, S., 110
Khan, W. A., 139
Khuddus, M. A., 161
Kienzle, F., 73, 118
Kilbourn, E., 175
Kimling, H., 9
Kimura, K., 86
King, S. W., 155, 163
Kinoshita, H., 146
Kinsman, R. G., 31
Kirschenbaum, H. D., 16
Kirschner, E., 193
Kirst, H. A., 3
Kishi, Y., 195
Kisić, A., 186, 197
Kiso, Y., 105
Kitahara, Y., 13
Kitayama, M., 61
Kitching, W., 70
Kite, G. F., 89
Klabunde, K. J., 82
Kleiman, R., 179
Klein, D. A., 112
Klein, J., 6, 35, 190
Klein, R. A., 188
Kleinstück, R., 109, 111
Klimov, E. M., 88
Kloster-Jenson, E., 7
Klumpp, G. W., 71
Knoess, H. P., 98
Knutson, K. K., 103
Kobata, H., 186
Kobayashi, S., 33
Kobayasi, M., 13
Kochetkov, N. K., 88
Kochi, J. K., 52, 56, 158
Köbrich, G., 6, 38
Köhler, W., 113
Koehn, W., 64
Koelliker, U., 139
König, W., 107
Kofron, W. G., 104
Koga, K., 104
Kohnle, J. F., 109
Kokes, R. J., 49
Kolewa, S., 100
Kolodny, N. H., 141
Kondo, K., 90, 108
Konen, D. A., 86
Koosterziel, H., 137
Kopchik, R. M., 26
Koppel, G. A., 138
Koritala, S., 193
Kornblum, N., 55, 173
Korobeinikova, S. A., 31
Korobitsyna, I. K., 176
Korth, T, 141
Kosma, S., 103
Kost, D., 190
Kotlyarevskii, I. L., 5
Koyama, H., 61
Kozhevnikova, A. N., 5

Kozhukhova, A. E., 147
Kramer, J. K. G., 186, 188
Kramer, K. E., 161
Kranz, E., 85
Kraus, J. L., 127
Kraus, M. A., 99
Krebs, A., 9
Kresge, A. J., 143
Kreuz, K. L., 173
Kricka, L. J., 13
Kronenberger, K., 141
Kropp, J. E., 69
Krou, L. C., 76
Krull, I. S., 92
Krusic, P. J., 52
Kubik, D., 165
Kučera, J., 120
Kühn, I., 172
Kuga, T., 111
Kuivila, H., 176
Kumar, V., 185
Kunau, W. H., 183
Kunstmann, R., 182
Kuo, Y.-N., 95
Kupchan, S. M., 92
Kurlansik, L., 7
Kurtev, B., 104
Kusamran, K., 71, 192
Kuwajima, I., 146
Kwiatkowska, S., 155

Laba, V. I., 17
Lachance, A., 102
Lack, D., 42
Lalancette, J. M., 102, 112, 151
Lalezari, I., 4
LaLonde, R. T., 109
Lamm, B., 119
Landor, P. D., 38, 178
Landor, S. R., 9, 178
Lands, W. E. M., 184
Lane, C. F., 69, 156, 160
Laneélle, G., 180
Laneélle, M. A., 180
Lansbury, P. T., 123
Lanser, A. C., 195
Lapidus, A. L., 3
Lapworth, A., 109
Larchevêque, M., 141
Lardicci, L., 139
Larkin, J. M., 173
Lastomirsky, R. R., 161
Lauer, R. F., 74, 118
Lauer, W. M., 188
Lavielle, G., 56, 94
Law, J. H., 199
Layton, R. B., 128
Leaffer, M. A., 182
Leaver, C., 32
Lebel, N. A., 43
Ledwith, A., 176

Lee, B., 90
Lee, C. V., 48
Lee, D. G., 124
Lee, J. B., 123
Lee, T. B. K., 155
Leeper, C. G., 165
Leffingwell, J. C., 145
Lehmann, H., 183
Lehmkuhl, H., 71
Lehnert, W., 111, 112
Lennarz, W. J., 188
Le Noble, W. J., 176
Leppik, R. A., 13
Leriverend, P., 62
Leroux, Y., 49
Le Tellier, P. R., 188
Levine, R., 139
Levitin, I., 157
Levitt, T. E., 85
Levy, A. B., 68
Levy, H. A., 78
Lewalter, J., 107
Lewis, A. J., 163
Ley, K., 33
Lie Ken Jie, M., 184, 187
Light, R. J., 185
Lin, H. C., 80
Lin, I.-H., 50
Lin, L.-C., 13
Lin, M.-S., 13
Lindert, A., 96, 97, 98
Linn, F., 190
Lion, C., 116
Lipinski, C. A., 3
Lipp, H. I., 153
List, G. R., 193
Litchfield, C., 181
Liu, K. T., 125
Lloyd, D., 143
Lloyd, D. J., 60
Löffler, A., 91
Lomas, J. S., 58
Lorber, M., 69
Louw, R., 132
Lowe, P. A., 175
Lown, J. W., 68, 125
Lowrey, E. R., 193
Lüttringhaus, A., 109
Lumma, W. C., jun., 148
Luong-Thi, N.-T., 116
Lutton, E. S., 187
Lyerla, R. O., 68

Ma, O. H., 148
Maasen, J. A., 174
Mabry, T. J., 181
McCarry, B. E., 24
McCarty, C. G., 165
McChesney, J. D., 137
McCloskey, J. A., 188, 197, 201
McCrae, W., 4, 185

McDonald, E., 139
McDonald, R. N., 93
McGillivray, G., 104
McIntosh, C. L., 92
McKenna, J. C., 91
McKenzie, D. A., 173
McKillop, A., 73, 118, 129
Mackor, A., 174
McManis, G. E., 187
MacMillan, J. H., 30
McMurry, J. E., 92, 122, 130
Macomber, R. S., 24, 48
McPherson, C. A., 67
McQuillin, F. J., 71, 154, 192
McWeeny, D. J., 194
Maddox, M. L., 100
Maekawa, K., 79
Maerker, G., 193, 201
Magnus, P. D., 135, 166
Mahadevan, V., 201
Maillard, A., 165
Makino, S., 145
Makosza, M., 138
Maldonado, L., 131
Malone, G. R., 126
Manabe, O., 170
Manchand, P. S., 103
Mandell, L., 145
Mangold, H. K., 186
Manitto, P., 149
Mann, M. E., 16
Mantione, R., 49
Markovetz, A. J., 201
Markowski, V., 65
Marquet, A., 144
Marr, P. W., 55
Marsh, W. C., 139
Marshall, J. A., 59, 88, 170, 171
Martin, J., 91
Martin, J. C., 57
Martin, K. A., 72
Martin, R. J. L., 140
Maruyama, M., 56
Maryanoff, B. E., 139
Masamune, S., 110
Masaracchia, J., 64
Maskens, F., 71
Maskens, K., 165, 180, 192
Massy, M., 98
Masui, M., 174
Mathiaparanam, P., 170
Mathieu, A., 187
Mathur, N. K., 161
Matl, V. G., 105
Matsueda, R., 105
Matsumura, N., 74, 89
Matuyama, H., 102
Maverick, E., 41
Mazur, U., 50

Medète, A., 115
Medvedeva, A. S., 32
Meerwein, H., 135
Meier, H., 4
Mellier, D., 173
Melton, J., 130
Menicagli, R., 139
Menzel, I., 4
Merkel, D., 6
Merzoni, S., 103
Metcalf, B. W., 9
Meyers, A. I., 84, 125, 126, 157
Miao, C. K., 186
Michael, U., 122
Michejda, C. J., 73
Midgley, J. M., 155
Midland, M. M., 152, 158, 161, 170
Miginiac, Ph., 98
Migita, T., 102
Mikhailov, B. M., 31
Mikolajczak, K. L., 179, 181
Mikolajczyk, M., 172
Miles, D. H., 24
Milewski, C. A., 139
Miller, A. W., 85
Miller, B. J., 9
Miller, D., 145
Miller, R. B., 91
Miller, R. L., 20
Miller, S. I., 3, 27, 29
Mills, R. W., 46
Mills, S. C., 196
Milne, G. M., 60
Milun, A., 190
Minale, L., 179
Minamini, K., 166
Minato, I., 86
Minnikin, D. E., 71, 184, 192
Misbach, P., 152
Mislow, K., 163
Mitani, M., 76, 147
Mitra, A., 16
Mitra, D. K., 153
Mitsudo, T., 117
Mitsunobu, O., 88
Mitsuyasu, T., 154
Miyano, S., 157
Miyaura, N., 57, 160, 161
Mizoguchi, T., 104, 105
Mladenova, M., 104
Mo, Y. K., 80
Mock, W. L., 101
Modari, B., 145
Möhrle, H., 123
Moersch, G. W., 95
Moffatt, J. G., 156
Mollet, P., 63
Molton, P. M., 201

Monahan, M. W., 67
Monson, R. S., 111, 159, 165
Montagné, M., 79
Montaigue, R., 63
Mooney, L. A., 199
Moore, G. G., 93
Moore, R. E., 170
Moore, W. R., 36, 39, 41, 50
Moralioglu, F., 163
Moran, H. W., 47
Moreau, C., 166
Morel, G., 116
Moreno-Mañas, M., 144
Mori, M., 172
Moriarty, R. M., 125
Moriconi, E. J., 15
Moritani, I., 61
Morley, J. R., 123
Moroz, A. A., 5
Morris, L. J., 199
Morris, S. G., 184
Morrison, A., 188
Morrison, G. F., 120
Morrison, J. D., 162
Morrison, W. H., 131, 201
Morrochi, S., 7, 149
Morrow, C. J., 134
Morschel, H., 135
Moser, H. A., 193
Mosher, H. S., 85
Moss, R. A., 53
Motoki, S., 147
Moulineau, C., 116
Moulton, K. J., 193
Mounts, T. L., 190, 194, 195
Müller, H., 20
Muir, D. M., 60
Mukaiyama, T., 79, 89, 105, 150
Mukasa, S., 178
Mukhametshin, F. M., 176
Murakami, M., 110, 128
Muraki, M., 105
Murray, N. G., 87
Murray, R. D. H., 46
Muscio, O. J., 41
Myers, R. F., 24

Nagata, W., 128
Nagendrappa, G., 46
Nakagawa, M., 6
Nakaido, S., 102
Nakajo, K., 147
Nambu, H., 104, 113
Nambudiry, M. E. N., 130
Narasaka, K., 79
Nasipuri, D., 140
Naudet, M., 185, 191, 194, 195, 201

Author Index

Nawar, W. W., 188
Nayak, U. R., 189
Naylor, R. D., 129
Needham, L. L., 163
Neergaard, J. R., 132
Nefedov, B. K., 3
Negishi, E., 128, 159, 160
Nelke, J. M., 141
Nelson, C. H., 74
Nesi, R., 58
Nestel, P. J., 200
Neuenschwander, M., 13
Newman, B. C., 168
Newman, H., 162
Newman, M. S., 36, 175
Nguyen, C. H., 172
Nichols, B. W., 200
Nichols, K. M., 33
Nicoud, J. F., 154
Niederhauser, A., 13
Niehaus, W. G., jun., 197
Niehouse, E. J., 123
Niewenhuyse, H., 132
Nishimura, J., 61
Nishimura, S., 86
Nivard, R. J. F., 93
Niznik, G. E., 131
Normant, H., 137, 175
Normant, J. F., 5, 18
Norris, F., 91
Northington, D. J., 37
Novikov, S. S., 176
Nowlan, V., 143
Noyori, R., 65, 131, 145
Nozaki, H., 13
Nozawa, S., 26, 128, 158
Nugteren, D. H., 183

O'Brien, D. H., 176
O'Brien, J. L., 155, 163
Ochrymowycz, L. A., 8
O'Connor, B. R., 33
O'Connor, C. J., 68
Odagi, T., 65
Odaira, Y., 86
Odham, G., 195
Öhler, E., 92
Oehlschlager, A. C., 157
Ogura, K., 132
Ohloff, G., 4, 90
Ohmori, H., 174
Ohno, K., 148
Oishi, T., 172
Okajima, T., 117
Okano, M., 68
Okorodudu, A. O. M., 36
Okumura, T., 128
Okushi, T., 170
Okuyama, H., 184
Okuyama, T., 45
Olah, G. A., 70, 80, 176
Olah, J. A., 80

Oldenziel, O. H., 118
Ollinger, J., 158
Olmstead, H. D., 136
Olmucki, M., 150
O'Neill, L. A., 187
Ono, N., 60
Oppolzer, W., 65
Orwig, B. A., 169
Osborn, C. L., 64
Osborn, J. A., 140
Otsuji, Y., 89
Ottenbrite, R. M., 165
Owen, C. R., 8
Owen, J. S., 186
Oxford, A. W., 85
Ozretich, T. M., 41, 50

Pabon, H. J. J., 183
Pace-Asciak, C., 178
Pachter, I. J., 170
Padwa, A., 64
Pässler, P., 114
Paetzold, P. I., 103
Page, C. B., 183
Palazzo, G., 6
Palenik, G. J., 62
Pandey, G. N., 86
Parham, W. E., 133
Parker, A. J., 52, 60
Parry, R. J., 24
Parsons, A. M., 201
Pasto, D. J., 44, 71, 120, 136
Patchornik, A., 99
Patterson, J. W., 92
Paulme, J. P., 50
Pavanaram, S. K., 186
Payling, D. W., 80
Pearce, P. J., 144
Pearson, C., 195
Pechnet, M. M., 110
Pelter, A., 85, 119
Penczek, St., 135
Penton, H. R., 57
Perkins, E. G., 185, 188, 194
Perkins, M. J., 174
Perron, R., 187
Peska, J., 3
Pète, J. P., 173
Peters, V. M., 159
Peterson, P. E., 22
Petit, J., 187
Pettit, R., 33
Pfeffer, P. E., 95
Pfeifer, W. D., 78
Pfeiffer, F. R., 186
Phillipe, J.-L., 4
Phillips, B. E., 179
Pidcock, A., 116
Pierri, F., 195
Pierson, C., 87

Pilgrim, W. R., 31
Pincock, J. A., 19
Pinnick, H. W., 55
Pinschmidt, R. K., 13
Pisareva, N. A., 181
Plé, G., 151
Pletcher, D., 81
Plonka, J. H., 81, 82, 83
Pober, K. B., 183
Poet, A., 172
Polgar, N., 71, 165, 180, 184, 192, 201
Polston, N. L., 6, 21
Pope, P., 157
Portella, C., 173
Portman, O. W., 200
Portnoy, N. A., 134
Posner, G. H., 3, 79, 116
Pouliquen, J., 38, 50
Poulter, C. D., 61
Power, A. J., 9
Powers, J. W., 158
Prager, R. H., 160
Pratt, R. J., 91
Priest, D. N., 111, 165
Prilezhaeva, E. N., 17
Prokipcek, J. M., 111
Prostenik, M., 186
Prout, C. K., 13
Proverb, R. J., 30
Pryde, E. H., 201
Pura, J. L., 58
Purcell, T. A., 85
Puzitskii, K. V., 3
Pyatin, B. M., 176

Quail, J. W., 125

Raber, D. J., 88
Rabjohn, N. A., 115
Rack, E. P., 73
Radatus, B., 166
Radziwill, A. N., 190
Ragonnet, B., 48
Raman, H., 130
Rampersad, M., 7, 149
Ranganathan, S., 130
Range, P., 182
Rankin, P. C., 188
Raphael, R. A., 46, 85
Raphalen, A., 127
Rapoport, H., 74, 100
Rao, G. S. K., 130
Rao, G. V., 194
Rao, V. V., 46
Rappoporte, Z., 48
Ratcliffe, R., 124
Rathke, M. W., 95, 96, 97, 98
Raunio, E. K., 17
Ravindranathan, T., 65
Rawls, H., 194

Ray, W. C., 75
Raymond, J. C., 44
Redfield, D. A., 72
Redshaw, D. J., 190
Rehman, Z., 23
Reichardt, C., 115
Reiffers, S., 97, 125
Reinehr, D., 71
Reininger, K., 92
Reist, E. J., 186
Reitz, E. C., 184
Reitz, R. R., 93
Renner, C. A., 32
Repič, O., 74, 118
Reske, E., 90, 122
Reucroft, J., 52, 185
Rhee, I., 114
Ricca, A., 7, 149
Richards, D. H., 144
Richey, H. G., 19, 153
Richheimer, S. L., 127
Richman, J. E., 129
Richmond, G. D., 27
Rickborn, B., 139
Rimmelin, P., 165
Ritchie, E., 122
Rivière, H., 116
Robbins, M. D., 173
Robey, R. L., 118
Rodeheaver, G. T., 118
Rodehorst, R., 124
Rodgers, M., 29
Rodin, J. O., 182
Roe, D. K., 141
Roe, R., 63
Roebke, H., 170
Röder, H., 117
Roelofsen, D. P., 89, 166
Rohwedder, W. K., 179, 197
Roller, P., 170
Roman, S. A., 3
Rone, A. M., 5
Roobeek, C. F., 23
Roocker, A. E., 69
Roomi, M. W., 191
Roosevelt, C. S., 133
Roscher, N. M., 165
Rose, J. G., 108
Rosenberg, D., 35
Rosenthal, A. F., 186
Ross, C. H., 65
Rosser, M. J., 68
Rostock, K., 103
Roth, M., 168
Rothman, E. S., 93, 195
Roumestant, M. L., 38
Rous, S., 201
Roussel, A., 63
Roussi, G., 143
Rozynov, B. V., 181
Rudler, H., 114, 151

Ruecker, G., 3
Rüesch, H. P., 91
Rühlmann, K., 176
Russell, G. A., 8
Russo, G., 149
Rutherford, K. G., 165
Rutledge, T. F., 3
Ryan, K. J., 13
Ryang, M., 114

Saegusa, T., 146
Safronova, L. P., 32
Sagatys, D. S., 58
Sagredos, A. N., 189, 190
Saha, J. G., 139
Said, A. I., 191, 192
Sainsbury, M., 31
Sakai, K., 105, 112
Sakakibara, T., 86
Sakurai, H., 26
Sallfrank, R. W., 100
Salomone, R. A., 15
Sammes, P. G., 52, 185
Samuelsson, B., 119, 188, 195, 200, 201
Samuelsson, K., 188, 201
Sandler, S. R., 156
Santelli, M., 39, 48, 51
Sasaki, O., 68
Sasaki, T., 15
Sasson, Y., 140
Satsumabayashi, S., 147
Savignac, P., 151
Scales, C. G., 13
Scanlon, B , 123
Schaaf, J V. D., 71
Schaaf, T. K., 139
Schäfer, H., 75, 155
Schäfer, W., 42
Scharf, D. J., 123
Schaumburg, K., 188
Scheer, W., 65
Scheeren, J. W., 93
Scheinbaum, M. L., 110, 170
Schelhorn, E., 37
Schell, R. A., 103
Schisla, R. M., 79
Schittenhelm, D., 123
Schlenk, H., 194
Schlessinger, R. H., 124
Schleyer, P. von R., 78
Schlosser, M., 52, 54, 61
Schmid, G. H., 77
Schmid, H. H. O., 180
Schmidt, G., 17
Schmidt, U., 7, 92
Schneller, S., 151
Schöllkopf, U., 98, 132
Scholfield, C. R., 187, 193, 195

Scholl, P. C., 80
Scholtz, H., 85
Schreiber, J., 4
Schreibmann, A. A. P., 142
Schrock, R. R., 140
Schroeder, J. P., 159
Schroepfer, G. J., jun., 197
Schue, F., 165
Schulte, K. E., 3
Schulte-Elte, K. H., 4
Schumaker, R. R., 144
Schutte, L., 166
Schvartsberg, M. S., 5
Schwab, A. W., 193
Schwartz, V., 100
Scilly, N. F., 144
Scott, A. I., 78
Scott, G. H., 186
Scott, M., 164
Scott, M. K., 114
Scott, T. W., 196
Scotton, M., 58
Seamen, F., 181
Seeley, D., 151
Seeliger, A., 106
Seigler, D., 181
Seigler, D. S., 181
Selke, E., 193
Selve, C., 151
Serafin, B., 138
Serebrennikovo, G. A., 186
Serota, S., 195
Servis, K. L., 165
Seux, R., 116
Seybold, G., 13
Sgoutas, D. S., 183, 184, 199
Shafiee, A., 4
Shah, D. O., 200
Shakhidayatov, Kh., 53
Sharpless, K. B., 74, 118, 162
Shaw, J. E., 103, 124
Shearing, D. J., 7, 60
Sheehan, J. C., 85
Sheehan, M. H., 178
Shekhtman, R. I., 17
Shen, J., 80
Shepard, K. L., 93
Shepherd, I. S., 190
Sheradsky, T., 16
Sherry, J. J., 124
Shevchuk, M. I., 113
Shevlin, P. B., 83
Shiavelli, M. D., 47
Shields, T. C., 103
Shimakawa, Y., 15
Shiner, V. J., jun., 175
Shinya, M., 33
Shô Itô, 178
Shone, T., 76, 147
Shostakovskii, M. F., 32

Author Index

Shulman, J. I., 54, 120, 121, 133
Sicher, J., 52
Siddall, J. B., 54
Siegel, J., 7
Silbert, L. S., 86, 95
Silveira, A., jun., 185
Silverstein, R. M., 182
Simeone, J. F., 141
Simonaitis, R., 29
Simpson, P., 29
Sims, C. L., 163
Singh, B., 108
Singh, S. P., 94
Singleton, D. M., 56
Sipos, J. C., 181
Sisido, K., 186
Skell, P. S., 81, 82, 83
Slaugh, L. S., 109
Slomkowski, St., 135
Slotboom, A. J., 201
Smalley, R. K., 175
Smith, C. R., jun., 179, 181, 201
Smith, D. B., 3
Smith, E. H., 54, 185
Smith, E. M., 125, 126
Smith, G., 39, 135
Smith, G. R., 193
Smith, H. A., 136
Smith, J. N., 105
Smith, K., 85, 119
Smith, P. J., 125
Smith, R. A. J., 141
Smith, R. G., 13, 55
Smith, T. N., 154
Snegotskii, V. I., 17
Snieckus, V., 13
Snow, J. T., 21, 39
Snyder, E. I., 43
Snyder, J. P., 174
Solomon, D. M., 183
Sommer, H. Z., 153
Sommer, J. M., 165
Soneda, R., 147
Sono, S., 152
Sonoda, N., 74, 90, 108
Sowerby, R. L., 136
Spencer, D. G., 102
Spencer, G. F., 179
Spencer, T. A., 141, 145
Spitzer, U. A., 124
Sprecher, H. W., 183
Stanacév, N. Ž., 186
Stanbury, P. F., 63
Stang, P. J., 48
Stang, P. S., 78
Starratt, A. N., 184
Steele, R. B., 162
Steenhoek, A., 183
Steglich, W., 83
Stein, O., 201

Stein, Y., 201
Steinbeck, W., 138
Stenberg, V. I., 89, 165
Stensiö, K.-E., 124
Stetter, H., 90, 122, 138
Steur, R., 50
Stevens, J. I., 93
Stevens, R. V., 109
Still, W. C., jun., 163
Stille, J. K., 148
Stirling, C. J. M., 39, 50, 85, 163
Stirton, A. J., 195
Stodola, F. H., 179
Stork, G., 122, 131, 157
Story, P. R., 75, 90
Strating, J., 97, 125
Streckert, G., 135, 166
Strelkov, T., 62
Stuchal, F. W., 173
Studzinkskii, O. P., 176
Stumpf, P. K., 199
Sturtz, G., 127
Suama, M., 138
Subbarao, R., 194
Suga, K., 95
Sugasawa, S., 111
Suginome, H., 110
Sugita, T., 138
Sugiura, S., 195
Sukh Dev, 189
Sullivan, A. B., 169
Sundt, E., 4
Sunko, D. E., 62
Surridge, J. H., 157
Suschitsky, H., 175
Suzuki, O., 187
Suzuki, A., 26, 57, 128, 152, 158, 160, 161
Suzuki, M., 105
Svanholt, K. L., 91
Swann, B. P., 118
Sweeley, C. C., 188
Swern, D., 53, 161, 187
Symons, E. A., 110
Szucs, S. S., 19

Taguchi, T., 150
Takagisi, H., 186
Takahashi, S., 93
Takamura, N., 104
Takaya, H., 65, 145
Takegami, Y., 117
Tamura, H., 186
Tanaka, M., 117
Tanaka, R., 3, 29
Tang, B. K., 165
Tang, R., 163
Taniguchi, H., 33
Tanino, H., 195
Tanizawa, K., 111
Tartivita, K., 180

Tatchell, A. R., 9
Taub, W., 91
Taylor, D. R., 41
Taylor, E. A., 57
Taylor, E. C., 73, 104, 118, 129
Taylor, J. A., 5
Taylor, J. B., 10
Taylor, W. C., 122
Temple, D. L., jun., 84
Teranishi, A. Y., 74, 118
Terashima, S., 65, 107
Tew, L. B., 159
Thakore, A. N., 157
Thaller, V., 183
Theissen, R. J., 127
Thiessen, W. E., 78
Thomas, A. M., 105
Thomas, F. L., 193
Tidwell, T. T., 72, 142
Tien, R. Y., 45
Timms, G. H., 130
Tindell, G. L., 38
Tjarks, L. W., 179, 181
Tobias, M. A., 62
Tocanne, G., 180
Tocanne, J. F., 180
Toda, F., 6
Tolstikov, G. A., 125
Tomaszewski, J. E., 162
Tomita, S., 146
Torkelson, A., 197
Torssell, K., 172
Toshima, N., 130
Toube, T. P., 52
Tove, S. B., 196
Townsend, J. M., 145
Trachtenberg, E. N., 74
Trahanovsky, W. S., 173
Trecker, D. J., 64, 108
Tronchet, J., 128
Tronchet, J. M. J., 128
Trost, B. M., 58
Truce, W. E., 25
Trueblood, K. N., 41
Tsuchihashi, G., 132
Tsuji, J., 148, 154
Tsutsumi, S., 74, 90, 108, 114
Tubul, A., 194
Tuccarbasu, S., 89
Tucker, W. P., 196
Tufariello, J. J., 69
Tullman, G. M., 89
Tulloch, A. P., 180
Tyler, W. E., 141

Uccella, N., 43
Ucciani, E., 191, 194, 195, 201
Ueki, M., 105
Uemura, S., 68

Umeda, H., 110
Ushakov, A. N., 181
Uyehara, T., 13

Vagelos, P. R., 182
Vaidyanathaswamy, R., 46
van Bekkum, H., 89, 166
van Deenen, L. L. M., 201
Vandenburg, L. E., 180
van den Elzen, R., 172
Vandenheuvel, F. A., 201
van der Baan, J. L., 116
Vander Kooi, J. P., 169
van der Ven, B., 180
van de Ven, L. J. M., 50
van Dongen, J. P. C. M., 50
van Dorp, D. A., 183
VanHorn, W. F., 169
van Leusen, A. M., 114
Vanlierde, H., 63
van Malick, J. E. W., 93
van Peppen, J. F., 193
van Senten, P. J., 194
van Tamelen, E. E., 60, 114, 151, 162
Vantillard, A., 191
Varma, R. K., 139
Vasilevskii, S. F., 5
Vaughan, W. R., 98
Vaver, V. A., 181
Vedejs, E., 56, 134
Veefkind, A. H., 71
Vela, F., 99
Veldink, G. A., 197
Vesley, G. F., 89, 165
Vesonder, R. F., 179
Vere Hodge, R. A., 183
Verheyden, J. P., 156
Vernon, J. M., 13
Verschoor, H. M., 89
Viau, R., 172
Vick, B. A., 197
Vida, J. A., 156
Viehe, H. G., 8, 13, 14, 28
Vijay, I. K., 199
Villaume, J. E., 83
Villieras, J., 56, 94
Vilsmaier, E., 100
Vinokur, V., 56, 122
Viola, A., 30, 122
Viviani, R., 200
Vliegenthart, J. F. G., 197, 198
Vogel, H. H., 79, 104
Volynskaya, E. M., 113
Vol'pin, M., 157
von Mikusch, J. D., 189, 190
von Rein, F. W., 19
Vo-Quang, L., 44
Vo-Quang, Y., 44

Voranov, V. K., 32
Vtorov, I. B., 186

Wagatsuma, M., 107
Wagner, E., 38
Wahl, G. H., 13
Wakil, S. J., 200
Walborsky, H. M., 131, 148
Walker, J. E., 184
Walker, L. E., 42
Walker, W. E., 103
Wall, D. K., 120
Wall, R. T., 184
Wallen, L. L., 197
Wantland, L. R., 194
Warburton, M. R., 41
Wasserman, H. H., 42, 101
Watanabe, S., 95
Watanabe, Y., 117
Waters, W. L., 45, 46
Watkins, R. J., 73
Watts, P., 150
Watts, P. C., 91
Weber, H. P., 65
Weber, W. P., 72, 142
Wedegaertner, D. K., 26
Weedon, B. C. L., 3, 52, 190
Weil, T. A., 185
Weiler, L., 102
Weingarten, H., 107
Weisbach, J. A., 186
Weisleder, D., 197
Weiss, K., 13
Welby-Gieusse, M., 180
Weller, J. W., 81
Wells, W. E., 175
Welvart, Z., 115
Wendler, P. A., 113
Wendt, G., 188
Wenkert, E., 142
Wentland, S. H., 101
Weslowski, T. J., 185
West, R., 35
Westwood, R., 10
Wetzel, B., 65
Whaley, H. A., 175
Whalley, W. B., 135
Wheeler, D. H., 190
Whereat, A. F., 201
Whipple, E. B., 150
White, A. M., 176
White, E. H., 155
White, J. D., 16, 94
White, J. G., 15
White, R. W., 155, 163, 196
Whitesides, G. M., 76
Whitfield, G. F., 194
Whitney, C. C., 20, 21
Whitten, C. E., 116

Wickerham, L. J., 179
Wiegandt, H., 201
Wieland, D. M., 162
Wieland, P., 4
Wieland, T., 106, 107
Wigfield, D. G., 62
Wilday, P. S., 72
Wilder, E. A., 180
Wildsmith, E., 130
Wilke, G., 20
Williams, A., 3
Williams, E. B., jun., 105
Williams, D. R., 59, 74, 118
Williams, J. C., 134
Williams, J. L., 183
Williams, N., 150
Williams, R. M., 201
Willich, R. K., 191
Willis, B. J., 54, 185
Wils, E. R. J., 166
Wilson, A. T., 195
Wilson, J. D., 107
Wilson, M. A., 68
Wilson, R., 85
Wilt, J. W., 113
Wineburg, J. P., 187
Winkler, H. J. S., 39
Winstein, S., 61
Winter, M., 4
Winterfeldt, E., 3, 17, 141
Wladislaw, B., 86
Wojtkowski, P. W., 120, 136
Wolf, A. P., 83
Wolf, G. C., 25
Wolfe, L. S., 178
Wolfe, S., 31, 77
Wolff, I. A., 179, 181
Wolff, V., 189
Wolters, E. T., 105
Wood, L. S., 81
Woodruff, R. A., 88
Woodward, R. B., 170
Woolsey, N. F., 138
Worsley, M., 125
Wright, D. A., 155
Wright, D. B., 41
Wrigley, A. N., 195
Wrixon, A. D., 78
Wu, Y. V., 197
Wudl, F., 155
Wulff, C. A., 155
Wunderlich, K., 135
Wyckoff, J. C., 87
Wycpalek, A. F., 137
Wynberg, H., 13, 97, 125, 135

Yahner, J. A., 95
Yajima, H., 105
Yalpani, M., 4

Author Index

Yamada, S., 104, 107
Yamagata, M., 26
Yamagishi, T., 166
Yamamoto, E., 111
Yamamoto, H., 53, 54, 121
Yamamoto, K., 117
Yamamoto, O., 187
Yamamoto, Y., 61
Yanase, R., 183
Yang, K.-W., 108
Yang, P. W., 13
Yanovskaya, L. A., 53
Yasuhara, T., 108
Yates, B. L., 30

Yates, K., 19
Yeo, A. N., 93
Yoder, J. E., 142
Yong, K. S., 134
Yoon, N. M., 175
Yoshida, Z., 170
Yoshimoto, T., 166
Yoshioka, M., 128
Young, A. T., 165
Yu, S. H., 140
Yukimoto, Y., 15
Yur'ev, V. P., 125

Zàhorszky, U.-I., 23
Zanarotti, A., 7, 149

Zarins, Z. M., 191
Ziege, E. A., 13
Ziegler, F. E., 113
Ziehn, K.-D., 109, 111
Zika, R. G., 22
Zimmer, H., 151
Zimmerman, D. C., 197
Zimmerman, J. P., 86
Zimmermann, H. E., 35
Zobova, N. N., 14
Zubiani, G., 56
Zurr, D., 135, 166
Zwanenburg, B., 101
Zweifel, G., 6, 20, 21, 39, 162
Zwiesler, M. L., 95